Electric Discharge Hybrid-Machining Processes

Electric Discharge Hybrid-Machining Processes

Fundamentals and Applications

Edited by
Basil Kuriachen, Jose Mathew and Uday Shanker Dixit

CRC Press
Taylor & Francis Group
Boca Raton London New York

CRC Press is an imprint of the
Taylor & Francis Group, an informa business

First edition published 2022
by CRC Press
6000 Broken Sound Parkway NW, Suite 300, Boca Raton, FL 33487-2742

and by CRC Press
4 Park Square, Milton Park, Abingdon, Oxon, OX14 4RN

© 2022 selection and editorial matter, Basil Kuriachen, Jose Mathew and Uday Shanker Dixit; individual chapters, the contributors

CRC Press is an imprint of Taylor & Francis Group, LLC

ISBN: 978-1-03-206432-1 (hbk)
ISBN: 978-1-03-206435-2 (pbk)
ISBN: 978-1-00-320230-1 (ebk)

DOI: 10.1201/9781003202301

Typeset in Times
by Newgen Publishing UK

Contents

Preface

Electrical discharge machining (EDM) is an advanced machining process based on thermal energy from electric sparks that causes localized melting and vaporization of the material. *Electrical Discharge Hybrid-Machining Processes: Fundamentals and Applications* presents in-depth knowledge of various processes and methods derived from EDM. In general, we can find EDM process applications in many engineering fields, thanks to a plethora of advantages of this technology. The process is no longer confined to academia; it has also become popular in industry. However, it has several limitations, too. In an attempt to overcome the limitations of EDM, researchers have invented a number of hybrid machining processes. Thus, a new area of research and development, called electrical discharge hybrid-machining processes, has appeared. The word "hybrid" denotes the combination of two or more methods to enhance the removal of material from a workpiece. In electrical discharge hybrid-machining processes, one or more sources of energy or methods, along with an electric spark, are utilized to remove material from the workpiece. The other energy sources can be chemical, electro-chemical, mechanical or thermal. Scientists have developed various methods to improve the process efficiency of electrical discharge machining processes with the assistance of magnetic fields, external powders in the dielectric and vibrations. Some researchers have developed a dry EDM process as part of the improvement of sustainability.

This book comprises 15 chapters. Chapter 1 by Basil introduces the reader to EDM. In Chapter 2, Jithin and Joshi present an elaborate review on dry and near-dry EDM processes. They have presented excellent discussion on the importance of choosing proper polarity in various EDM processes. Sharma et al. have described a hybrid machining process, called electrochemical spark machining in Chapter 3. This process is motivated by EDM as well as Electro-Chemical Machining (ECM). Subsequently in Chapter 4, Tayade and Doloi have described in detail how EDM and ECM can be carried out sequentially to obtain a desired product. Ahmed et al. have lucidly presented the latest trends in arc machining processes in Chapter 5, e.g., how plasma arc machining can be used with EDM.

In Chapter 6, George et al. describe electric discharge hybrid-turning processes, and Setti presents a detailed description of electrical discharge grinding and electrical discharge abrasive grinding processes in Chapter 7. In Chapter 8, Vipindas et al. have described the electric discharge-assisted milling process. This chapter highlights how the performance of conventional machining processes can be enhanced by using EDM as an assistive process. In Chapter 9, Mertiya and Unune have described vibration-assisted EDM processes, including discussion on micro-EDM processes. Singh et al. have described the magnetic field-assisted EDM process in Chapter 10.

In Chapter 11, Suvin and Sahu have described laser-assisted EDM. Philip and Kuriachen have presented state-of-the-art powder mixed electrical discharge machining in Chapter 12. A focused discussion on various types of micro-EDM processes has been presented by Dilip and Mathew in Chapter 13. In the near future, modelling and simulation of various EDM-based processes will gain prominence.

With this in mind, Chapter 14 by Das et al. introduces the readers to modelling and optimization of EDM-based processes. Finally, Skoczypiec describes applications of EDM-based processes with practical examples.

All contributors are active researchers in various aspects of EDM-based processes. The information provided by them has been presented in an organized manner, such that this book not only acts as a reference book for researchers and professionals, but it can also be used as a textbook in undergraduate- and postgraduate-level courses. While maintaining the continuity of the entire treatment, an attempt has been made to preserve the independence and completeness of each chapter. Hence, some overlap between the chapters has been deliberately retained. We also feel that key concepts should be explained in various ways by different experts. Editors thank all the authors for their efforts in presenting the latest information in the area of EDM in a lucid form.

We would especially like to thank friends and colleagues who reviewed the early drafts of the book and provided constructive criticism. A group of early reviewers includes Dr. Vipindas K., Dr. Jiju V. Elias and Dr. Anand Krishnan. In addition, we express our sincere gratitude to the institutes, with whom at least one of us is affiliated, namely, the Indian Institute of Technology Guwahati, and the National Institute of Technology, Calicut for providing a proper ambience for the development of this book. Finally, thanks to the editorial staff of CRC Press for their excellent cooperation. We welcome constructive feedback from the readers.

Basil Kuriachen
Jose Mathew
Uday Shanker Dixit

Editors

Basil Kuriachen is serving as an assistant professor in the Advanced Manufacturing Centre, Department of Mechanical Engineering, National Institute of Technology Calicut, Kerala, India, from 2020. His vivacity and dexterity toward an abiding commitment to a sublime work ethic conferred him with PhD and MTech degrees from NIT Calicut (2015) and Mahatma Gandhi University, Kottayam (2011), respectively. His specific research niches are in the fields of metal additive manufacturing, micro/nano-machining processes, precision and ultra-precision machining, modeling and analysis in the machining of "difficult-to-machine" materials, tribology of manufacturing processes, nano-tribology, surface coatings, nano-lubricants, advanced machining processes, friction stir welding and processing and surface characterization. He is the principal investigator (PI) for two currently active projects from DST-SERB and ARDB, DRDO, Govt. of India. Excellency in academics and research enabled him to be the recipient of several prestigious national awards inclusive of IEI Young Engineers Award 2019–20, SERB-International Travel Fellowship 2017, and the Early Career Research Award of DST-SERB 2016. He has organized five GIAN programs (in collaboration with foreign faculties), and is the reviewer for several international peer-reviewed journals of repute associated with various publishing houses such as ASME, Elsevier, Springer, Taylor & Francis, and more. He has to his credit, 90 research publications in international refereed journals and conferences alongside 2 filed patents. Under his guidance, one PhD thesis was awarded successfully. Presently, he is guiding six PhD scholars (one thesis submitted) and six MTech (PG) students for carrying out research in the distinct fields of electrical discharge machining-assisted alloying, friction stir welding and processing, laser surface texturing and more. He has organized many national and state-level programs such as 1 conference, 3 short-term courses, and 4 state-level seminars, and he has participated in 17 short-term courses in his career so far.

Jose Mathew is a professor (HAG) and HOD (ME) and Ex-Dean (R&C) of the National Institute of Technology Calicut, Kerala. He received his MTech and PhD from IIT Kanpur (1990), and IIT Bombay (1999), respectively. His interests are micro- and nano-machining process, precision- and ultra-precision machining, modelling and analysis of machining of "difficult-to-machine" materials, and more. He has published more than 125 research papers in international journals and international conferences. Several MTech and PhD theses have been completed under his guidance. He also has worked on a number of industry-sponsored research and development projects. He was awarded with: the Best All-Around Performance Award among NITC faculties during 2005–06 (instituted by NIT Calicut and was the first award winner); ISTE – Certificate of Achievement 2008 National Award (first prize) for having guided the Best MTech Thesis in Mechanical Engineering for the thesis, "Theoretical and Experimental Investigations on Mechanical Micro Machining (M^3)" submitted by Sooraj V. S.; ISTE – Certificate of Achievement 2012 National Award (first prize) for having guided the Best MTech Thesis in Mechanical Engineering for the thesis, "Multi-Objective Optimization of Process Parameters in Micro Drilling Using Vibration Assisted Micro EDM" submitted by Deepak G. Dilip; and DST International Travel Support (full amount) Scheme Award for attending the ASME International Conference at Chicago, US in 2012. Also, he has developed many state-of-the-art research facilities such as the DST-FIST-sponsored Centre for Precision Measurements and Nanomechanical Testing, Micromachining Centre, Additive Manufacturing Centre, CNC Centre, Sophisticated Instruments Centre, CAD/CAM Centre, and more.

Uday Shanker Dixit received a BE in mechanical engineering from University of Roorkee (now Indian Institute of Technology Roorkee) in 1987, an MTech in mechanical engineering from the Indian Institute of Technology Kanpur in 1993, and a PhD in mechanical engineering from IIT Kanpur in 1998. He has worked in two industries: HMT, Pinjore and INDOMAG Steel Technology, New Delhi, where his main responsibility was designing various machines. Dr. Dixit joined the Department of Mechanical Engineering, Indian Institute of Technology Guwahati in 1998 where he is currently a professor. He was also the Officiating Director of Central Institute of Technology, Kokrajhar from 2014–15. Dr. Dixit has been actively engaged in research in various areas of design and manufacturing for the last 33 years. He has authored/co-authored about 125 journal papers, about 115 conference papers, 31 book chapters and 7 books in mechanical engineering. He has also co-edited 8 books related to manufacturing. Out of these 15 books, 2 have been published by CRC Press. He has guest-edited 11 special issues of journals. Presently, he is an associate editor of the *Journal of Institution of Engineers (India), Series C* and regional editor Asia of *International Journal of Mechatronics and Manufacturing Systems*. He has guided 14 doctoral and 44 master's students. Dr. Dixit has investigated a number of sponsored projects and developed several courses. Presently, he is the vice president of AIMTDR conference.

Contributors

Afzaal Ahmed, Assistant Professor, Indian Institute of Technology Palakkad, Kerala, India, afzaal@iitpkd.ac.in.

Jibin Boban, Department of Mechanical Engineering, Indian Institute of Technology Palakkad, Kerala, India.

Sanghamitra Das, Department of Mechanical Engineering, Indian Institute of Technology Guwahati, Assam, India.

Deepak G. Dilip, Assistant Professor, Mar Baselios College of Engineering and Technology, Kerala, India, deepakgd1@yahoo.co.in, deepak.dilip@mbcet.ac.in.

Uday Shanker Dixit, Professor, Department of Mechanical Engineering, Indian Institute of Technology Guwahati, Assam, India.

B. Doloi, Professor, Department of Production Engineering, Jadavpur University, Kolkata, India.

Jees George, Amal Jyothi College of Engineering, Kanjirappally, Kerala, India, jeesgeorge@gmail.com.

S. Jithin, Department of Mechanical Engineering, Indian Institute of Technology Mumbai, India, jithins.iitb@gmail.com.

Shrikrishna Nandkishor Joshi, Associate Professor, Department of Mechanical Engineering, Indian Institute of Technology Guwahati, Assam, India, snj@iitg.ac.in.

Suhas S. Joshi, Professor, Indian Institute of Technology Bombay, ssjoshi@iitb.ac.in.

Vipindas K., Assistant Professor, Department of Mechanical Engineering, Indian Institute of Information Technology, Design and Manufacturing, Kurnool, India, vipindas.k@iiitk.ac.in.

Basil Kuriachen, Assistant Professor, Advanced Manufacturing Centre, Department of Mechanical Engineering, National Institute of Technology Calicut, Kerala, India, and National Institute of Technology Mizoram, Aizawl, Mizoram, India, bk@nitc.ac.in.

R. Manu, Professor, National Institute of Technology Calicut, Kerala, India.

Jose Mathew, Professor (HAG), Department of Mechanical Engineering, National Institute of Technology Calicut, Kerala, India.

Abhimanyu Singh Mertiya, Department of Mechanical-Mechatronics Engineering, The LNM Institute of Information Technology, Jaipur, India.

Jibin T. Philip, Assistant Professor, Amal Jyothi College of Engineering, Kanjirappally, Kerala, India, philip.jibin07@gmail.com.

M. Azizur Rahman, Department of Mechanical and Production Engineering, AUST Dhaka, Bangladesh.

Janakarajan Ramkumar, Professor, Department of Mechanical Engineering, Indian Institute of Technology Kanpur, Uttar Pradesh, India, jrkumar@iitk.ac.in.

Ranjeet Kumar Sahu, Department of Mechanical Engineering, National Institute of Technology Karnataka, Mangalore, India.

Dinesh Setti, Assistant Professor, Department of Mechanical Engineering, Indian Institute of Technology Palakkad, Kerala, India, dinesh@iitpkd.ac.in.

Vyom Sharma, Department of Mechanical Engineering, Indian Institute of Technology Kanpur, Uttar Pradesh, India.

Mahavir Singh, Department of Mechanical Engineering, Indian Institute of Technology Kanpur, Uttar Pradesh, India.

Sebastian Skoczypiec, Professor, Chair of Production Engineering, Faculty of Mechanical Engineering, Cracow University of Technology, Kraków, Poland, sebastian.skoczypiec@pk.edu.pl.

P. S. Suvin, Assistant Professor, Department of Mechanical Engineering, National Institute of Technology Karnataka, Mangalore, India, suvin@nitk.edu.in.

R. M. Tayade, Department of Production Engineering, Jadavpur University, Kolkata, rmtayade@vjti.org.in.

Deepak Rajendra Unune, Department of Materials Science and Engineering, INSIGNEO Institute for *in silico* Medicine, The University of Sheffield, UK, deepunune@gmail.com.

1 Electrical Discharge Machining (EDM)

Basil Kuriachen
Advanced Manufacturing Centre, Department
of Mechanical Engineering, National Institute of
Technology, Calicut, Kerala, India, and National Institute of
Technology, Mizoram, Aizawl, Mizoram, India

CONTENTS

1.1 INTRODUCTION

Electric discharge machining (EDM) is a modern machining process that has become a well-established alternative for machining advanced materials throughout the world regardless of each country's level of academic or industrial advancement. Several research projects in EDM improved its machining capabilities and extended its application from working with only metallic materials to being able to work with non-conducting materials. These research projects have also developed the hybrid electric discharge machining process. At present, there are many variants of the hybrid electric discharge machining process in addition to Wire EDM, Die Sinking EDM, and Micro EDM, and so on. This chapter provides an elementary introduction to the history, material removal mechanism, essential process parameters, and the influences of the major process parameters, by considering all of the available EDM processes.

1.2 HISTORY OF EDM

The first recorded attempt to use electrical energy to remove material, a 'disintegrator', was reported by Matulaitis and Harding in 1930 [1]. Later, during World

DOI: 10.1201/9781003202301-1

War II, physicists B. R. and N. I. Lazarenko in Moscow observed the effect of metal removal in electric circuit breakers and maximized it for the purpose of metal machining [1]–[3]. In the early years of development, relaxation type generators, in other words, charging condensers, were used to store the discharge energy. In 1960, semiconductor switched static pulse generators achieved a duty cycle of up to 99% and increased the material removal rate [4]. With the invention of computer numerical control in the 1980s, maintaining the constant spark gap between the electrodes provided more stable sparks.

In due course, an electronic servo control system was introduced to the EDM machines, which automatically provided the required spark gap [5]. Since then, the improvement of EDM has greatly benefited the manufacturing industry, aiding in the processing of difficult-to-machine materials and it has generated specific research areas. At the same time, electrical discharge machining drew the attention of the research community who were interested in improving its process efficiency by developing a number of different variants, including WEDM, EDM drilling, EDM milling, μEDM, and the like. However, the unique nature of material removal through melting and vaporization and the problems arising around the small inter-electrode gap make the process more complex. In addition, the discharge phenomena occur for a small time interval, of the order of microseconds, making it more complicated. Therefore, both theoretical and experimental investigations of the electric spark in particular, and electrical discharge machining in general, are extremely difficult.

1.3 WORKING PRINCIPLE

The basic working principle of the electrical discharge machining process is the localized melting and vaporization of electrodes through controlled electric sparks. In other words, the electrical energy is converted into thermal energy through sparks, and it is effectively utilized to remove the materials. It is a non-contact material removal process wherein no mechanical interaction between the tool and workpiece occurs. Hence, the mechanical properties of tools and work pieces haven't got a significant role in the process. The removal of material takes place between the tool and workpiece during the pulse 'on' time of a pulsed DC power supply.

A basic scheme of the EDM process is shown in Figure 1.1. The electrodes are connected to a pulsed DC power supply, and a suitable inter-electrode gap (IEG) is maintained between tool and workpiece inside a dielectric fluid. Once the appropriate power supply is established between the electrodes, the cold emission of electrons (also known as 'field emission') starts from the cathode surface due to the strong electric field. The cold emission of electrons begins where the local distance between the tool and workpiece is the narrowest due to the roughness in the bottom surface of the tool and top surface of the workpiece (Figure 1.1). For example, point A has the smaller distance between the tool and workpiece compared to point B in Figure 1.1. In the absence of an electric field, the electrons need to acquire a minimum energy level to escape from the surface of the electrodes, which is referred to as the work function. However, under the influence of an electric field, the work function required to release the electrons is reduced, and electrons start to be emitted

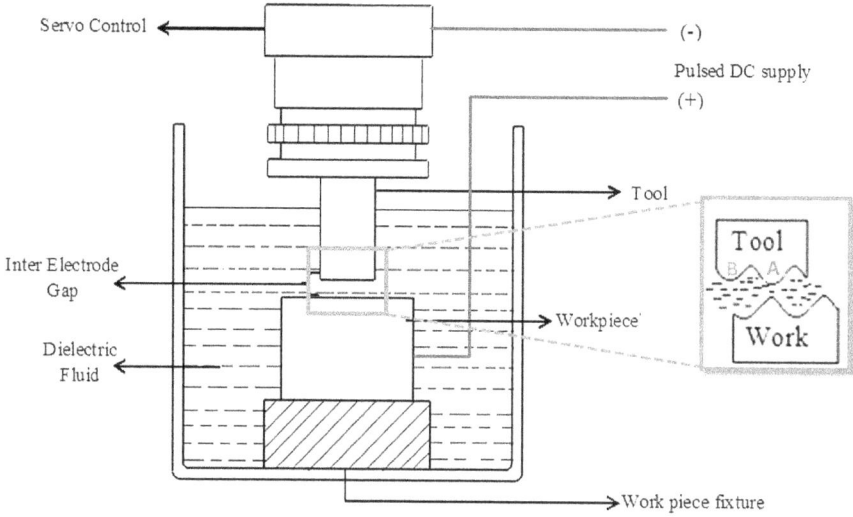

FIGURE 1.1 Basic schematic diagram of EDM.

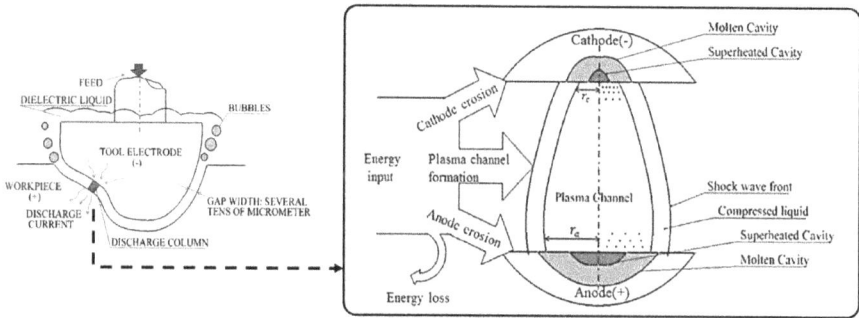

FIGURE 1.2 Formation of plasma channel in the IEG [1], [6]. With permission from Springer Nature.

from the cathode surface due to the cold emission. These liberated electrons are accelerated towards the surface of the anode. Once the electrons acquire enough velocity, they strike the molecules of the dielectric fluid present in the inter-electrode gap (IEG), in other words, the gap between two electrodes. Accelerated electrons divide the dielectric molecules into ions and electrons. Electrons produced in this way are referred to as 'secondary electrons'. These electrons and ions accelerate towards the workpiece- (+) and tool-electrode (−) surfaces, respectively, within the IEG. The process repeats in the IEG and an avalanche of electrons and ions forms there. In this way, a plasma channel of very high conductivity is established between the electrodes, as shown in Figure 1.2. These electrons strike the anode surface with very high kinetic energy, thus an large amount of heat is produced locally on the

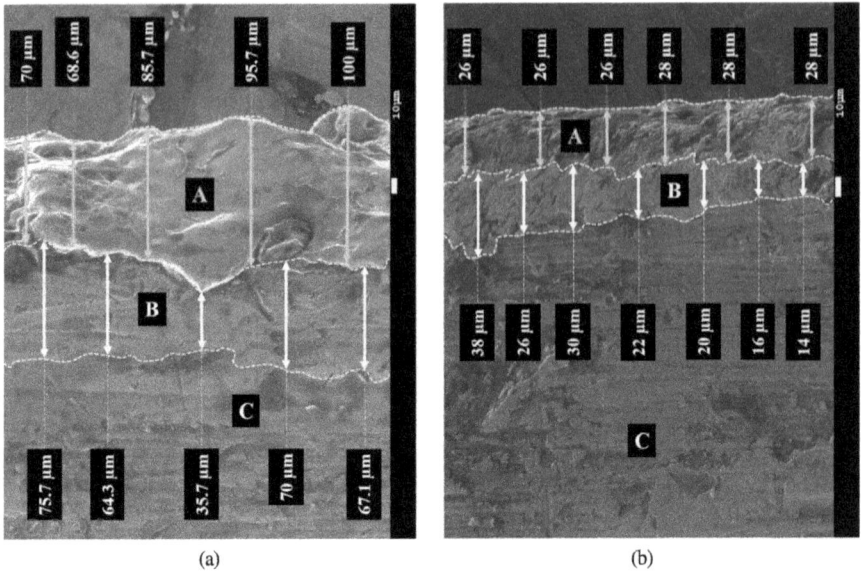

FIGURE 1.3 Cross-sectional images of electrical discharge machined surfaces (a) ETi64 and (b) WETi64. RL (zone A), HAZ (zone B), substrate (zone C) [7]. With permission from Elsevier.

anode surface. It is also understood that out of the total heat generated, two-thirds is at the anode surface and one-third is at the cathode surface. This is due to the difference in the kinetic energies of the ions and the electrons. The mass of electrons is comparatively less than that of the ions but their velocities are higher, therefore the discharge energy of the electron stream is higher.

Upon analysis, researchers [6] found that the temperature generated was in the range of 8,000–12,000 °C. The heat is conducted into the workpiece and tool followed by melting, and vaporization. At the end of the spark, the plasma channel disappears, and the dielectric fluid present, gushes back into the space previously occupied by the plasma channel. Hence, the dielectric removes the melted portion of the electrodes in the form of debris that is semi-spherical in shape. Moreover, the dielectric acts as a cooling medium and some of the melted portion gets re-solidified on the electrodes (both workpiece and tool). It is reported [7] that the average thickness of the recast layer is of the order of 70 to 100 μm. Figure 1.3 shows the re-solidified layer formed during EDM and WEDM.

1.4 PULSE GENERATORS

Even though many pulse generators are available, transistor-type and relaxation-type (RC type) generators are mostly used in EDM machines at present. Hence, more attention is given to these pulse generators. The schematic representation of both transistor-type and RC-type generators are depicted in Figures 1.4 and 1.5.

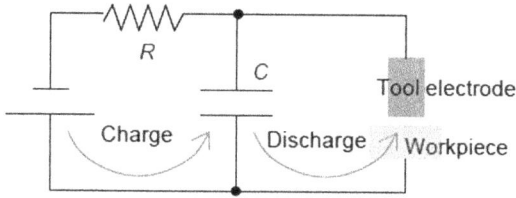

FIGURE 1.4   RC type circuit [1]. With permission from Elsevier.

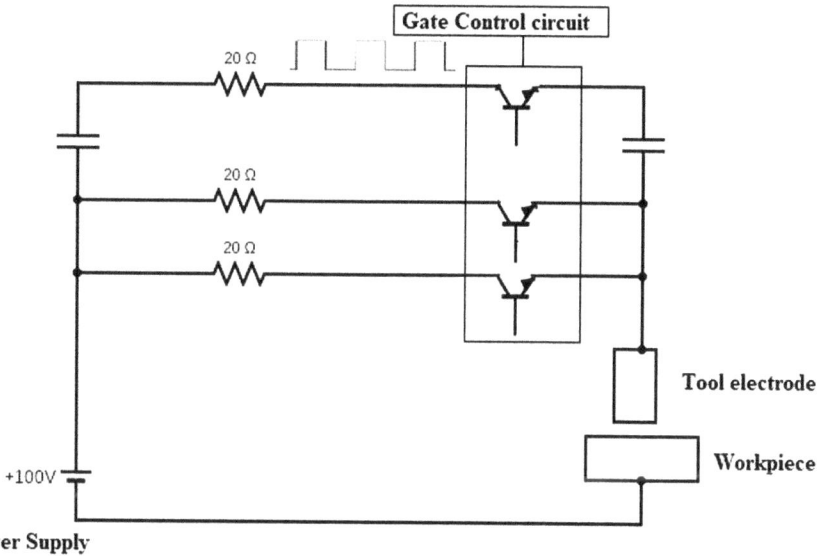

FIGURE 1.5   Transistor circuit [1]. With permission from Elsevier.

Relaxation-type pulse generators with a capacitor, as shown in Figure 1.4, were used in EDM equipment in the early stages of its invention. These were replaced by power transistors, as shown in Figure 1.5, that can handle large currents with a good response time. Even though transistor-type pulse generators are used in conventional EDM, relaxation-type pulse generators are still being used in finishing and micromachining because of their ability to obtain short pulse duration with constant energy, which is very difficult to attain in transistor pulse generators [8]. The relaxation circuit is composed of variable resistors and capacitors, connected to the tool electrode and work piece, as shown in Figure 1.4. The capacitor gets charged by the DC power supply through a variable resistor. Later, it is discharged into the interelectrode gap. The voltage across the IEG keeps on increasing from the time when the capacitor starts to get a charge from the DC power source, and it tries to become equal to the open-circuit voltage 'V$_0$' if allowed. However, discharge of the capacitor occurs once the voltage across the IEG reaches a value equal to the discharge voltage 'V$_d$' and the dielectric medium and IEG are selected to achieve this. The maximum

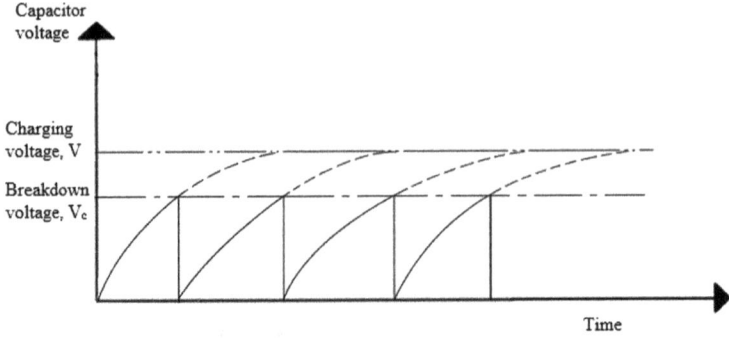

FIGURE 1.6 Variation of voltage against time in RC circuit [11]. With permission from Elsevier.

power delivered to the IEG is at $V_d = 0.72 V_0$ appriximately. The discharge energy per spark is given by,

$$E_d = \frac{1}{2}CV_d^2 \tag{1.1}$$

where E_d is the discharge energy per spark, C is the capacitance and V_d is the discharge voltage [6]. In transistor type generators, the discharge energy per unit time can be calculated as follows:

$$E_d = V_p I_p T_{ON} \frac{1}{T_{ON} + T_{OFF}} \tag{1.2}$$

where, $V_p, I_p, T_{ON}, and\, T_{OFF}$ are the voltage of the pulse, current of the pulse, pulse ON time, and pulse OFF time, respectively [9]. A comparison of both relaxation circuit-based resistance capacitance and transistor-based pulse generators is reported while machining tungsten carbide with μEDM. It is established that RC-type circuits are more suitable for the reduction in the discharge energy in IEG, thereby achieving a more stable removal process for μEDM applications [10]. Moreover, the construction and working of RC-type pulse generators is simple. During the process of machining, the major portion of the machining time is used in charging of the capacitor as shown in Figure 1.6. Figure 1.6 shows the variation of capacitor voltage with time in an RC circuit.

1.5 EDM PROCESS PARAMETERS

The electrical discharge machining process has a large number of process parameters that significantly influence its machining performance. These can be divided into two broad categories: electrical and non-electrical parameters, as shown in Figure 1.7.

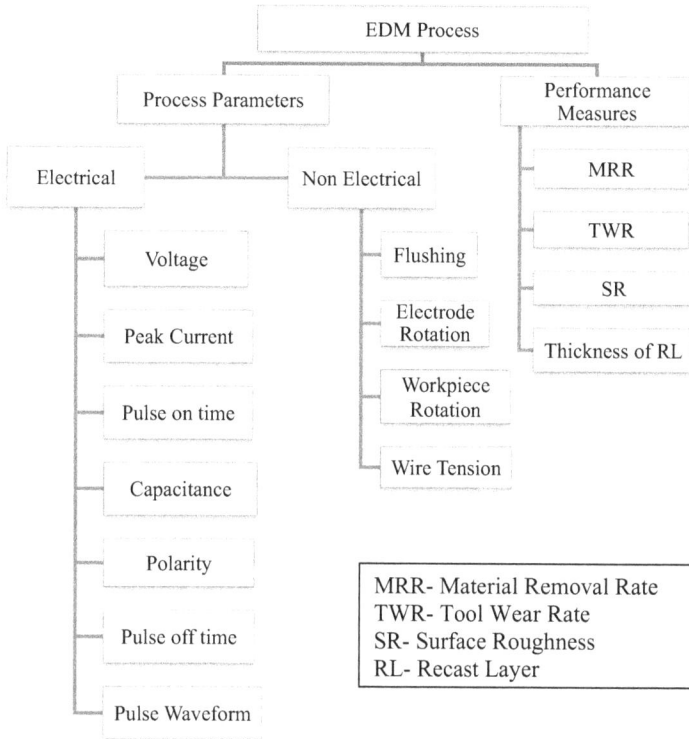

FIGURE 1.7 Electrical and non-electrical parameters in EDM.

1.5.1 ELECTRICAL PARAMETERS

Voltage: In EDM operation, generally, there are two types of voltages specified: open-circuit voltage and discharge voltage (or working voltage). Open circuit voltage is the voltage from the power supply to the electrodes or the voltage measured across the electrodes before actual machining or sparking starts. In contrast, the discharge voltage is the level of the voltage in the IEG during sparking. It is lesser than that of the open-circuit voltage by 35–38 %. The discharge voltage (or machining voltage) is controlled by the servo voltage set in the servo control unit that is available in most commercial EDM machines. During the actual machining, the tool is operated along with the feed motion, and it moves downwards and stops at a point very close to the workpiece. The downward movement of the tool increases the voltage between the tool and the workpiece. Once the voltage between the electrodes reaches a preset servo voltage, the downward movement of the tool stops, and a series of sparks take place. These sparks remove the material, thereby increasing the IEG and voltage between the tool and workpiece. The servo control feeds the tool downwards to maintain the preset servo voltage and the sparks continue. During sparking, ionization or plasma channel formation in the IEG takes place. As a result, current starts to flow, thereby the voltage drops and stabilizes the IEG [12]. On this basis, the servo voltage

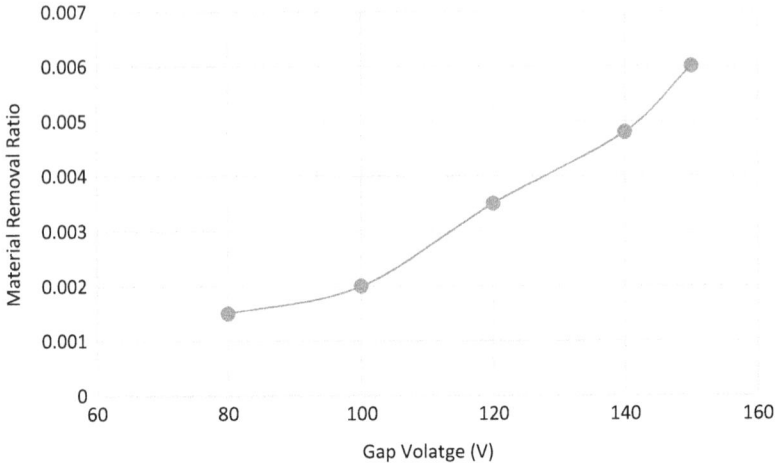

FIGURE 1.8 Influence of voltage on Material removal ratio while machining tungsten carbide [14].

has a direct liner relationship to the inter-electrode gap. This is that the larger the servo voltage, the higher the IEG [13].

Figure 1.8 shows the effect of the gap voltage on the amount of material removed while machining cemented carbide and it is found that the gap voltage raises the material removal rate (MRR). This is due to the fact that a rise in gap voltage, in turn, increases IEG, thereby effective removal of debris or flushing of the IEG takes place. In addition, the increase in the voltage raises the discharge energy, thereby increasing the MRR and R_a.

Peak Current: In the transistor-based circuit, the amount of power delivered in the IEG mainly depends on the amount of current supplied. Hence, it is important to discuss its influence. During the discharge time, current increases to reach a maximum value normally called the peak current. The level of peak current is generally decided by the area that needs to be machined. The increase in peak current enhances the amount of material removed from the workpiece. The amount of discharge energy directly depends on the current in the transistor-based machines, as depicted in Figure 1.9. Obviously, there is an increase in the MRR due to an upsurge in the current whereas it reduces surface finish (in other words, it increases the surface roughness), as evident from Figure 1.10. During the roughing operation, a high current level is used to enhance the MRR (Figure 1.11), causing rougher surface finish and an increase in the tool wear rate. Therefore, it is necessary to carefully decide on the current settings based on the particular applications and process response value.

Pulse Duration: is the amount of time during which the actual discharge occurs in the IEG and is measured in microseconds. Generally, it is also referred to as the pulse on time and is expressed in microseconds. The discharge energy directly depends upon the number of sparks and their duration. Hence, the pulse on time is important for

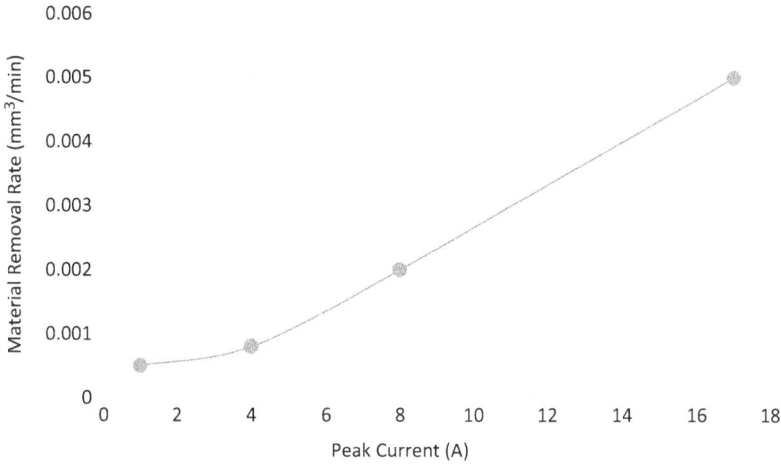

FIGURE 1.9 Influence of current during μEDM of WC-Co [14].

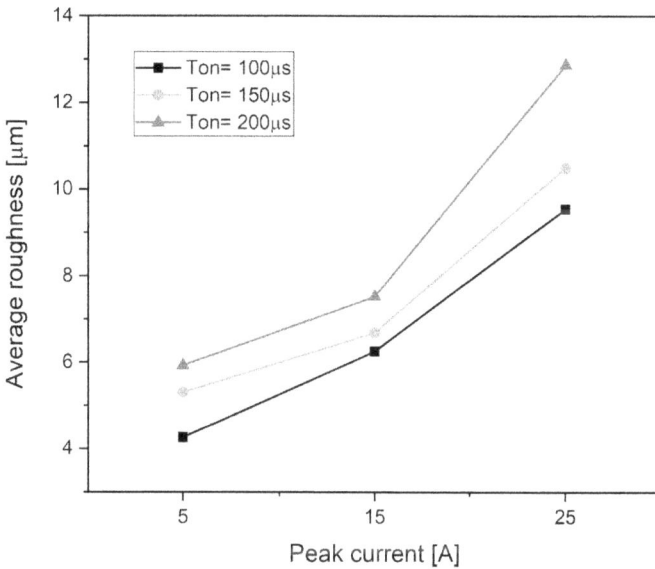

FIGURE 1.10 Influence of current and pulse ON time on the R_a during EDM of Ti6Al4V.

EDM / μEDM applications because the material removal occurs during the pulse on-time (in other words, whilst the sparks are occurring). The discharge energy supplied is higher for a longer pulse on time, therefore more workpiece material gets melted away and the size of the crater (both the diameter and depth) formed is broader and deeper as shown in Figures 1.12 and 1.13. The Ra value increases with an increase in the pulse on time as shown in Figure 1.14. Moreover, the increase in discharge

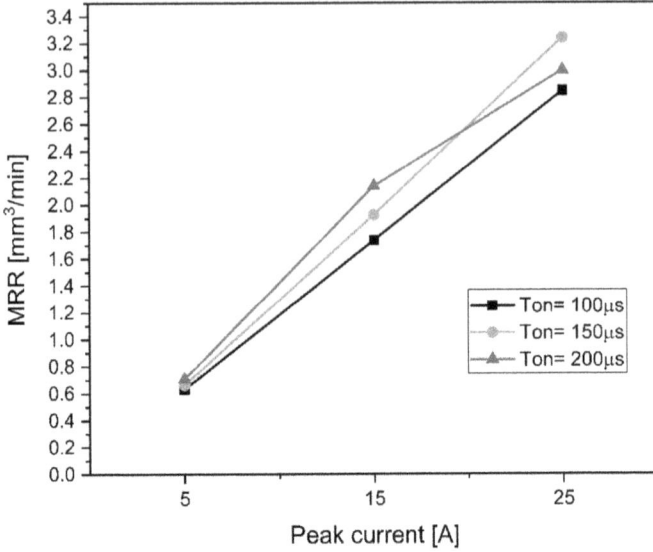

FIGURE 1.11   Influence of current & pulse ON time on material removal rate while machining Ti6Al4V with copper tool electrode.

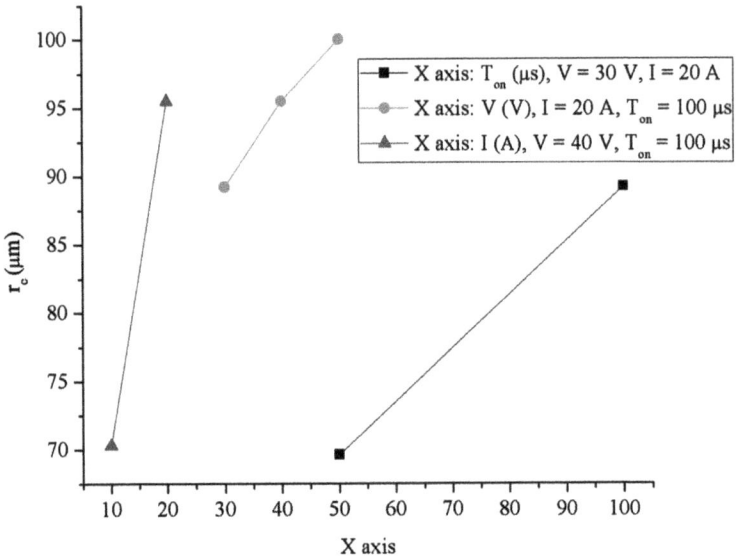

FIGURE 1.12   Effect of single crater radius (r_c) concerning Current, Voltage, and Pulse ON time [15], [16]. With permission from Springer Nature.

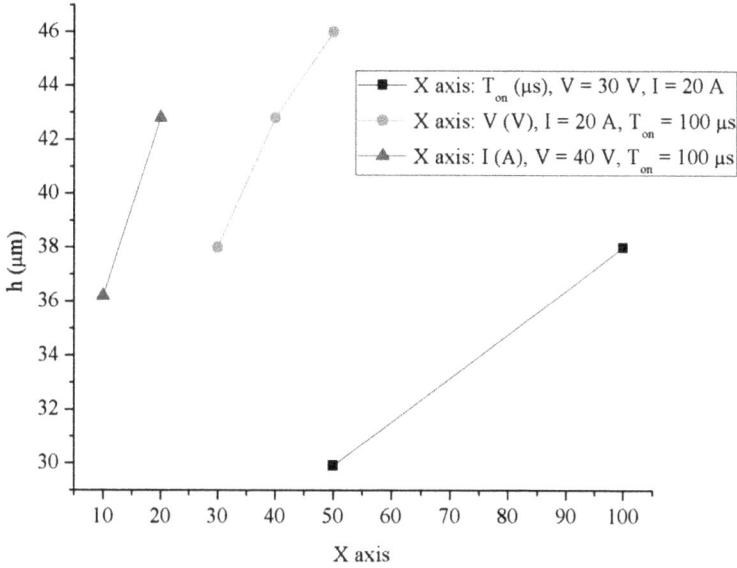

FIGURE 1.13 Effect of single crater depth (r_c) concerning Current, Voltage, and Pulse ON time [15], [16].

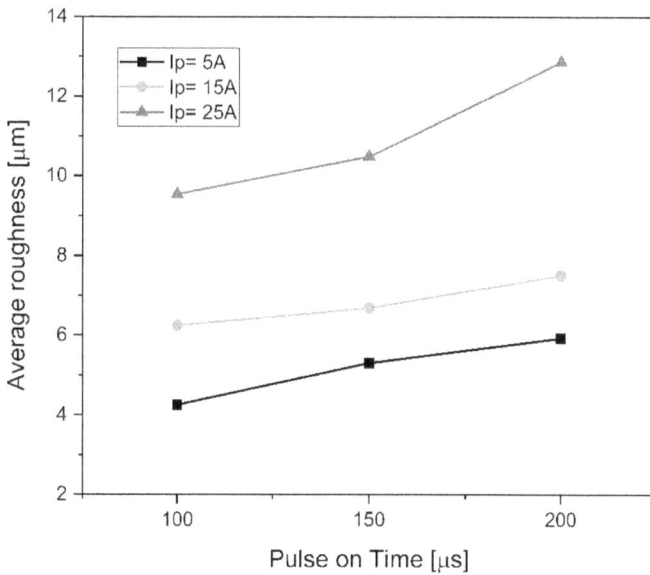

FIGURE 1.14 Influence of pulse ON time on R_a while machining Ti6Al4V with copper tool electrode.

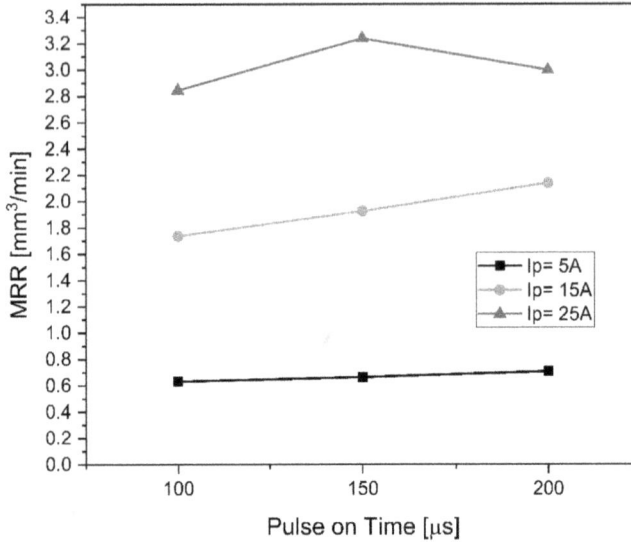

FIGURE 1.15   Influence of pulse ON time on material removal rate while machining Ti6Al4V with copper tool electrode.

energy and time duration increases the amount of heat dissipated into the workpiece, resulting in a greater MRR as shown in Figure 1.15.

Capacitance: is an essential parameter in the resistance-capacitance-based relaxation circuit (Figure 1.4). In the RC circuit, the value of the capacitance affects the amount of discharge energy in the IEG. The electric charge becomes stored in the capacitor during the charging cycle and gets discharged into the IEG during the discharge cycle. Hence, the magnitude of the MRR increases with the surge in capacitance value and it gives a poorer surface finish. Moreover, the pulse on time and off time, in other words, the duration of spark on and spark off time, respectively, are critical parameters based on the capacitance value. Charging time (spark off time) is comparatively larger than the sparking time (that is, the spark on time), and the discharge time is found to be approximately 10% of the charging time.

Polarity: Another important electrical parameter in EDM is electrode polarity. In the EDM process, the employment of a pulsed DC power supply and its controls are essential to achieve better process effectiveness. During material removal, the electrodes can be connected as anode or cathode. To enhance the MRR during the process, it is advisable to keep the workpiece as an anode. The maximum heat is generated in the anode as explained in Section 1.3, thereby causing the maximum material removal through melting and vaporization.

Lesser material removal takes place at the cathode surface. Hence, the tool wear rate can be minimized by keeping the tool electrode as the cathode. In some studies, more material was removed while keeping the workpiece as the cathode and tool as the anode even when tool and workpiece were made of the same material. In contrast, for the minimization of material removal from the workpiece and maximizing electric discharge deposition, it is advisable to keep the tool electrode as the anode and the workpiece as the cathode. The amount of material removed from the tool can be enhanced, thereby enhancing the deposition on the workpiece.

Pulse Off Time: This can be expressed as the duration, measured in microseconds, of time when sparking is not taking place due to the nature of the pulsed power supply. The increase in pulse off time increases the MRR to an optimum level. Thereafter the MRR decreases with a rise in machining time. The initial rise in off time enhances the flushing efficiency of the dielectrics in the IEG, thereby maximizing effective sparking by avoiding secondary sparking and arcing. A further increase in the off time reduces the MRR.

1.5.2 NON-ELECTRICAL PARAMETERS

In addition to the electrical parameters that directly influence the discharge energy, there is another category of process parameter called non-electrical parameters. These process parameters don't directly influence the discharge energy into the IEG, but they can significantly influence the process performances. As explained in Figure 1.7, flushing conditions, electrode rotation speed, workpiece rotation, and wire tension in WEDM, and the like, are some of these parameters. Flushing conditions have a substantial effect on machining performance, especially on the MRR and surface integrity of the machined surface. Generally, two types of flushing conditions: the submerged type and the jet type dielectric supply, are provided. In the submerged type, the IEG is fully submerged inside the dielectric fluid, and there is no forced circulation of the dielectric, as shown in Figure 1.1. This adversely affects the effective removal of debris from the IEG and leads to poor machining performance. On the other hand, a nozzle-based pressurized dielectric supply system, as shown in Figure 1.16 can result in washing away debris more effectively from the IEG to increase machining performance and reduce the deposition of re-solidified material.

The second most important non-electrical parameter is the electrode rotation speed (ERS). It is the speed at which the tool electrode is rotating during the EDM process. It is irrelevant during the die-sinking process. However, it significantly influences the other variants of EDM such as EDM drilling, milling, and the like. The ERS can produce a centrifugal force on the dielectric present in the IEG, which enhances the effective removal or washing away of the debris present in the IEG thereby improving process efficiency in terms of the MRR and surface integrity.

1.6   SURFACE INTEGRITY OF THE MACHINED SURFACE

Surface integrity of the machined surface refers to the surface topography and it encompasses the hardness of the surface, changes in surface composition, formation of the deposited layer, microhardness, residual stress formation, and the like. Electric discharge machining has the inherent ability to modify the machined surface with a redeposited layer. The surface roughness value of machined surface ranges from 2 to 5 μm, whereas it can be controlled in the range of 0.1 to 3 μm. The comparison of the surface roughness parameters (R_a and S_a) on electric discharge modified and wire

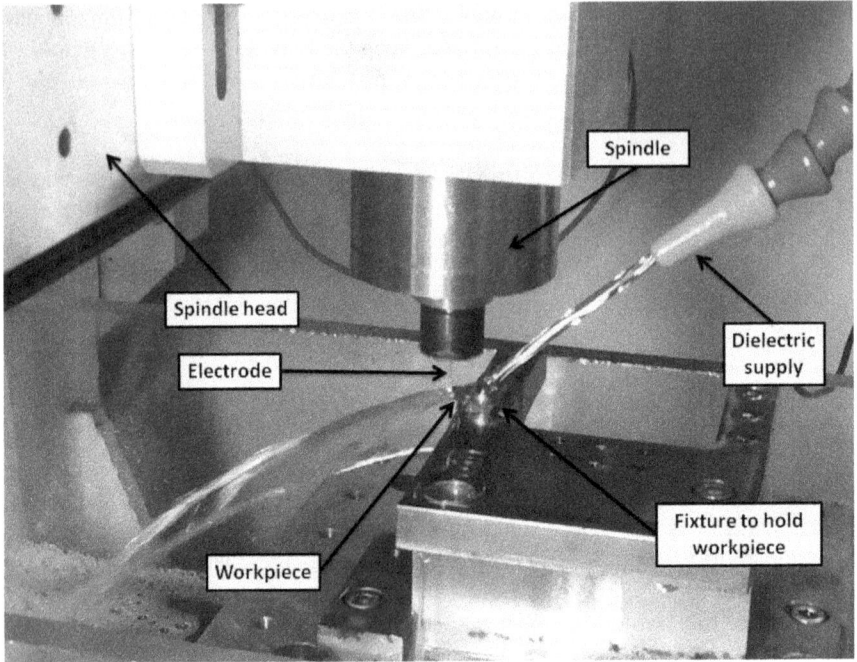

FIGURE 1.16   Jet dielectric supply system [8].

electric discharge modified Ti6Al4V is shown in Figure 1.17, and it is evident that R_a and S_a are much higher in electric discharge modified surfaces compared to other surfaces and can be optimized. Hence, surfaces produced through the EDM or micro EDM processes can be adopted for many functional applications, including the manufacturing of bio-implants, associated with accelerated bone and tissue growth. The studies recommend that the EDM-treated Ti6Al4V is a suitable material for the production of bone implants due to its adhesion and good integration with surrounding bone tissue [17]. The nature of the surface produced through electric discharge surface modification is comprised of several micro-pores, which in turn enhances adhesion and cell growth. The modified layer is non-uniform in its thickness, and it varies from 36 to 117 μm on an EDM-modified Ti6Al4V alloy [18]. The surface microhardness of the modified re-solidified layer is found to be increased more than two-fold [18]. The modified layer comprises of two distinct regions, the re-solidified layer and the heat-affected layer, apart from the base material. The variation of the hardness in the electric discharge modified Ti6Al4V, wire electric discharge modified Ti6Al4V, and bare Ti6Al4V is shown in Figure 1.18. It can be understood that surface hardness increases substantially, after the electric discharge surface modification, thereby increasing wear resistance.

In addition, the modified layer may be affected by the migration of various elements from tool and dielectric fluid, as explained in Section 1.3. As per Figure 1.19 that the tool material, copper, and the presence of oxygen and carbon from the dielectric on the modified layer while processing the Ti6Al4V, with the copper electrode

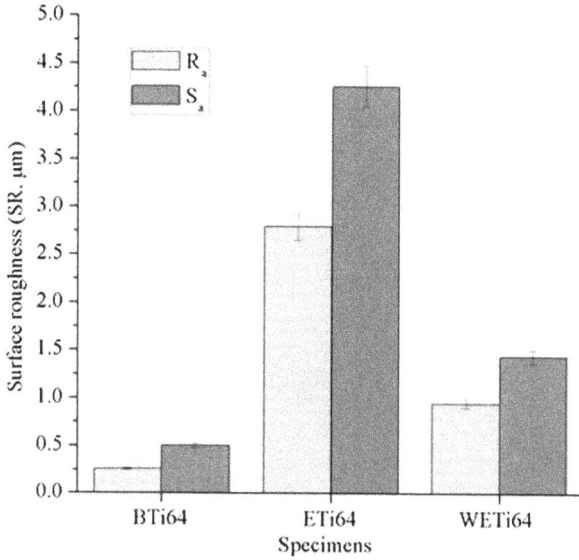

FIGURE 1.17 Variation in SR (R_a and S_a) of the electric discharge modified Ti64, and Wire electric discharge modified Ti64 [7]. With permission from Elsevier.

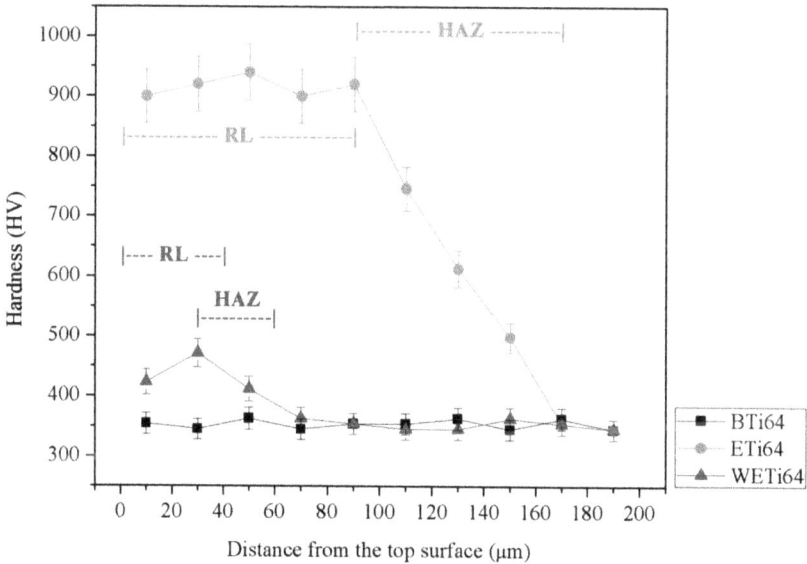

FIGURE 1.18 Difference in surfaces hardness of electric discharge modified Ti64, and Wire electric discharge modified Ti64 [7]. With permission from Elsevier.

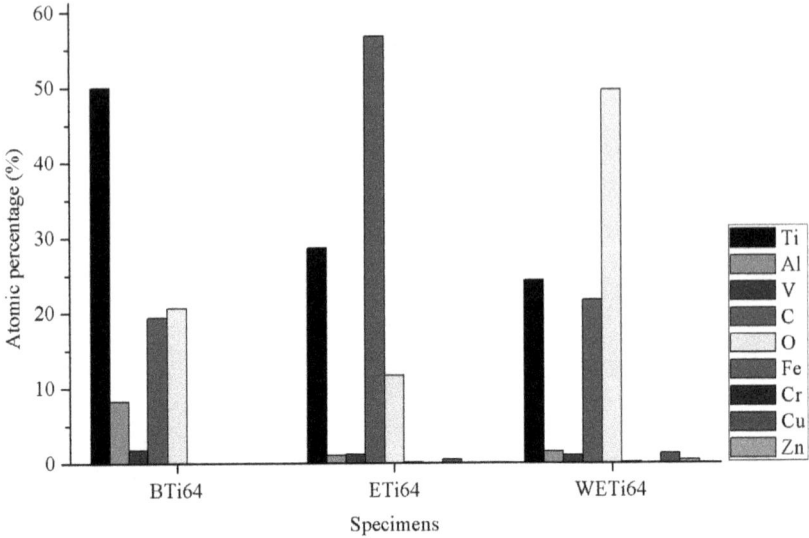

FIGURE 1.19 Difference in surface elemental composition of BTi64, ETi64, WETi64 modified with copper electrode and EDM oil as dielctric [7].

and EDM3 oil as the dielectric, have considerable impact. Through X-ray diffraction spectroscopy (XRD), it is understood that a rise in oxygen, as well as carbon content on the machined surface, leads to the formation of various carbide faces (TiC, and the like) and oxides (TiO, and so forth).

1.7 ADVANTAGES AND DISADVANTAGES

EDM is one of the extensively used modern machining or non-conventional machining processes used in industry because of several of its advantages. Firstly, it can machine or remove materials from all electrically conductive materials regardless of their mechanical properties, unlike in the conventional machining process. Any electrically conductive materials also can act as the tool electrode to complete the electric circuit. However, it is preferable to select the tool electrode based on its thermal properties like thermal conductivity, melting and vaporization temperatures, specific heat capacity, and so forth, to improve the machining performance and decrease tool wear. Secondly, a wide range of variants of the EDM process can quickly produce complex 3D shapes.

On the other hand, EDM has several disadvantages which need to be addressed carefully to improve its machining performance. Among these, tool wear is one of the critical problems that influence dimensional accuracy. Along with the workpiece removal, a portion of the tool also gets removed and wears off. Hence, the required dimensions might not be achieved. Even though some tool wear compensation models were developed, none of them is integrated with commercial machines yet. Secondly, due to the melting and vaporization and to the presence of the dielectric fluid, a small portion of the molten metal is re-deposited on the machined surface. The re-solidified layer may be composed of various elements from the tool electrode and dielectric. In addition, to these re-solidified layers, a thermally modified layer is also formed. These thermally modified and re-solidified layers have different mechanical properties than the base material, including improved hardness and wear resistance.

Thirdly, the MRR is very low compared to conventional processes and many advanced machining processes. Moreover, the processing of the electrically non-conductive material is not possible in EDM. To overcome such difficulties, researchers have tried various methods and developed multiple technologies. Among these developed technologies, electrical discharge hybrid machining processes offer an important area that will gain popularity in the future. The hybrid machining process is when two or more different energy sources are effectively used to remove the material from the workpiece. Hence, other types of energy are combined with the electric discharge to improve the process efficiency in electrical discharge hybrid machining processes. Based on these facts, the upcoming chapters are designed and arranged in such a way as to provide an insight for the reader into the available electrical discharge hybrid machining processes and their fundamental working principles along with the latest research trends in the respective areas.

1.8   CONCLUSION

In this chapter, the basic mechanism of material removal in the electrical discharge machining process is explained. The critical electrical and non-electrical process parameters on MRR, Ra, surface integrity, and material migration phenomena are explained. Moreover, the chapter highlights the electrical discharge hybrid machining process as one of the solutions to overcome EDM's several disadvantages in addition to various EDM variants.

REFERENCES

[1] M. Kunieda, B. Lauwers, K. P. Rajurkar, and B. M. Schumacher, "Advancing EDM through fundamental insight into the process," *CIRP Ann. – Manuf. Technol.*, vol. 54, no. 2, pp. 64–87, Jan. 2005, doi: 10.1016/s0007-8506(07)60020-1.

[2] B. M. Schumacher, R. Krampitz, and J. P. Kruth, "Historical phases of EDM development driven by the dual influence of 'Market Pull' and 'Science Push,'" in *Procedia CIRP*, vol. 6, pp. 5–12, Jan. 2013, doi: 10.1016/j.procir.2013.03.001.

[3] B. R. Lazarenko, "To Invert the Effect of Wear on Electric Power Contacts," The All-Union Institute for Electro Technique, Moscow/CCCP (In Russian), 1943.

[4] B. M. Schumacher, "After 60 years of EDM the discharge process remains still disputed," in *J. Mater. Process. Technol.*, vol. 149, no. 1–3, pp. 376–381, Jun. 2004, doi: 10.1016/j.jmatprotec.2003.11.060.

[5] S. Singh and A. Bhardwaj, "Review to EDM by Using Water and Powder-Mixed Dielectric Fluid," 2011, www.scirp.org/pdf/JMMCE20110200007_13006610.pdf.

[6] B. Kuriachen, A. Varghese, K. P. Somashekhar, S. Panda, and J. Mathew, "Three-dimensional numerical simulation of microelectric discharge machining of Ti-6Al-4V," *Int. J. Adv. Manuf. Technol.*, vol. 79, no. 1–4, pp. 147–160, Jul. 2015, doi: 10.1007/s00170-015-6794-y.

[7] J. T. Philip, D. Kumar, J. Mathew, and B. Kuriachen, "Tribological investigations of wear resistant layers developed through EDA and WEDA techniques on Ti6Al4V surfaces: Part I – Ambient temperature," *Wear*, vol. 458–459, p. 203409, Oct. 2020, doi: 10.1016/j.wear.2020.203409.

[8] B. Kuriachen and J. Mathew, "Experimental investigations into the effects of microelectric-discharge milling process parameters on processing Ti-6Al-4V," *Mater. Manuf. Process.*, vol. 30, no. 8, pp. 983–990, 2015, doi: 10.1080/10426914.2014.984206.

[9] S. M. Son, H. S. Lim, A. S. Kumar, and M. Rahman, "Influences of pulsed power condition on the machining properties in micro EDM," *J. Mater. Process. Technol.*, vol. 190, no. 1–3, pp. 73–76, 2007, doi: 10.1016/j.jmatprotec.2007.03.108.

[10] M. P. Jahan, Y. S. Wong, and M. Rahman, "A study on the quality micro-hole machining of tungsten carbide by micro-EDM process using transistor and RC-type pulse generator," *J. Mater. Process. Technol.*, vol. 209, no. 4, pp. 1706–1716, Feb. 2009, doi: 10.1016/j.jmatprotec.2008.04.029.

[11] S. Kumar, R. Singh, T. P. Singh, and B. L. Sethi, "Surface modification by electrical discharge machining: A review," *J. Mater. Process. Technol.*, vol. 209, no. 8. Elsevier, pp. 3675–3687, Apr. 21, 2009, doi: 10.1016/j.jmatprotec.2008.09.032.

[12] "Gap Voltage – an overview | ScienceDirect Topics." www.sciencedirect.com/topics/engineering/gap-voltage (accessed Apr. 25, 2021).

[13] M. Zhou, X. Mu, L. He, and Q. Ye, "Improving EDM performance by adapting gap servo-voltage to machining state," *J. Manuf. Process.*, vol. 37, pp. 101–113, Jan. 2019, doi: 10.1016/j.jmapro.2018.11.013.

[14] M. P. Jahan, Y. S. Wong, and M. Rahman, "Experimental investigations into the influence of major operating parameters during micro-electro discharge drilling of cemented carbide," *Mach. Sci. Technol.*, vol. 16, no. 1, pp. 131–156, Jan. 2012, doi: 10.1080/10910344.2012.648575.

[15] S. N. Joshi and S. S. Pande, "Development of an intelligent process model for EDM," doi: 10.1007/s00170-009-1972-4.

[16] J. T. Philip, J. Mathew, and Basil Kuriachen, "Numerical simulation of the effect of crater morphology for the prediction of surface roughness on electrical discharge textured Ti6Al4V," *J. Brazilian Soc. Mech. Sci. Eng.*, vol. 42, no. 3, p. 248, 2020, doi: 10.1007/s40430-020-02321-6.

[17] P. Harcuba, L. Bačáková, J. Stráský, M. Bačáková, K. Novotná, and M. Janeček, "Surface treatment by electric discharge machining of Ti – 6Al – 4V alloy for potential application in orthopaedics," *Journal of the Mechanical Behavior of Biomedical Materials*, vol. 7, pp. 96–105, 2012, doi: 10.1016/j.jmbbm.2011.07.001.

[18] J. T. Philip, D. Kumar, J. Mathew, and B. Kuriachen, "Experimental investigations on the tribological performance of electric discharge alloyed Ti–6Al–4V at 200–600 °C," *J. Tribol.*, vol. 142, no. 6, 2020, doi: 10.1115/1.4046016.

2 Dry and Near-Dry Electrical Discharge Machining

S. Jithin[1] and Suhas S. Joshi[2]
[1]Indian Institute of Technology Mumbai, India
[2]Indian Institute of Technology Bombay, India

CONTENTS

DOI: 10.1201/9781003202301-2

2.1   INTRODUCTION

Conventional die-sinking EDM, though it offers many advantages, is not envir-
onmentally friendly. It produces harmful gases and waste. Researchers, therefore,
have come up with many solutions for this issue, one of the prominent ones among
them being the use of a gaseous dielectric or a liquid mixed gaseous dielectric.
Here, the quantity of liquid dielectric used is zero to very little. Consequently,
fewer harmful process wastes are generated. The EDM variant using a gas dielectric is widely known as dry EDM, whereas the one using a liquid-gas mixture is
known as near-dry EDM. This chapter aims to elucidate working mechanisms,
influencing parameters, and the advantages and disadvantages of dry and near-
dry EDM.

2.2   PRINCIPLES OF WET EDM, DRY EDM, AND NEAR-DRY EDM

2.2.1   Principle of Wet EDM

The working principle of the conventional wet EDM process is represented in Figure 2.1. The workpiece and the tool are connected to a DC power supply so that they act as electrodes. These electrodes are placed with a small gap between them, called the spark gap. In the case of wet EDM, the spark gap is filled with a liquid dielectric. The dielectric initially behaves as an insulating material and prevents current flow. A servo control mechanism moves the tool electrode towards the work. When the work and tool electrodes come sufficiently close, the electric potential between them crosses a critical value, and the dielectric in that region breaks down to form a plasma comprised of electrons and positively charged ions. This region becomes highly conductive, and a large current passes through the plasma channel in a small amount of time, known as the pulse duration (t_{on}). Thus, the ions and electrons are accelerated, and they impinge upon the cathode and anode, respectively, resulting in local high temperatures. These high temperatures cause a small amount of material removal from both workpiece and tool surfaces due to melting and evaporation. The eroded material is flushed away from the spark gap by the flowing liquid dielectric medium. During the EDM process, a portion of molten material from both the anode and the cathode surfaces resolidifies to form a recast or white layer (see Figure 2.1). Below the recast layer, the high temperatures result in a heat-affected zone (HAZ). The cathodic craters are usually larger compared with anodic craters, at pulse durations greater than 0.5 µs [1]. This is due to the higher erosion rate of the cathode compared with that of the anode during the EDM process at pulse durations higher than 0.5 µs [1] (see Figure 2.2). Moreover, a number of important research

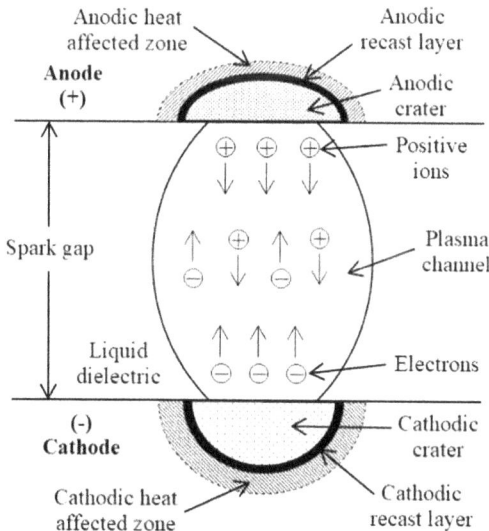

FIGURE 2.1   Working of conventional wet EDM [14]. With permission from Elsevier.

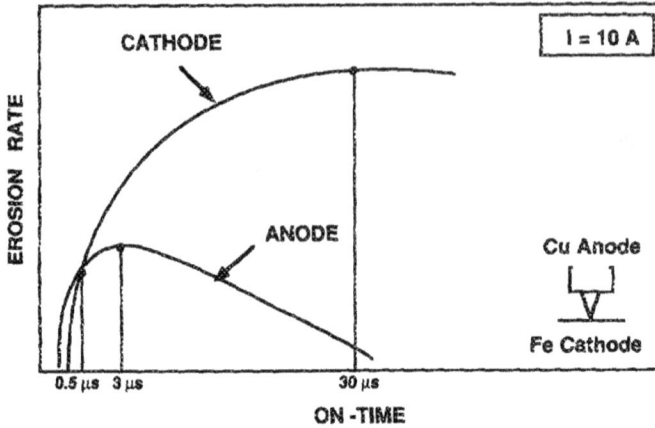

FIGURE 2.2 Difference between anode and cathode erosion rates with different pulse on-times [1]. With permission from AIP.

works, including simulation [1–7] and experimentation [8–10], consider the work-piece as the cathode for normal polarity. Major review works on conventional die-sinking EDM by Kunieda et al. [11] and Schumacher [12] also consider the same polarity for the workpiece. However, a good number of researchers have also assumed the workpiece to be the anode for normal polarity. Recent research by Holsten et al. [13] states that although conventionally, the workpiece is selected as the cathode for a higher material removal rate (MRR), some workpiece-tool material combinations show higher MRR for an anodic workpiece. Thus, researchers should select polarity, based on a number of factors including the variant of EDM, electric parameter levels, and materials used. In this chapter, we assume that the workpiece is the cathode as the normal polarity for conventional wet EDM.

A schematic for a wet EDM setup is given in Figure 2.3 (a).

It is understood that there is a dedicated plasma flushing system required to ensure the continuous working of the wet EDM. This flushing system makes the whole machine very bulky. An experimental setup of conventional wet EDM for texturing experiments is shown in Figure 2.3 (b).

This whole setup will be submerged in dielectric during the machining process.

2.2.2 Principle of Dry EDM

Dry electrical discharge machining uses a gaseous dielectric in the place of the liquid dielectric used by conventional wet EDM. Figure 2.4 represents the working principle of dry EDM. Contrary to conventional wet EDM, a majority of research works adopt a negative polarity of tool electrode for dry EDM. This is because of the low tool wear at this polarity [17]. When a plasma channel is formed in a gas dielectric, the pressure exerted by the surrounding gas on the plasma channel is much less in comparison to the pressure in a liquid dielectric. This leads to an uncontrolled lateral expansion of the plasma region and thereby, a larger discharge channel. The use of a shield around

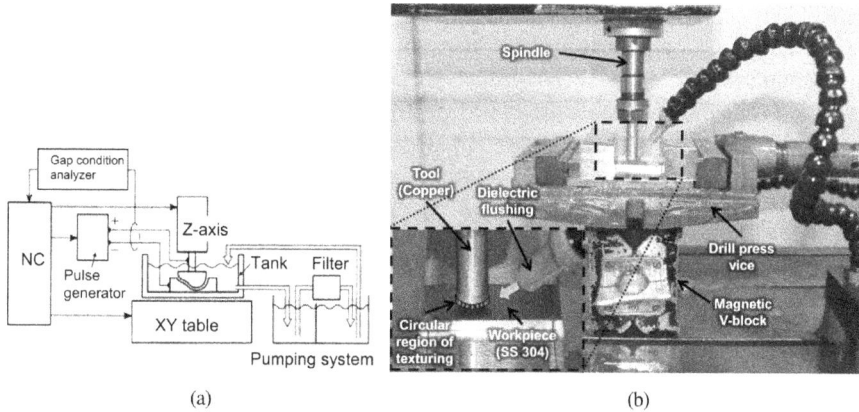

FIGURE 2.3 Wet EDM (a) schematic [11] and (b) experimental setup for texturing [15]. With permission from Elsevier.

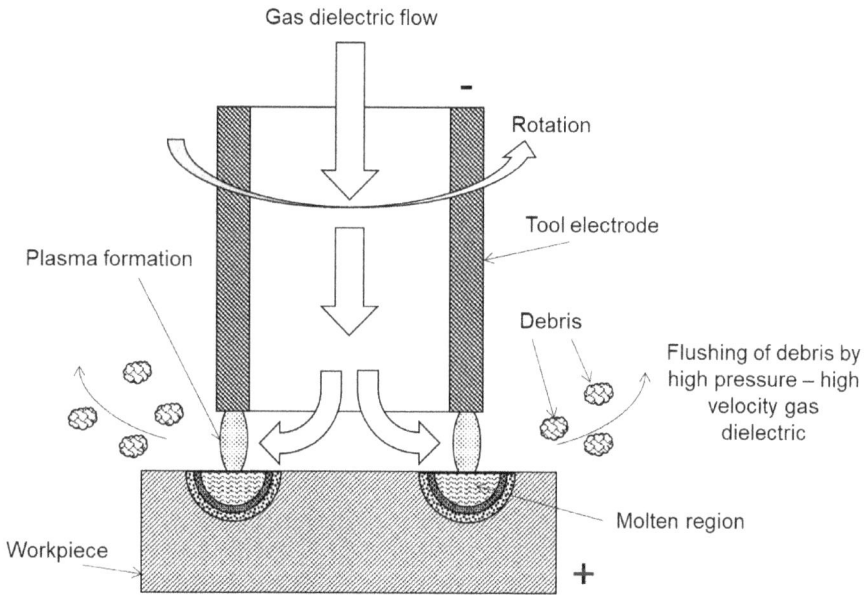

FIGURE 2.4 Working mechanism of dry EDM.

the sparking region is identified as one of the solutions [16]. Sparks occur in the gas medium leading to the melting of workpiece material in the form of resolidified debris. The high-velocity gas flow aids in the molten material removal. However, the molten material flushing is not as effective as that in wet EDM due to the gaseous medium being less dense. This leads to significant molten material deposition on the tool electrode. Suitable operating parameters need to be selected so as not to hinder the EDM process.

Kunieda et al. [17], in one of the first works on dry EDM, identified that the biggest performance advantage of dry EDM over conventional wet EDM is its relatively negligible tool electrode wear. They also reported that the dry variant of EDM is also capable of machining 3D shapes with reasonable accuracy. An important performance indicator, material removal rate (MRR), was considerably less with a dry EDM process with air dielectric than that with wet EDM. However, when oxygen was used as the gas dielectric material, a higher MRR was obtained for dry EDM than that for wet EDM. In their previous work [18] on single discharges performed in a dielectric liquid and air, they found that molten material sticks to the electrode surface and accumulates in the discharge gap in the case of an air dielectric. The debris accumulation is undesirable as it leads to the short-circuiting and hindrance of machining. This showed that the debris deposition problem needs to be solved for effective EDM with gas dielectric. This motivated Kunieda et al. [17] to use high-velocity gas flow in their dry EDM setup. Shen et al. [19] investigated high-speed dry EDM milling. They developed 'quasi-explosion' and 'intake method' to significantly improve the MRR. Lee and Oh [20] developed a 3-axis tabletop dry EDM machine and performed experiments under no air, air blowing, and air suction conditions. They found air suction to be more suitable for machining aluminium by giving a higher MRR and a better accuracy.

In dry EDM, due to the low viscosity of gas dielectric, the lateral expansion of spark plasma is more than that in the case of a liquid dielectric. Govindan and Joshi [16] proposed using a shield to cover the sparking region of dry EDM drilling to reduce the lateral expansion of spark plasma (see Figure 2.5). They found that a higher radial clearance of the shield with the hollow tool electrode results in lesser machining oversize. However, this comes at the cost of a higher amount of debris

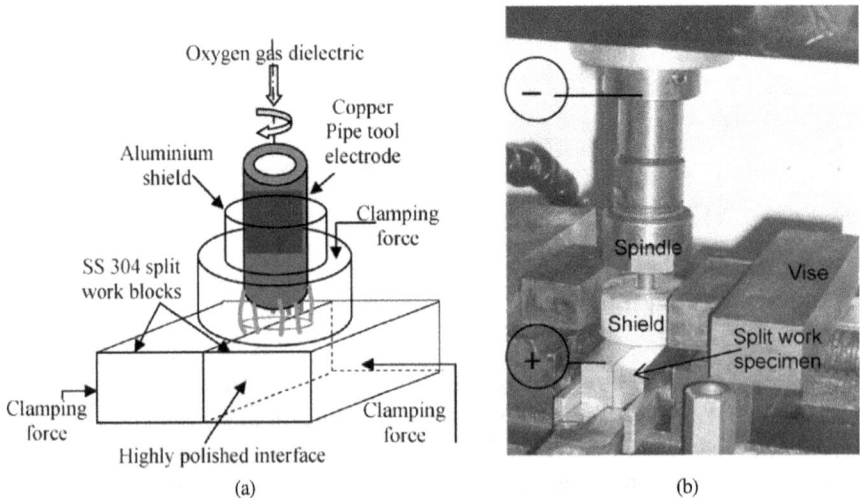

FIGURE 2.5 Dry EDM with a shield to reduce the lateral expansion of spark plasma (a) schematic and (b) photograph [16]. With permission from Elsevier.

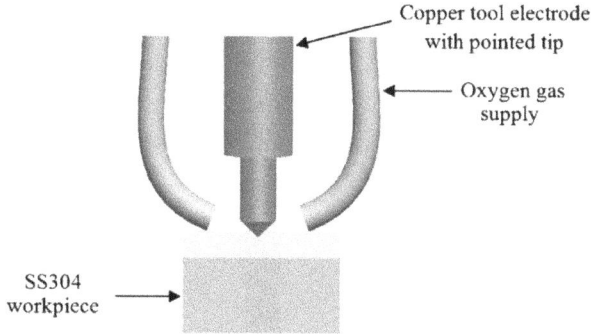

FIGURE 2.6 Single-spark experiment setup [21]. With permission from Elsevier.

deposition on the tool. Their EDAX analysis showed the deposition of shield material (aluminium) used on the machined work (SS304) surface. In another work [21], they performed single spark experiments in dry EDM conditions (see Figure 2.6). They reported that at low discharge energies, the single-spark crater dimensions of dry EDM are larger than those obtained with conventional wet EDM. This leads to a higher MRR in dry EDM than that in wet EDM at low discharge energies. Joshi et al. [22] investigated a hybrid dry EDM process with the presence of a pulsating magnetic field. The magnetic field was found to enhance the performance of the dry EDM process by increasing the electron movement, the degree of ionization in plasma, and constraining the spread of plasma in the gaseous medium. The use of the magnetic field also resulted in an improvement in MRR and a reduction in TWR. Govindan and Joshi [16] analyzed micro-cracks generated during dry EDM and studied the influence of different parameters on the average crack length. They found that the average crack length and the number of micro-cracks are less on dry EDM surfaces than those on wet EDM surfaces. Liqing and Yingjie [23] investigated dry EDM performance improvement using two methods: an oxygen gas dielectric and a cryogenically cooled workpiece. They reported that the use of an oxygen dielectric improved MRR by more than 200 % compared to other non-oxygen gas dielectrics. The use of cryogenically cooled dielectric resulted in an improvement in MRR of 30–50 %. It also resulted in a better surface finish with a reduction (1–10 %) in surface roughness (R_a).

2.2.3 PRINCIPLE OF NEAR-DRY EDM

Though dry EDM gives an advantage of operation without a liquid dielectric, over wet EDM, it has a very low material removal rate (*MRR*) and results in high debris deposition, thus making it less desirable. In such a scenario, the use of a liquid and gas mixture as dielectric gives the advantages of comparatively high MRR and significantly lower debris deposition while requiring only a very small quantity of liquid dielectric. Such a variant of EDM using a liquid and gas mixture as the dielectric is known as near-dry electrical discharge machining. The way in which near-dry EDM works is represented in Figure 2.7. The mist supplied acts as the dielectric in the near-dry EDM process. In the same way as with the cases of wet and dry EDMs, a

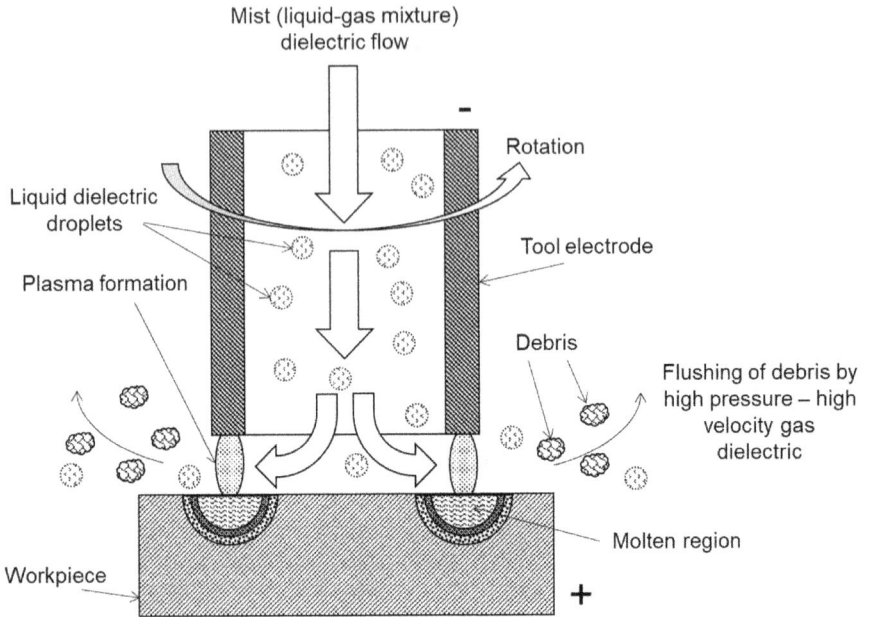

FIGURE 2.7   Working mechanism of near-dry EDM.

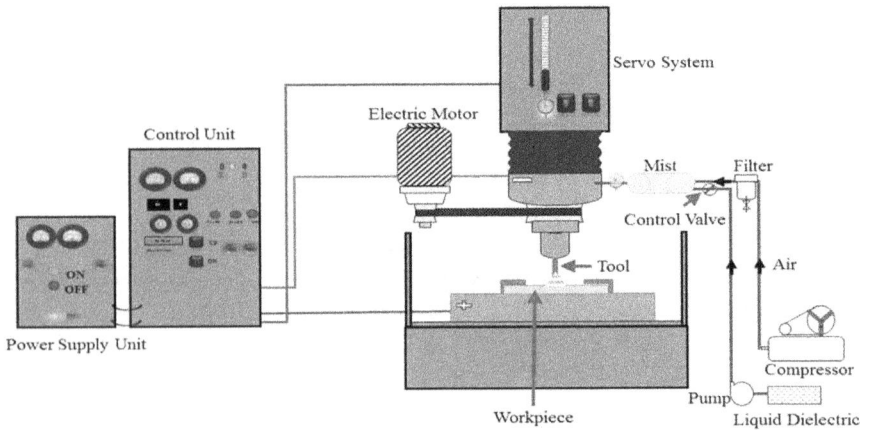

FIGURE 2.8   Schematic of near-dry EDM [24]. With permission from Taylor & Francis.

plasma is formed under suitable electrical conditions and a consequent passing of high current results in material erosion from both anode and cathode. The liquid-gas mist has better dielectric properties in comparison with gas medium and results in a higher MRR and improved debris flushing. As in the case of dry EDM, most of the research work in near-dry EDM adopts negative polarity for the tool electrode. A schematic for a near-dry EDM is shown in Figure 2.8.

The feasibility of EDM with liquid-gas mixtures was first investigated by Tanimura et al. [25]. They performed EDM with water mist as the dielectric in various gases such as air, nitrogen, and argon. Kao et al. [26] performed wire EDM and EDM drilling experiments under dry, near-dry, and wet conditions. They reported that the use of this dielectric mixture enables the adoption of liquid-gas compositions, which give optimized performance results. They used a water-air mixture as the dielectric in their work. Their experiments revealed near-dry EDM to be advantageous over dry EDM by providing a higher MRR, lower wear of tool edges, and lower deposition of debris. They also reported near-dry EDM to have certain advantages, such as more MRR at low discharge energy and lesser gap distance (which improves machining accuracy), over wet EDM. Their EDM drilling experiments revealed that holes with the lowest taper were obtained with the near-dry condition as opposed to those of dry and wet conditions (see Figure 2.9).

Shen et al. [27] investigated high speed near-dry EDM milling. They were able to achieve a higher MRR, lower R_a, and lesser overcut in comparison with those obtained with high-speed dry EDM. In other research, Dhakar et al. [28] found that near-dry EDM resulted in 97 % lower gas emissions as compared to those of wet EDM (see Figure 2.10). Thus, near-dry EDM falls into the category of green machining processes. They also reported that the near-dry EDM gave a higher MRR than that of wet EDM. Yadav et al. [24] studied the influence of different operating parameters on MRR, surface roughness, and hole overcut of rotary tool near-dry EDM (RT-ND-EDM). They reported that tool rotation is beneficial in flushing debris from the spark gap during the near-dry EDM process. Moreover, RT-ND-EDM surfaces had lesser

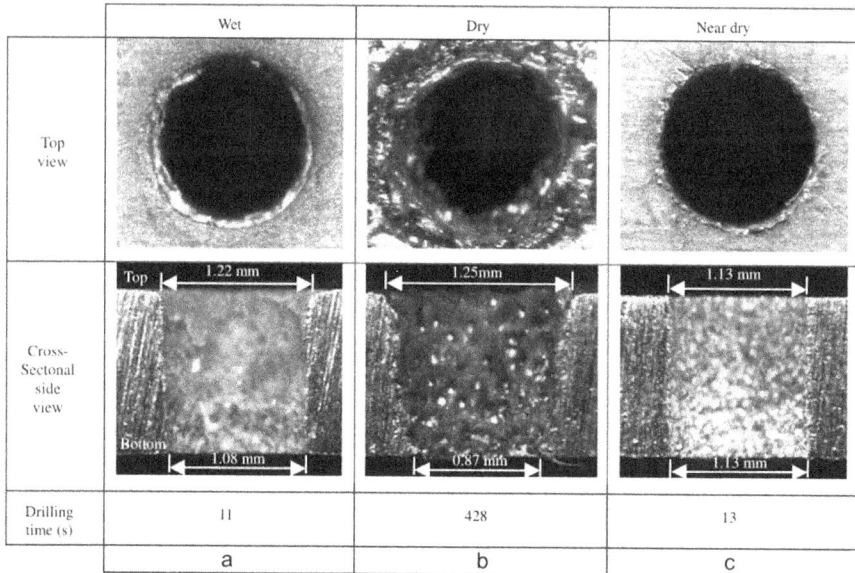

FIGURE 2.9   Comparison of wet, dry, and near-dry EDM drilled holes [26]. With permission from Elsevier.

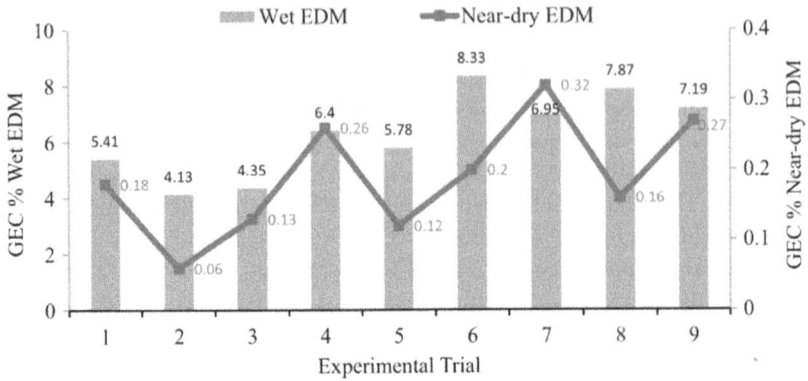

FIGURE 2.10 Comparison of gas emission concentration of wet and near-dry EDM [28]. With permission from Taylor & Francis.

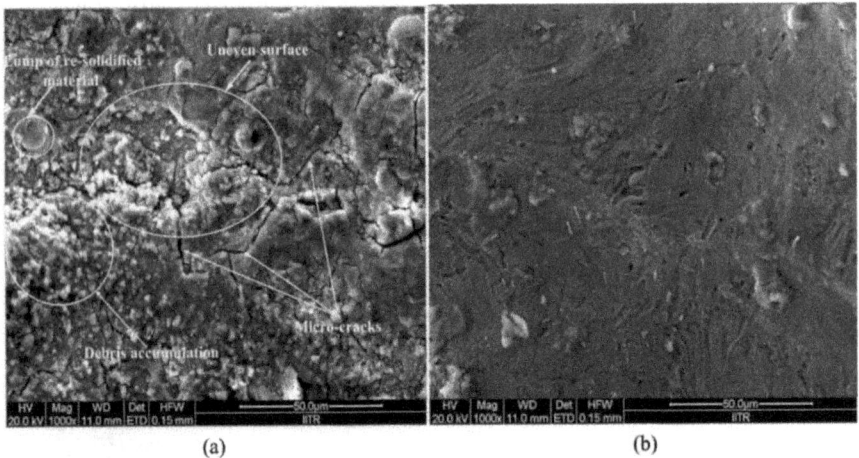

FIGURE 2.11 SEM images of ND-EDM and RT-ND-EDM samples [24]. With permission from Taylor & Francis.

surface irregularities and damage than those of near-dry EDM (ND-EDM) surfaces (see Figure 2.11).

2.3 PARAMETRIC STUDY FOR DRY AND NEAR-DRY EDM

The parametric study of dry and near-dry EDM is essential to understanding the influence of different operating parameters on machining performance. The important machining performance indicators are material removal rate (MRR), tool wear rate (TWR), and surface roughness (R_a).

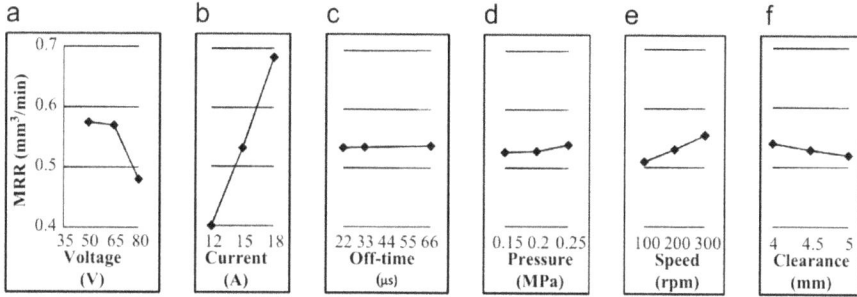

FIGURE 2.12   Main effect plots of different parameters on MRR [21]. With permission from Elsevier.

2.3.1   MATERIAL REMOVAL RATE (MRR) OF DRY EDM

Material removal rate (MRR) is one of the important performance indicators for machining processes. Higher MRR is desired for higher productivity. One of the disadvantages of the dry EDM process is its low MRR. Hence, it is of critical importance to increase the MRR of the process by different means. Therefore, a study of different parameters influencing the MRR of dry EDM is essential.

2.3.1.1   Effect of Discharge Current

Discharge current is one of the significant electrical parameters influencing the EDM process. Govindan and Joshi [21] studied the influence of different parameters on MRR of dry EDM and found that discharge current has the most significant influence among different parameters studied, see Figure 2.12 (b). This is because discharge current directly influences the amount of discharge energy and thereby, the size of the crater formed. As discharge current increases, the resultant spark crater size increases, leading to an increase in MRR.

2.3.1.2   Effect of Gap Voltage

Gap voltage is another significant parameter influencing MRR in dry EDM. An increase in gap voltage results in a decrease in MRR (see Figure 2.12 (a)).

2.3.1.3   Effect of Pulse Off-time

Pulse off-time does not show any significant influence on the MRR of EDM (see Figure 2.12 (c)). However, this parameter does play a significant role in the quasi-explosion mode of dry EDM.

2.3.1.4   Effect of Polarity

Polarity influences the EDM process by deciding the percentage of spark energy reaching the workpiece at different operating conditions. Kunieda et al. [17] studied the influence of polarity on tool wear ratio and MRR. They found that the negative polarity of the tool electrode is more desirable in dry EDM as it results in significantly lesser tool wear and

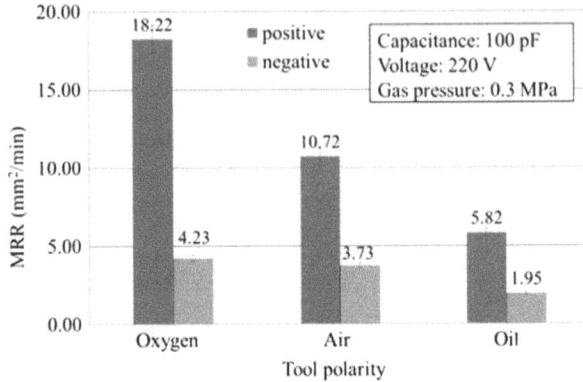

FIGURE 2.13   Effect of tool polarity on MRR of dry EDM [29]. With permission from Springer Nature.

higher material removal. However, dry EDM studies by Islam et al. [29] have shown that higher MRR is observed with a positive polarity for the tool (see Figure 2.13).

2.3.1.5   Effect of Tool Electrode Wall Thickness

Kunieda et al. [17] studied the influence of wall thickness of tool electrodes on dry EDM performance. They found that the wall thickness of the tool significantly influences dry EDM material removal as opposed to minor influence in the case of wet EDM, as shown in Figure 2.14. It is understood that above a threshold value of wall thickness, the material removal reduces significantly.

2.3.1.6   Effect of Dielectric Gas Material

Dielectric gas influences the share of spark energy transferred to the workpiece for erosion. Kunieda et al. [17] reported that the amount of material machined during dry EDM doubles when pure oxygen is used as a gas dielectric instead of air. Liqing and Yingjie [23] found that MRR was increased at least by 200 % when oxygen gas was used as dielectric as compared to other gas dielectric materials (see Figure 2.15 (a)). This is believed to be due to oxidation of molten work material leading to heat generation and resulting in further material erosion. Liqing and Yingjie [23] also reported that an increase in oxygen content in gaseous dielectrics such as compressed air, nitrogen, and argon resulted in an improvement in MRR (see Figure 2.15 (b)).

2.3.1.7   Effect of Magnetic Field

The provision of a pulsating magnetic field improves the machining performance of dry EDM in terms of MRR [22] (see Figure 2.16). This is due to the higher ionization and plasma confinement in the presence of a magnetic field, which leads to larger individual crater volume. The larger the individual crater volume, the larger will be the MRR. It was also reported that MRR increases with an increase in the intensity of the magnetic field (see Figure 2.16(c)).

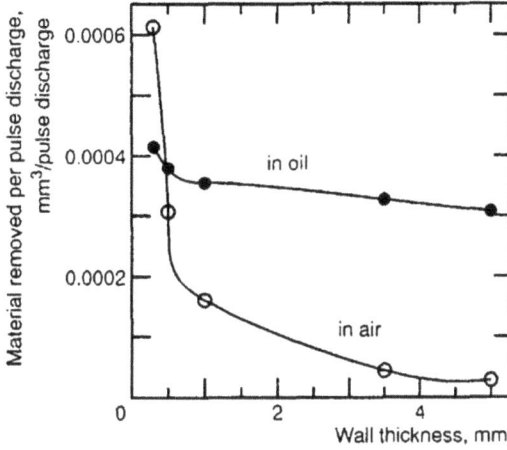

FIGURE 2.14 Variation of MRR with wall thickness of tubular electrodes [17]. With permission from Elsevier.

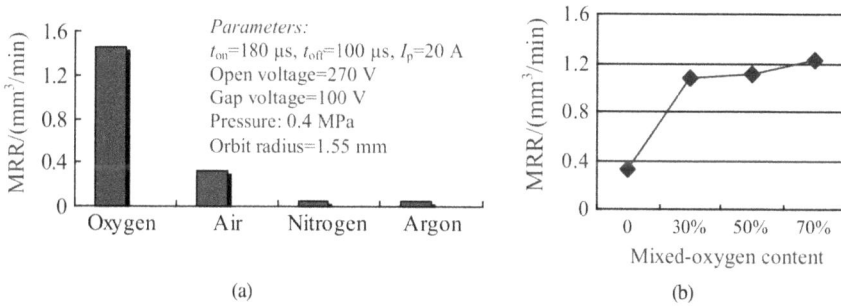

FIGURE 2.15 MRR for different gas dielectrics [23]. With permission from Elsevier.

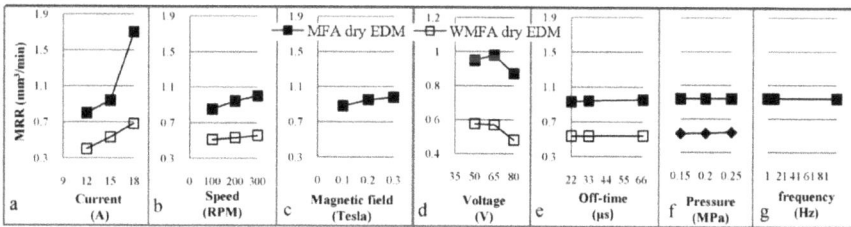

FIGURE 2.16 Comparison of MRR of magnetic field assisted (MFA) dry EDM and without magnetic field (WMFA) dry EDM [22]. With permission from Elsevier.

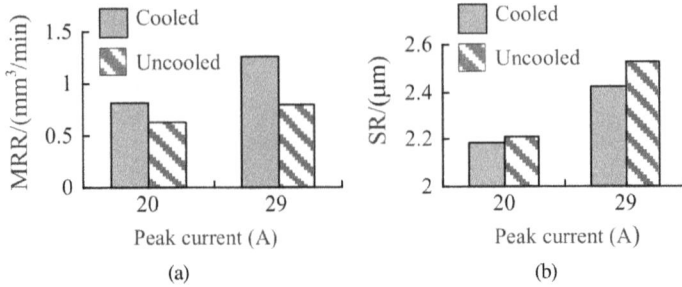

FIGURE 2.17 MRR and surface roughness of cryogenically cooled and non-cooled workpieces [23]. With permission from Elsevier.

2.3.1.8 Effect of Cryogenic Cooling of the Workpiece

Cryogenic cooling of the workpiece is another parameter that has a positive influence on the material removal rate of the surface. Liqing and Yingjie [23] reported that the MRR obtained was higher with the cryogenically cooled workpiece as compared to that with the uncooled workpiece (see Figure 2.17 (a)). Moreover, cryogenic cooling of the workpiece also improved the surface finish of the machined surfaces (see Figure 2.17 (b)).

2.3.2 MATERIAL REMOVAL RATE (MRR) OF NEAR-DRY EDM

Kao et al. [26] found that the MRR of near-dry EDM is higher than that of dry EDM but slightly lesser than that of wet EDM. Researchers have discussed the influence of various factors in the near-dry EDM process as follows.

2.3.2.1 Effect of Discharge Current

The variation of MRR with discharge current was investigated by Kao et al. [26]. They reported that in the experimental range of discharge current, MRR showed an initial increase, and then a decrease (see Figure 2.18). They reported that this trend is owing to debris removal problems at high discharge power.

2.3.2.2 Effect of Flow Rate

As per the investigations of Kao et al. [26], MRR shows a continuous increase with an increase in the flow rate of the dielectric at all discharge current values (see Figure 2.18). This can be attributed to the improved flushing of molten materials at a higher dielectric flow rate.

2.3.2.3 Effect of Pulse On-time

MRR of near-dry EDM shows a continuous increase as the pulse on-time increases from 4 to 18 μs [26] (see Figure 2.19). This can be attributed to the higher discharge energy supplied to forming the crater as the pulse on-time increases.

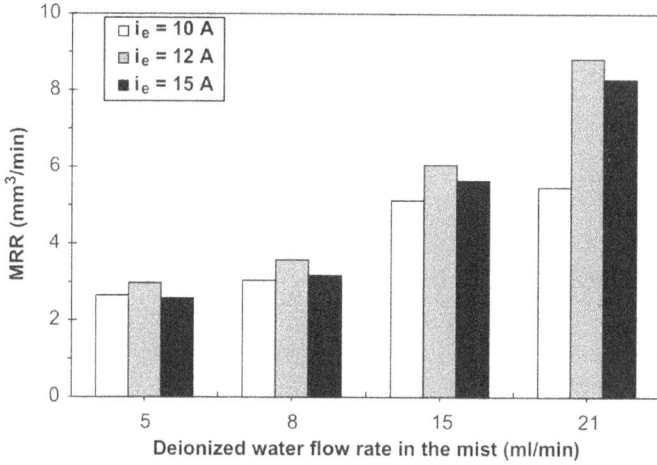

FIGURE 2.18   MRR at different discharge currents and flow rates [26]. With permission from Elsevier.

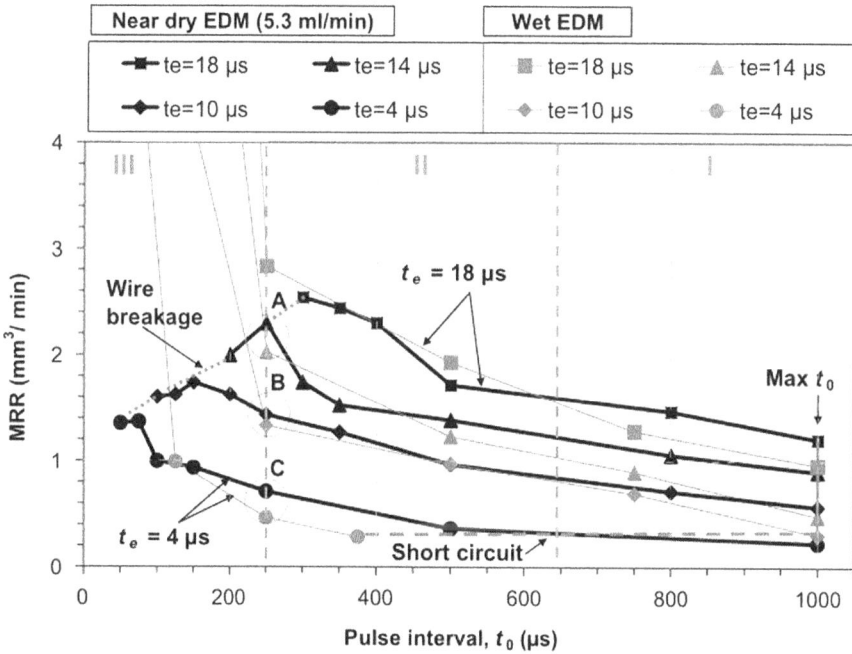

FIGURE 2.19   Comparison of near dry and wet EDM in terms of MRR at different pulse intervals [26]. With permission from Elsevier.

2.3.2.4 Effect of Pulse Off-time

Kao et al. [26] reported that as the pulse off-time or pulse interval increases, the MRR tends to decrease (see Figure 2.19). However, very low pulse off-time is also not desirable due to occurrences of wire breakage in wire EDM. The MRR of wet EDM is much higher than that of near-dry EDM at low pulse off-times. However, at high pulse off-time values, the MRR of near-dry EDM is larger than that of wet EDM.

2.3.2.5 Effect of Dielectric Material

Tao et al. [30] investigated combinations of different liquid gas dielectrics on EDM performance. They used gases such as air, oxygen, nitrogen, and helium with deionized water for near-dry EDM. This study found that for copper tool electrodes, a water-oxygen dielectric gave the highest MRR for near-dry EDM at both high and low discharge energy settings. They observed that EDM action fails to occur normally in the majority of high discharge energy experiments. Only a few near-dry EDM experiments with copper electrodes were successful at high discharge energy. They also investigated use of gases such as air, nitrogen, and helium, mixed with kerosene as the dielectric. At low discharge energy experiments, in near-dry EDM with a kerosene-gas dielectric, only those with graphite tools were successful. Among these, the highest MRR was observed for a kerosene-air mixture dielectric.

2.3.2.6 Effect of Tool Electrode Material

Tao et al. [30] also investigated the effect of different electrode materials, such as copper and graphite, on near-dry EDM performance. A graphite electrode was found not to be favorable for successful EDM action in high discharge energy experiments. However, in low discharge energy experiments, the graphite tool gave a higher MRR in comparison with the copper tool. Moreover, the use of a copper tool with kerosene-gas mixture dielectrics was found to be not suitable for the EDM cutting action.

2.3.2.7 Effect of Input Pressure

An increase in input pressure of the dielectric spray was found to positively influence the MRR of the near-dry EDM process [30].

2.3.3 Tool Wear Rate (TWR) of Dry EDM

Tool wear rate (TWR) is another important performance indicator of dry EDM. One of the advantages of dry EDM over wet EDM is its low tool wear rate.

2.3.3.1 Effect of Discharge Current

Govindan and Joshi [21] reported that none of the major parameters, including discharge current, showed any significant influence on tool wear rate, see Figure 2.20. This is due to almost negligible variation in tool wear rate due to molten material deposition from tool and work electrodes.

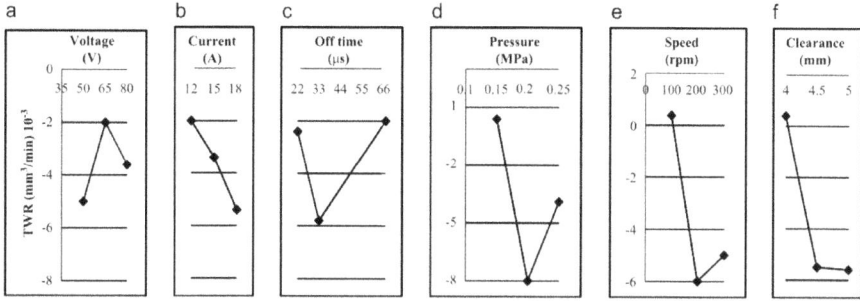

FIGURE 2.20   Main effect plots of different parameters on TWR [21]. With permission from Elsevier.

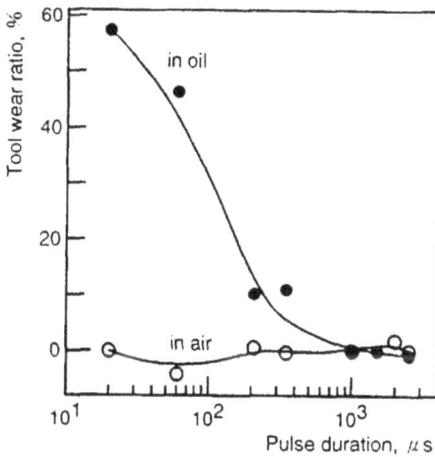

FIGURE 2.21   Variation of tool wear ratio (%) with pulse duration (μs) [17]. With permission.

2.3.3.2   Effect of Pulse Duration

Kunieda et al. [17] investigated the influence of pulse duration on the TWR of EDM in oil and air. It was found that contrary to wet EDM in oil, dry EDM with air had almost zero tool wear (see Figure 2.21).

2.3.4   Tool Wear Rate (TWR) of Near-dry EDM

Tool wear rate (TWR) in near-dry EDM is found to be less than that in dry EDM [31]. Thus, the tool change required is less frequent.

2.3.4.1   Effect of Discharge Current

Rajabinasab et al. [32] reported that an increase in discharge current led to an increase in TWR for tool electrode materials such as Cu and Cu-Cr. Dhakar and Dvivedi

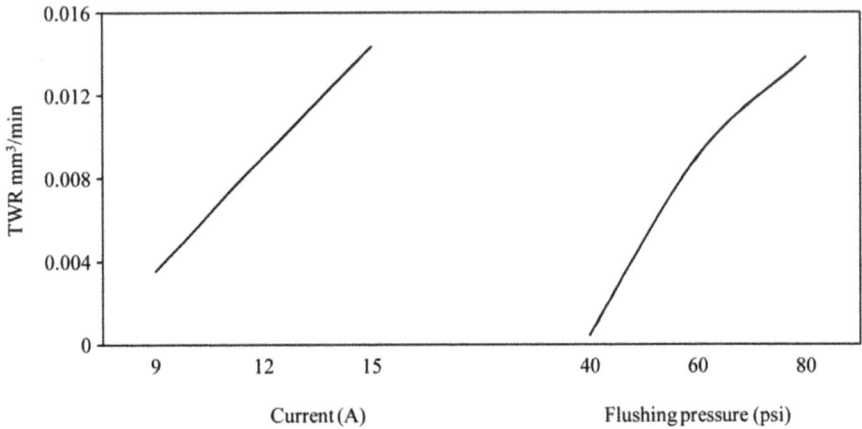

FIGURE 2.22   Effect of current and flushing pressure on TWR [33]. With permission from Taylor and Francis.

[33] also reported an increase in TWR with an increase in discharge current (see Figure 2.22).

2.3.4.2   Effect of Tool Material

Rajabinasab et al. [32] reported that at low discharge currents, Cu-Sn displays a higher TWR, whereas, at high discharge currents, Cu displays a higher TWR among tool materials such as Cu, Cu-Cr, and Cu-Sn.

2.3.4.3   Effect of Flushing Pressure

Dhakar and Dvivedi [33] reported that an increase in flushing pressure is detrimental in terms of tool wear with an increase in the TWR (see Figure 2.22).

2.3.4.4   Effect of Dielectric

The use of a mist dielectric with a higher viscosity than a gas dielectric results in a better flushing efficiency and a thinner recast layer in near-dry EDM than those in dry EDM [31].

2.3.5   Parametric Influences on Surface Roughness (R_a) of Dry EDM

Surface roughness (R_a) of the machined surface is another important performance indicator for a machining process.

2.3.5.1   Effect of Discharge Energy

Gholipoor et al. [34] found that the surface roughness of dry EDM surfaces increases with an increase in discharge energy (see Figure 2.23).

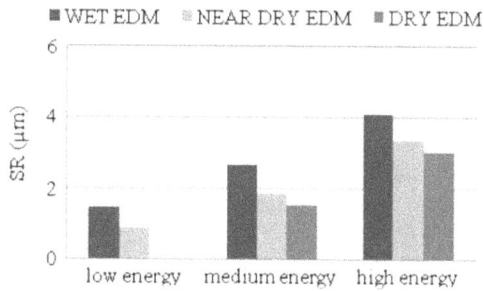

FIGURE 2.23 Effect of discharge energy on surface roughness of wet, near-dry, and dry EDM [34]. With permission from Springer Nature.

2.3.6 PARAMETRIC INFLUENCES ON SURFACE ROUGHNESS (R_a) OF NEAR-DRY EDM

2.3.6.1 Effect of Dielectric Material

Among the different water-gas mixture dielectrics, the highest R_a was generally observed when a water-oxygen mixture dielectric was used for the EDM process [30]. A kerosene-air mixture dielectric gave a higher R_a compared with that of other kerosene-gas mixture dielectrics.

2.3.6.2 Effect of Tool Electrode Material

A copper tool electrode gives a better surface finish in terms of lower R_a in comparison with a graphite tool electrode [30]. This trend is also supported by the findings of Tao et al. [35].

2.3.6.3 Effect of Input Pressure

R_a does not show a significant variation with an increase in the input dielectric pressure [30].

2.4 ISSUES FACED BY DRY AND NEAR-DRY EDM PROCESSES

2.4.1 ISSUES WITH DRY EDM PROCESSES

2.4.1.1 Probability of Shorting

The probability of shorting is the ratio between the number of occurrences of shorting and the total number of pulses. It is very high for dry EDM as compared to that of wet EDM. Kunieda et al. [17] experimented with different electrode motions to find their influence on the probability of shorting (see Figure 2.24). They found out that the highest probability of shorting is observed for dry EDM with neither rotation nor planetary motion, whereas the lowest probability of shorting was observed for dry EDM with combined rotation and planetary motion.

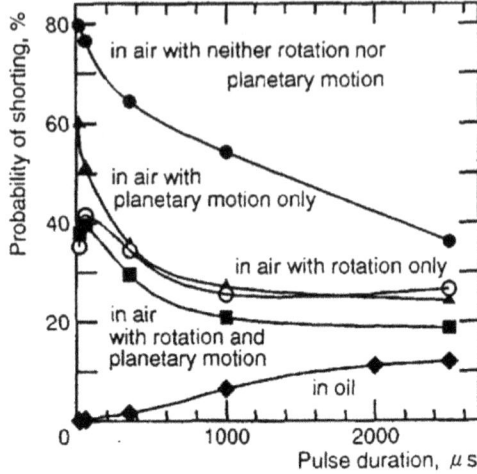

FIGURE 2.24 Influence of rotary and planetary motion on shorting probability [17]. With permission from Elsevier.

2.4.1.2   3D Machining

Kunieda et al. [17] investigated different tool electrode movements for 3D machining. They found out that machining in thin layers helps achieve 3D machining.

2.4.2   ISSUES WITH NEAR-DRY EDM PROCESSES

2.4.2.1   Wire Breakage

Kao et al. [26] reported that the probability of wire breakage is high for near-dry EDM due to the lower capacity of the liquid-gas mixture to remove heat from the spark gap.

2.5   SIMULATIONS IN DRY AND NEAR-DRY EDM

2.5.1   SINGLE-SPARK SIMULATION IN DRY EDM

The majority of simulation research works in the domain of EDM are on the single-spark of conventional wet EDM. Single-spark simulation in dry EDM offers challenges due to changes in process physics. Wang et al. [36] performed a single-spark simulation in dry EDM to evaluate the resultant temperature distribution. They compared the single-spark craters generated under similar conditions in wet and dry EDMs and found that a larger work material region is heated in dry EDM compared to that of wet EDM. In later research [37], they were also able to simulate the raised edges of dry EDM craters through fluid field simulation in ANSYS. Single-spark crater simulations can be further improved by considering values for aspects of the model such as cathode energy fraction, spark radius, plasma flushing

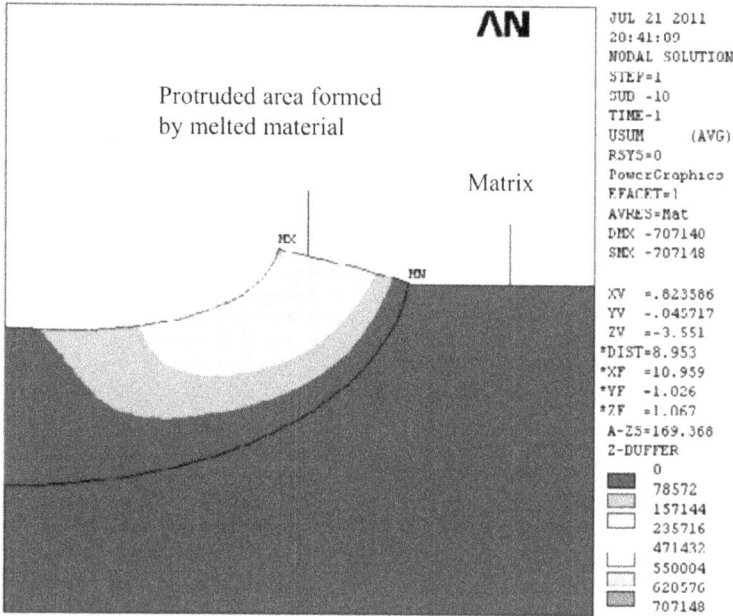

FIGURE 2.25 Single-spark simulation in dry EDM with protruded area formation [37]. With permission from Elsevier.

efficiency, heat flux distribution, and the heat convection coefficient, specific to dry EDM.

2.5.2 SINGLE-SPARK SIMULATION IN NEAR-DRY EDM

Single-spark simulations in near-dry EDM are scarce in the available literature. Ganachari et al. [31] studied the tool wear in near-dry EDM through single-spark simulation. While simulating the near-dry EDM condition, they adopted the heat convection coefficient (h) of an air-water mist in the convection region outside of the spark plasma, in place of the heat convection coefficient of water considered in the simulation of wet EDM. There is great scope for improving single-crater simulations in near-dry EDM by considering the change in cathode energy fraction, spark radius, heat flux distribution, and plasma flushing efficiency as the EDM condition changes from wet to near-dry.

2.6 CRITICAL COMPARISON BETWEEN WET EDM, DRY EDM, AND NEAR-DRY EDM

A critical comparison between wet, dry, and near-dry EDM technologies in terms of their advantages, disadvantages, and applications is made in Table 2.1.

TABLE 2.1
Comparison of wet, dry, and near-dry EDM

EDM technology	Dielectric medium	Advantages	Limitations	Applications
Wet EDM	Liquid	High MRR	Hazardous wastes Bulky machine	Die and mould making
Dry EDM	Compressed air	Helps achieve near-zero TWR	Short circuiting between work and tool electrodes	Precise machining of 3-D surfaces [17]
	Oxygen	Increase in MRR by 100% than air due to thermally activated oxidation	Poor surface finish and surface appearance	Roughing dry EDM operation [30]
	Inert gases	Fine surface finish, upto 0.80 μm R_a (Helium) and 0.85 μm R_a (Nitrogen).	Very low MRR at all machining conditions	Pocket machining with mirror finish [35]
Near-dry EDM	Liquid-air mixture	Increase in MRR at low energy input	Extremely high TWR	Finish machining [30]

2.6.1 ADVANTAGES AND DISADVANTAGES OF DRY EDM

2.6.1.1 Advantages of Dry EDM over Wet EDM

- Dry EDM has characteristic low tool wear as compared to that of wet EDM. This enables less frequent tool change due to tool wear. Moreover, tool wear in dry EDM is independent of pulse duration.
- Since dry EDM does not require a dielectric flushing system and a dielectric tank, the corresponding EDM machines can be made compact compared to wet EDM machines. Even desktop dry EDM is feasible and has already been fabricated by some researchers.
- In wet EDM, machining direction needs to be chosen to be in the direction of gravity owing to the large quantity of liquid dielectric used. However, dry EDM machining does not impose such restrictions.
- The danger of fire is significantly reduced due to the absence of flammable liquid dielectric.
- Process wastes that are harmful to the environment are greatly reduced.

2.6.1.2 Disadvantages of Dry EDM

- High debris deposition is observed in dry EDM in comparison to that in wet EDM due to the inability of the less viscous gaseous dielectric to carry away molten material from the discharge gap.

- The MRR is dependent on electrode wall thickness. It becomes very low as wall thickness crosses a threshold.

2.6.2 ADVANTAGES AND DISADVANTAGES OF NEAR-DRY EDM

2.6.2.1 Advantages of Near-dry EDM over Dry EDM
- Near-dry EDM gives a higher material removal rate (MRR) in comparison to dry EDM.
- Tool cutting edges maintain their sharpness in near-dry EDM in comparison to dry EDM.
- Debris deposition in near-dry EDM is significantly reduced in comparison to that in dry EDM.
- Tool wear in near-dry EDM is 25 % less than that in dry EDM, thus reducing the frequency of the requirement for tool change [31].

2.6.2.2 Advantages of Near-dry EDM over Wet EDM
- Near-dry EDM gives a higher MRR than that of wet EDM at low discharge energies. In another research by Dhakar et al. [28], near-dry EDM was observed to give almost five times higher MRR than that of wet EDM.
- Near-dry EDM generates a smaller gap width between electrodes in comparison to that of wet EDM, thus reducing the machining overcut.

2.6.2.3 Disadvantages of Near-dry EDM
- There is a high thermal load acting on the electrode during the near-dry EDM process. This is highly detrimental to the machining process by causing frequent wire breaks in wire EDM and higher tool erosion in EDM drilling.

2.7 FUTURE AVENUES OF RESEARCH

The following are the major future avenues of research for dry and near-dry EDM:

1 Surfaces textured using EDM are finding applications in the generation of ortho-implant [38], self-cleaning [39], antifouling [40], and antibacterial [41] surfaces. Dry and near-dry EDM surfaces need to be investigated for their feasibility in these applications.
2 The surfaces generated using EDM are investigated for different topographical aspects in terms of areal texture parameters [42]. Such analysis imparts deeper knowledge about the developed surface topography. Researchers need to perform similar topographical analyses for dry and near-dry EDM generated surfaces.
3 Analytical [43] and numerical [44] multi-spark models are currently available to predict different EDM performance indicators. Multi-spark models also need to be developed for dry and near-dry EDM variants.

2.8 CONCLUSION

Towards countering the high liquid dielectric cost and hazardous waste production in wet EDM, two new EDM technologies were proposed by researchers: dry EDM using gas dielectric and near-dry EDM using a liquid-gas mixture dielectric. Among different dry EDM technologies, the use of oxygen gas was found to give the highest MRR. In general, the dry EDM MRR was only one-fifth of that of wet EDM. Hence, researchers began using liquid-gas mixture dielectrics for EDM. Among the liquid-gas dielectric combinations, water-oxygen and kerosene-air combinations were found to give high MRRs. Further research is necessary in this domain to apply these processes in meso to micro domains and expand their applications for various tool and work combinations.

REFERENCES

[1] DiBitonto D, Eubank P, Patel MR, Barrufet MA. Theoretical models of the electrical discharge machining process. I. A simple cathode erosion model. J Appl Phys 1989;66:4095–103. https://doi.org/10.1063/1.343994.

[2] Singh A, Ghosh A. A thermo-electric model of material removal during electric discharge machining. Int J Mach Tools Manuf 1999;39:669–82. https://doi.org/10.1016/S0890-6955(98)00047-9.

[3] Kunieda M, Kowaguchi W, Takita T. Reverse simulation of die-sinking EDM. CIRP Ann – Manuf Technol 1999;48:115–8. https://doi.org/10.1016/S0007-8506(07)63144-8.

[4] Joshi SN, Pande SS. Thermo-physical modeling of die-sinking EDM process. J Manuf Process 2010;12:45–56. https://doi.org/10.1016/j.jmapro.2010.02.001.

[5] Izquierdo B, Sánchez J a., Plaza S, Pombo I, Ortega N. A numerical model of the EDM process considering the effect of multiple discharges. Int J Mach Tools Manuf 2009;49:220–9. https://doi.org/10.1016/j.ijmachtools.2008.11.003.

[6] Salah N Ben, Ghanem F, Atig K Ben. Numerical study of thermal aspects of electric discharge machining process. Int J Mach Tools Manuf 2006;46:908–11. https://doi.org/10.1016/j.ijmachtools.2005.04.022.

[7] Tlili A, Ghanem F, Salah N Ben. A contribution in EDM simulation field. Int J Adv Manuf Technol 2015;79:921–35. https://doi.org/10.1007/s00170-015-6880-1.

[8] Soni JS, Chakraverti G. Experimental investigation on migration of material during EDM of die steel (T215 Cr12). J Mater Process Technol 1996;56:439–51. https://doi.org/10.1016/0924-0136(95)01858-1.

[9] Singh S, Maheshwari S, Pandey PC. Some investigations into the electric discharge machining of hardened tool steel using different electrode materials. J Mater Process Technol 2004;149:272–7. https://doi.org/10.1016/j.jmatprotec.2003.11.046.

[10] Kitamura T, Kunieda M. Clarification of EDM gap phenomena using transparent electrodes. CIRP Ann – Manuf Technol 2014;63:213–6. https://doi.org/10.1016/j.cirp.2014.03.059.

[11] Kunieda M, Lauwers B, Rajurkar KP, Schumacher BM. Advancing EDM through fundamental insight into the process. CIRP Ann – Manuf Technol 2005;54:64–87. https://doi.org/10.1016/S0007-8506(07)60020-1.

[12] Schumacher BM. After 60 years of EDM the discharge process remains still disputed. J Mater Process Technol 2004;149:376–81. https://doi.org/10.1016/j.jmatprotec.2003.11.060.

[13] Holsten M, Koshy P, Klink A, Schwedt A. Anomalous influence of polarity in sink EDM of titanium alloys. CIRP Ann 2018;67:221–4. https://doi.org/10.1016/j.cirp.2018.04.069.

[14] Jithin S, Raut A, Bhandarkar U V, Joshi SS. FE modeling for single spark in EDM considering plasma flushing efficiency. Procedia Manuf 2018;26:617–28. https://doi.org/10.1016/j.promfg.2018.07.072.

[15] Jithin S, Bhandarkar U V, Joshi SS. Three-dimensional topography analysis of electrical discharge textured SS304 surfaces. J Manuf Process 2020;60:384–99. https://doi.org/10.1016/j.jmapro.2020.10.066.

[16] Govindan P, Joshi SS. Analysis of micro-cracks on machined surfaces in dry electrical discharge machining. J Manuf Process 2012;14:277–88. https://doi.org/10.1016/j.jmapro.2012.05.003.

[17] Kunieda M, Yoshida M, Taniguchi N. Electrical discharge machining in gas. CIRP Ann 1997;46:143–6. https://doi.org/10.1016/S0007-8506(07)60794-X.

[18] Yoshida M, Kunieda M. Study on the distribution of scattered debris generated by a single pulse discharge in EDM process. J Japan Soc Electr Mach Eng 1996;30:27–36. https://doi.org/10.2526/jseme.30.64_27.

[19] Shen Y, Liu Y, Zhang Y, Dong H, Sun W, Wang X, et al. High-speed dry electrical discharge machining. Int J Mach Tools Manuf 2015;93:19–25. https://doi.org/10.1016/j.ijmachtools.2015.03.004.

[20] Lee SW, Oh YS. A study on dry electrical discharge machining process. Int. Conf. Smart Manuf. Appl., IEEE; 2008, p. 234–8. https://doi.org/10.1109/ICSMA.2008.4505648.

[21] Govindan P, Joshi SS. Experimental characterization of material removal in dry electrical discharge drilling. Int J Mach Tools Manuf 2010;50:431–43. https://doi.org/10.1016/j.ijmachtools.2010.02.004.

[22] Joshi S, Govindan P, Malshe A, Rajurkar K. Experimental characterization of dry EDM performed in a pulsating magnetic field. CIRP Ann – Manuf Technol 2011;60:239–42. https://doi.org/10.1016/j.cirp.2011.03.114.

[23] Liqing L, Yingjie S. Study of dry EDM with oxygen-mixed and cryogenic cooling approaches. Procedia CIRP 2013;6:344–50. https://doi.org/10.1016/j.procir.2013.03.055.

[24] Yadav VK, Kumar P, Dvivedi A. Effect of tool rotation in near-dry EDM process on machining characteristics of HSS. Mater Manuf Process 2019;34:779–90. https://doi.org/10.1080/10426914.2019.1605171.

[25] Tanimura, T. Development of EDM in the mist. Proc IJEM, 1989 1989;9:313–6.

[26] Kao CC, Tao J, Shih AJ. Near dry electrical discharge machining. Int J Mach Tools Manuf 2007;47:2273–81. https://doi.org/10.1016/j.ijmachtools.2007.06.001.

[27] Shen Y, Liu Y, Sun W, Zhang Y, Dong H, Zheng C, et al. High-speed near dry electrical discharge machining. J Mater Process Technol 2016;233:9–18. https://doi.org/10.1016/j.jmatprotec.2016.02.008.

[28] Dhakar K, Chaudhary K, Dvivedi A, Bembalge O. An environment-friendly and sustainable machining method: near-dry EDM. Mater Manuf Process 2019;34:1307–15. https://doi.org/10.1080/10426914.2019.1643471.

[29] Islam MM, Li CP, Ko TJ. Dry electrical discharge machining for deburring drilled holes in CFRP composite. Int J Precis Eng Manuf – Green Technol 2017;4:149–54. https://doi.org/10.1007/s40684-017-0018-x.

[30] Tao J, Shih AJ, Ni J. Experimental study of the dry and near-dry electrical discharge milling processes. J Manuf Sci Eng 2008;130:0110021–9. https://doi.org/10.1115/1.2784276.

[31] Ganachari VS, Chate U, Waghmode L, Jadhav P, Mullya S. Simulation and experimental investigation of tool wear rate in dry and near-dry EDM process. World J Eng 2021;18(5):701–9. https://doi.org/10.1108/WJE-07-2020-0322.

[32] Rajabinasab F, Abedini V, Hadad M, Hajighorbani R. Experimental investigation of the effect of tool material on the performance of AISI 4140 steel in the rotary near dry electrical discharge machining. Proc Inst Mech Eng Part E J Process Mech Eng 2020;234:308–17. https://doi.org/10.1177/0954408920922102.

[33] Dhakar K, Dvivedi A. Parametric evaluation on near-dry electric discharge machining. Mater Manuf Process 2016;31:413–21. https://doi.org/10.1080/10426914.2015.1037 905.

[34] Gholipoor A, Baseri H, Shabgard MR. Investigation of near dry EDM compared with wet and dry EDM processes. J Mech Sci Technol 2015;29:2213–8. https://doi.org/10.1007/s12206-015-0441-2.

[35] Tao J, Shih AJ, Ni J. Near-Dry EDM Milling of Mirror-Like Surface Finish. Int J Electr Mach 2008;13:29–33. https://doi.org/10.2526/ijem.13.29.

[36] Wang T, Xie SQ, Xu XC. Thermal field analysis of single pulse EDM in gas. Key Eng Mater 2010;426–427:633–7. https://doi.org/10.4028/www.scientific.net/KEM.426-427.633.

[37] Wang T, Zhe J, Zhang YQ, Li YL, Wen XR. Thermal and fluid field simulation of single pulse discharge in dry EDM. Procedia CIRP 2013;6:427–31. https://doi.org/10.1016/j.procir.2013.03.032.

[38] Harcuba P, Bačáková L, Stráský J, Bačáková M, Novotná K, Janeček M. Surface treatment by electric discharge machining of Ti-6Al-4V alloy for potential application in orthopaedics. J Mech Behav Biomed Mater 2012;7:96–105. https://doi.org/10.1016/j.jmbbm.2011.07.001.

[39] Wang H, Chi G, Jia Y, Yu F, Wang Z, Wang Y. A novel combination of electrical discharge machining and electrodeposition for superamphiphobic metallic surface fabrication. Appl Surf Sci 2020;504:144285. https://doi.org/10.1016/j.apsusc.2019.144 285.

[40] He ZR, Luo ST, Liu CS, Jie XH, Lian WQ. Hierarchical micro/nano structure surface fabricated by electrical discharge machining for anti-fouling application. J Mater Res Technol 2019;8:3878–90. https://doi.org/10.1016/j.jmrt.2019.06.051.

[41] Bui VD, Mwangi JW, Schubert A. Powder mixed electrical discharge machining for antibacterial coating on titanium implant surfaces. J Manuf Process 2019;44:261–70. https://doi.org/10.1016/j.jmapro.2019.05.032.

[42] Jithin S, Bhandarkar U V, Joshi SS. Characterization of surface topographies generated using circular- and cylindrical-face EDT. Surf Topogr Metrol Prop 2020;8:045018. https://doi.org/10.1088/2051-672X/abc320.

[43] Jithin S, Bhandarkar U, Joshi SS. Analytical simulation of random textures generated in electrical discharge texturing. J Manuf Sci Eng 2017;139:111002 (1–12). https://doi.org/10.1115/1.4037322.

[44] Jithin S, Raut A, Bhandarkar U V, Joshi SS. Finite element model for topography prediction of electrical discharge textured surfaces considering multi-discharge phenomenon. Int J Mech Sci 2020;177:105604. https://doi.org/10.1016/j.ijmecsci.2020.105 604.

3 Electrochemical Spark Machining Process

*Vyom Sharma, Mahavir Singh and Janakarajan Ramkumar**

Department of Mechanical Engineering, Indian Institute of Technology Kanpur, Uttar Pradesh, India

*Corresponding author

CONTENTS

DOI: 10.1201/9781003202301-3

45

3.1   INTRODUCTION

In the past few decades, the introduction of new engineering materials has led manufacturing engineers to develop new methods for processing. Machining of ceramics,
pyrex, and silicon-based materials such as quartz and glass is not very convenient
because these materials are hard, brittle, and non-conductive in nature. The conventional machining processes such as turning, drilling, milling, and grinding are
incapable of fabricating complex micro features on these materials within required
tolerance levels. It is because these processes involve mechanical forces (due to physical contact between tool and workpiece) and due to their brittle nature, glass and
quartz usually fail during their processing. Non-conventional machining processes
such as EDM and ECM are the most widely used options for the machining of hard
materials. However, these processes are incapable of machining non-conductive
materials. Other processes such as laser beam machining (LSM) and electron beam
machining (EBM) are capable of machining non-conductive materials. But these
processes often result in a generation of a heat affected zone (HAZ). Also, these
methods are not capable of machining high aspect ratio features and their equipment
cost is also relatively high. Abrasive jet machining (AJM), ultrasonic machining
(USM), and abrasive water jet machining (AWJM) [1] are other options for the
machining of non-conductive materials. But, these processes results in poor surface
quality and dimensional inaccuracy. Also, these processes are not suitable for the
machining of ductile materials. This has prompted researchers to develop a process
that is capable of overcoming these challenges and can prove to be an industrially
viable option for the machining. ECSM is a hybrid material processing technique
in which material gets removed through two processes: (1) melting and subsequent

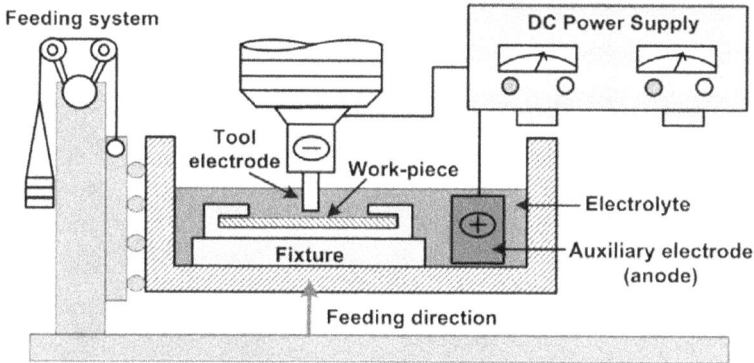

FIGURE 3.1 Diagram showing one of the possible arrangements for ECSM process [2].

vaporization (2) accelerated chemical etching under the high electrical energy discharged on the electrode tip during electrolysis.

3.1.1 INTRODUCTION TO ECSM TECHNOLOGY

ECSM is a hybrid machining process that is developed especially for the machining of electrically nonconductive materials. As in EDM, the primary mode of material removal in this process is melting and vaporization of workpiece material due to sparking. The sparks are generated due to the ionization of a gas film which forms on the tool surface due to the electrolysis of an electrolyte solution, similar to the ECM process. The secondary mode of material removal in this process is through chemical etching. Since ECSM is a result of the combination of two processes, it overcomes some of the drawbacks of its constituent processes. Although the particular value of this process lies in the machining of non-conductive materials, it can also be used for the machining of conductive materials as well. Figure 3.1 is a schematic diagram of a typical ECSM machining setup.

The material removal rate (MRR) in ECSM is appreciably higher than that of its constituent processes for identical machining conditions. Over a period of time, researchers have proposed different names for this process. These include: discharge machining (DM) for non-conductors, electro-chemical arc machining (ECAM), spark-assisted etching (SAE), electro erosion-dissolution machining (EEDM), spark-assisted chemical engraving (SACE), electro-chemical discharge machining (ECDM), and electrochemical spark machining (ECSM). Amongst these, ECDM and ECSM are the most commonly used names and therefore ECSM is used throughout this chapter to refer to this process and to maintain uniformity in the text.

3.1.2 COMPARISON OF ECSM WITH EDM AND ECM

Highlighted in Table 3.1 are the key differences between the ECM, EDM, and ECSM processes based on various parameters.

TABLE 3.1
A comparison between ECM, EDM, and ECSM process based on different factors (compiled from Ref. [2, 3])

Basis of comparison	ECM	EDM	ECSM
Mechanism of material removal	Anodic dissolution at ionic level	Melting and vaporization	Melting and vaporization
Material conductivity dependency	Yes	Yes	No
Medium required	Electrolyte	Dielectric	Electrolyte
Polarity required	Negative	Both negative and positive	Both negative and positive
Power supply type	DC power (pulse and continuous)	DC power (pulse)	DC power (pulse and continuous)
Power rating	V= 2- 30 V I= 50–10000 A	V= 20- 120 V I= 5–120 A	V= 20- 30 V I≈ 1 A
Typical Interelectrode gap in the process	200 μm–600 μm	10 μm–500 μm	Approx. 20 μm–25 μm
Capability to machine high aspect ratio features	Yes	Yes	No
Process complexity	Low	Low	High
Thermal induced defects (HAZ, recast layer, micro cracks)	No	Yes	Yes
Dependence on workpiece hardness	No	No	No
Commercially used	Yes	Yes	No
Eco-friendly	No	Yes	No

3.2 THEORIES OF THE ECSM PROCESS

Over a period of time, researchers have proposed different theories to explain the theory and the mechanism of spark formation in ECSM which results in material removal from the workpiece. The pioneer work in ECSM was carried out in 1968 when Kurafuji and Suda [4] drilled a micro hole in glass. However, their report didn't explain the reason for spark generation. In 1984, Crichton and McGeough [5] demonstrated the different stages of the discharge process using the technique of streak photography. They suspected that the gas film formation over the tool surface was responsible for spark formation. However, the actual mechanism of spark generation in the presence of a gas film was not understood at that time. Basak and Ghosh [6] 1996 presented a theory for spark generation in ECSM. It was proposed that Ohmic heating of the electrolyte lead to the generation of vapour bubbles which covered the tool surface and insulated it from the electrolyte. It is analogous to the circuit being temporarily switched off and current in the circuit dropping to zero. Later, in 1999, Jain et al. [7] proposed an '*arc discharge valve theory*' to explain the mechanism of spark formation in the process. They proposed that the gas bubbles are generated near the tool due to an electrochemical reaction (hydrolysis of water and reduction of hydrogen ions near the tool) and act as a valve for the spark. In the presence of a high intensity electric field, breakdown of these gas bubbles (gas acting as a dielectric medium) takes place, resulting in the generation of a spark. In 2002, Kulkarni et al. [8] presented a graph showing the variation of current with time during machining in ECSM. Finally, in 2004, Fascio et al. [9] proposed a percolation theory to explain the mechanism of spark generation in ECSM. It was proposed that smaller bubbles coalesce to form bigger bubbles. These bubbles grow further and form an insulating film of gas over the tool surface which is in contact with the electrolyte. In the presence of an electric field, this gas film breaks down and leads to the generation of the spark [10].

3.3   FUNDAMENTALS OF SPARK FORMATION

In the ECSM process, a DC power source (constant or pulsating) is used to apply potential between the tool and the auxiliary electrode. Electrolyte solution (an alkaline solution is commonly used) completes the circuit and the process of electrolysis starts. Anodic material starts dissolving into the electrolyte according to the following reaction:

$$N \rightarrow N^{z+} + Z\,e^-$$

where, 'N' represents anodic material (in case of machining non-conductive materials, it is the auxiliary electrode) and 'Z' represents the dissolution valency of anodic material. Metal ions combine with hydroxyl ions in the solution (generated due to hydrolysis of water) to form metal hydroxide according to the following reaction:

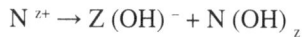

$$N^{z+} \rightarrow Z\,(OH)^- + N\,(OH)_z$$

Hydrogen ions drift towards the negatively polarized tool and are reduced to form hydrogen gas according to the following reaction:

$$2H^+ + 2\,e^- \rightarrow H_2 \uparrow$$

The positively charged metal ions get neutralized to form metal which further forms metal hydroxide according to the following reactions:

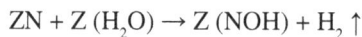

$$N^{z+} + Z\,e^- \rightarrow N\,(Metal)$$

$$ZN + Z\,(H_2O) \rightarrow Z\,(NOH) + H_2 \uparrow$$

It is important to note that above a certain critical voltage (depending upon electrolyte type and workpiece material), the rate of generation of hydrogen gas bubbles near the tool (cathode) is more than the rate of dispersion of these bubbles in the bulk electrolyte. As a result, gas bubbles surround the tool surface and form a film. This gas film insulates the tool surface from the electrolyte and results in a drop in circuit current.

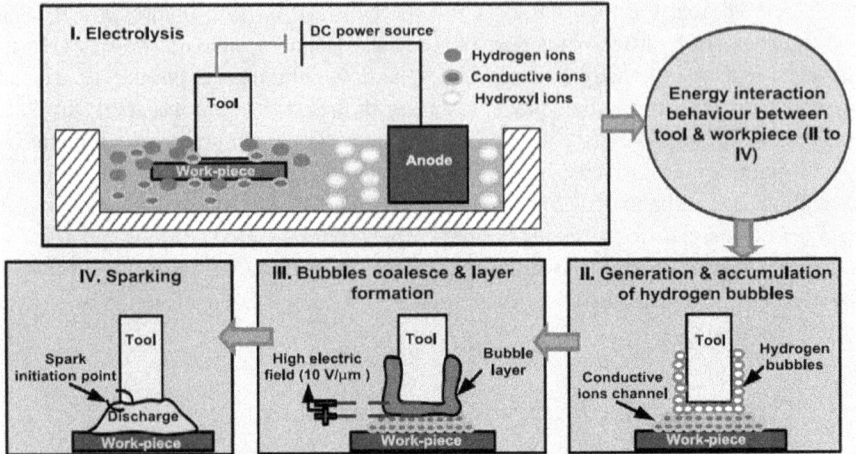

FIGURE 3.2   Schematic diagram showing the phenomenon of spark generation in ECSM [2].

Also, this gas film behaves as a dielectric medium and consequently, an electric field of high intensity (of the order of $10V / \mu m$) is established across it [2]. Near the sharp corners of the tool, current density becomes very high and causes the breakdown of this dielectric medium, resulting in sparking. After sparks initiate from the tool edges, its location changes along the complete tool surface. During sparking, a large number of high velocity electrons bombard the workpiece surface leading to its melting and vaporization. Figure 3.2 shows schematically, the phenomenon of spark generation in the ECSM process.

3.4   PROCESS CAPABILITIES AND AREAS OF APPLICATION

Since the applicability of the ECSM process does not depend upon the electrical conductivity of the material to be machined, it can be conveniently employed for the processing of a large spectrum of materials including metals, alloys, semi-conductors, and non-conductors. With an appropriate arrangement for electrolyte flushing in the sparking region, the process is capable of machining high aspect ratio micro features at a rate superior to that of its constituents [11]. Some of the key advantages of this process are as follows:

1   The process does not depend upon degree of hardness or brittleness in the workpiece. [12, 13].
2   Tool wear in the process is significantly less than that in the EDM process [14–16].
3   ECSM can be hybridized with other machining processes easily [17].
4   Due to chemical etching, the machined surface generated in ECSM is of superior quality [16].
5   Machining efficiency, accuracy, and quality can easily be controlled by varying different process parameters.

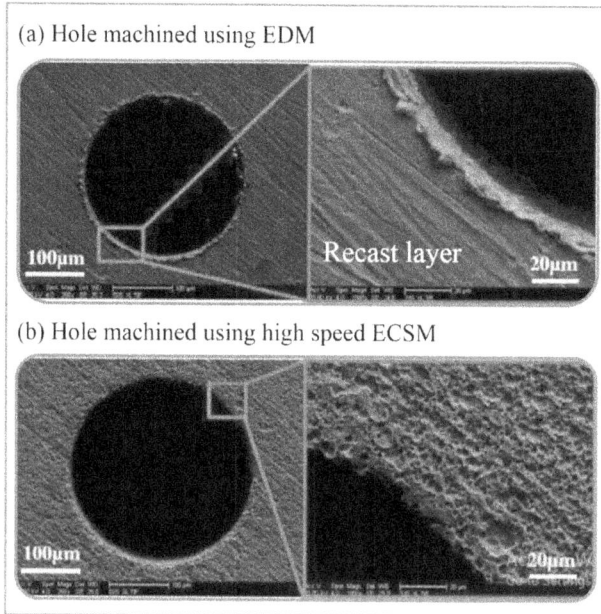

(a) Hole machined using EDM

100μm   Recast layer   20μm

(b) Hole machined using high speed ECSM

100μm   20μm

FIGURE 3.3 SEM image of hole machined using: (a) EDM, and (b) high speed Electrochemical Spark/Discharge Drilling process (modified from Ref. [18]).

Figure 3.3 shows an SEM image of a hole machined using an EDM and a high speed ECSM process. It can be seen from this figure that no appreciable recast layer is formed in the latter case as it gets removed due to chemical etching.

Ever since its inception in 1962, the process has grown in leaps and bounds. However, due to the lack of knowledge transfer, ECSM still remains a lab-based technique. Process capabilities and their prospective areas of application can be showcased at different levels by providing detailed research and data sheets to the practising engineers. Parts and features machined using ECSM may find their application in different fields of science and technology which include: bio-medical engineering, micro electromechanical systems, the energy sector for the slicing of silicon (a semiconductor) ingots to make solar cells, the electronic industry for machining complex circuits on silicon chips, and micro-assemblies [17, 19, 21]. Figure 3.4 shows images of features machined using ECSM.

3.5 PROCESS AND PERFORMANCE PARAMETERS IN ECSM

In ECSM, there are various parameters that can be controlled by the operator and which have a significant influence on its performance. Such parameters are generally termed as process parameters and their classification is shown in Figure 3.5. The parameters on which the performance of the process is based are termed performance parameters as shown in Figure 3.6. This section highlights all such parameters in ECSM and a brief discussion of each is presented.

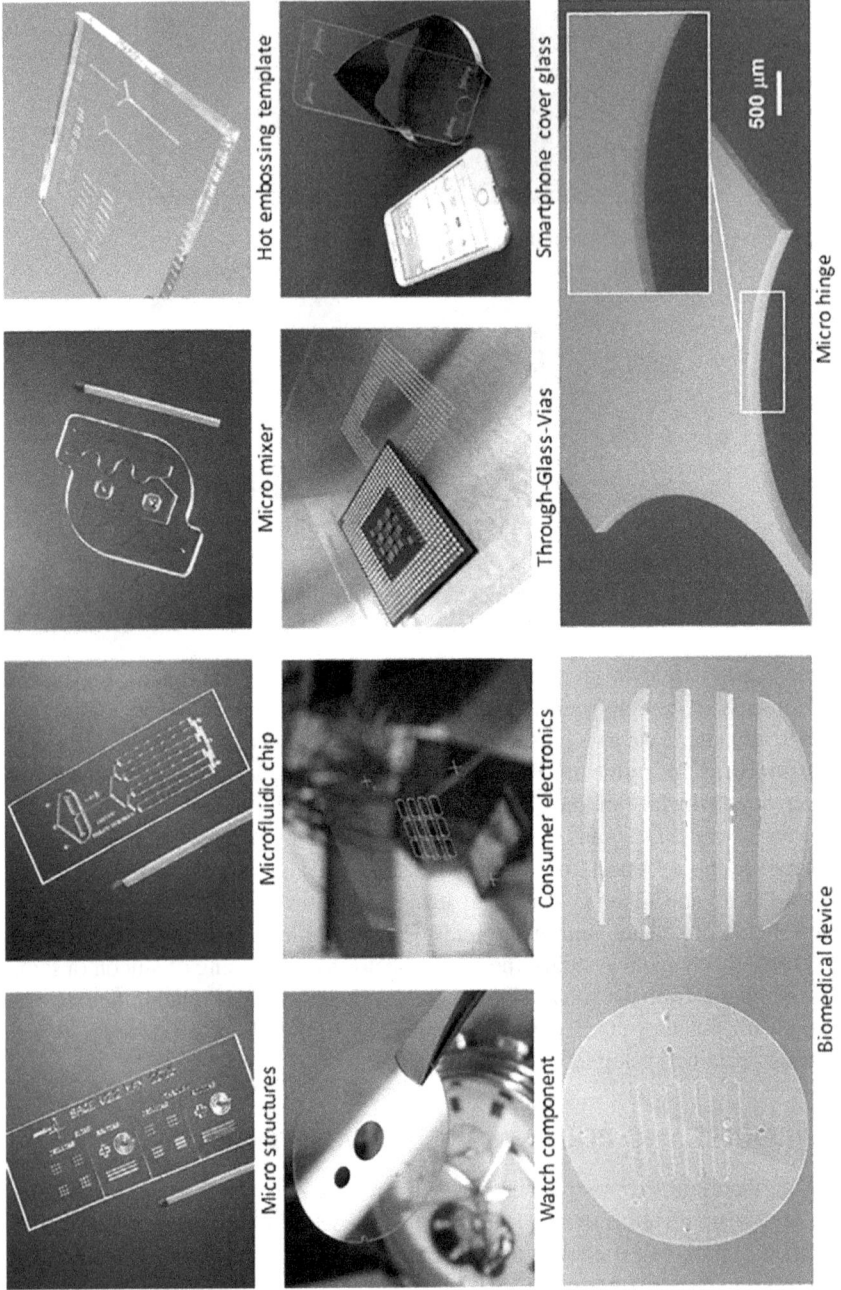

FIGURE 3.4 Images of different micro features machined on a glass workpiece using ECSM process [20].

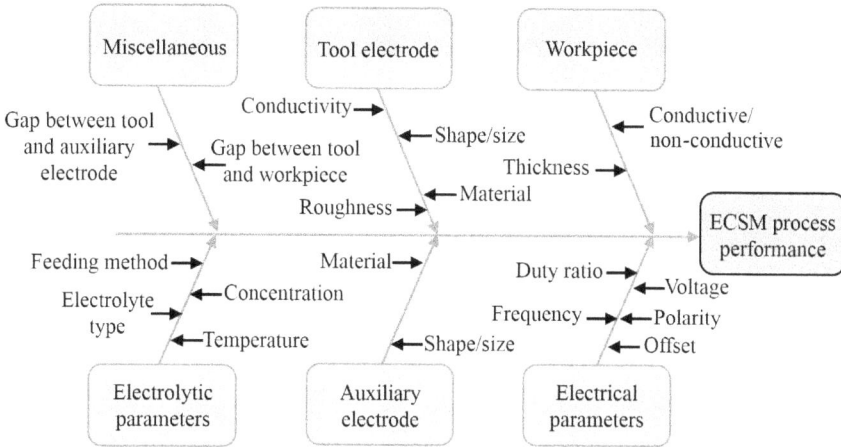

FIGURE 3.5 Cause and effect diagram showing different process parameters in ECSM.

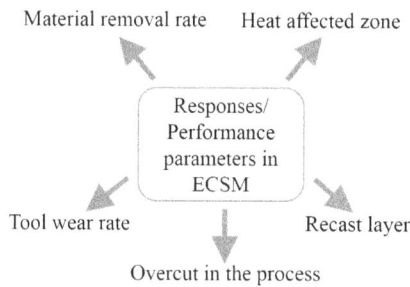

FIGURE 3.6 Performance parameters in ECSM process.

3.5.1 PROCESS PARAMETERS

3.5.1.1 Applied Voltage

Voltage plays an important role in the generation of sparks in the ECSM process. When the voltage is below a critical value, sparking does not take place and machining takes place only due to electrochemical dissolution (only when the workpiece is conductive). Different factors which influence the critical voltage in the process are: conductivity of the tool material, tool diameter [6], electrolyte type and concentration, and the distance between the anode and cathode. At a high value of applied voltage, the intensity of sparks is also high which results in an increase in the MRR. This, however, adversely affects the quality of the machined surface and this often occurs particularly in the vicinity of the machined feature.

3.5.1.2 Duty Ratio and Pulse Frequency

The ratio of pulse on-time to the pulse period is called duty ratio or alternatively, the duty factor. On increasing the duty ratio and keeping the pulse frequency fixed, pulse on-time increases. This means the sparking time increases and consequently, the MRR, heat affected zone (HAZ), and depth of penetration also increases. However, the duty ratio does not affect the aspect ratio of the machined feature significantly as the increase in input thermal energy due to the increase in duty ratio is compensated by the increased rate of dissipation of this energy from both the curved and flat surface area of the tool and the machined feature. Pulse on and off-time also influences the process performance in ECSM. As the pulse on-time increases, energy is supplied to the workpiece for a longer duration. As a result, a larger volume of material gets machined by melting and subsequent vaporisation from the workpiece and the MRR tends to increase. As a result of this, surface roughness of the machined feature also increases and thermal damage on the workpiece surface becomes more evident [15]. The pulse off-time of a duty cycle is also an important parameter in determining the process performance. In the pulse off-time, no sparking takes place and both tool and workpiece cool down during this period by losing their thermal energy. In this period, the machining products get flushed out of the sparking zone and a gas layer around the tool gets re-established. However, if the pulse off-time is too small in comparison to the pulse on-time, the thermal energy supplied during pulse on-time might not be sufficient to melt and vaporize the workpiece, thereby, resulting in no machining [22].

3.5.1.3   Tool Polarity

Generally, ECSM is performed with negative polarity. This implies that the tool is connected to the negative terminal of the voltage source. However, researchers have also studied the process performance with positive polarity (tool is connected with the positive terminal of the voltage source) [23]. The MRR, tool wear, and dimension inaccuracy increase when positive polarity is used. Surface roughness of the machined feature also increases in this case due to an increase in the material removed due to chemical etching. It was observed that with the same tool shape and dimensions, the machined hole was of spherical shape with positive polarity and of conical shape with reverse polarity [24].

3.5.1.4   Electrolyte Types and Concentration

Types: Electrolytes in ECSM are generally categorised on the basis of their pH value as acidic (HCl, H_2SO_4), basic (NaOH, KOH), and neutral (NaNO$_3$, NaCl). Selection of an electrolyte type depends upon the material to be machined, required MRR, and surface finish. For most of the workpiece materials, acidic electrolytes result in unstable sparking, low MRR, and dimensional inaccuracy. Although neutral electrolytes are not hazardous to the health of the operator and environment, their use often results in high tool wear rate and low dimensional accuracy. Basic electrolytes are most widely used in ECSM as they offer high MMR, surface finish, and good dimensional accuracy.

Concentration: Electrolyte concentration significantly influences the machining quality in ECSM [25]. As the concentration increases, spark intensity also increases, resulting in an increase in the MRR. The critical voltage needed for sparking also decreases as the concentration of electrolyte increases. Low concentration electrolytes are preferred for machining dimensionally accurate features and also for the machining of thin and fragile materials. But, with low concentration electrolytes, surface roughness of the machined feature is high.

3.5.1.5   Electrolyte Flow and Temperature

Electrolyte flow: During machining, debris is generated and gets accumulated in the narrow sparking zone. This debris interferes with the formation of a gas film and often stops the generation of sparks. To overcome this problem, electrolyte with some pressure must be injected into the gap. This also ensures the supply of fresh electrolyte in the gap which is necessary for electrolysis to initiate. Researchers have shown that the aspect ratio of the machined feature can be increased when pressurized electrolyte jet is used for machining. Also, due to the removal of debris and quick renewal of fresh electrolyte in the machining region, the MRR in the process increases significantly (by 40 %) in comparison to the situation where stagnant electrolyte is used [26].

Electrolyte temperature: The temperature of the electrolyte solution has a significant effect on the machining efficiency of the process. At elevated temperatures, the conductivity of the solution also increases. As a result, the rate of electrolysis increases and more gas bubbles are generated per unit time. The MRR and aspect ratio of the machined feature also increases with electrolyte temperature. The magnitude of critical voltage required for spark initiation decreases with an increase in electrolyte temperature and sparks can be created at a relatively lower voltage [13]. Generally, the temperature of the electrolyte in ECSM varies between 30° and 70°. However, it is also important to note that at elevated electrolyte temperature, machining accuracy gets compromised.

3.5.1.6   Surfactants

In ECSM, surfactants such as sodium dodecyl sulphate (SDS) are added to the electrolyte solution for enhancing the process performance. In the presence of surfactants, gas bubbles are produced in a larger volume (due to the increase in the concentration of hydroxyl ions in the solution) and the magnitude of current density over the tool surface also increases [27]. Surfactants reduce the critical voltage needed for the sparking to take place and improve the tool wettability, resulting in the generation of a thin film over the tool surface. Researchers have shown that by adding the surfactant in the electrolyte solution, the quality of the machined surface improves in addition to which the size of the HAZ decreases [28].

3.5.1.7   Feed Rate

Feed rate is a vital factor for controlling the gap between the tool and the workpiece. Feed motion can be given either to the tool or to the workpiece. An optimum value for the feed rate should be selected to ensure that the tool-workpiece gap remains small. This is because when the gap remains small, the spark does not get dispersed in the electrolyte and a large proportion of the spark energy reaches the workpiece [29]. As a result, the MRR in the process increases. However, when the feed rate is kept at more than the actual linear MRR (especially in a drilling operation), there is a physical contact between the tool and the workpiece. In such a situation, sparking occurs only from the curved surface of the tool, which eventually hampers the dimensional accuracy of the process. Tool breakage may also occur in this case.

3.5.1.8 Gas Film

During ECSM, gas bubbles cover the tool surface which is immersed in the electrolyte solution and form a film. This film comprises of hydrogen gas and vapours of electrolyte generated due to Ohmic heating [30]. The thickness of this gas film determines the rate and intensity of sparking during machining. Low film thickness is ideal for achieving a high degree of accuracy in the machined feature. Film thickness can be reduced by increasing the concentration of hydroxyl ions in the solution [30]. A low value of applied voltage is useful for the generation of a dense and compact gas film. However, when high frequency voltage pulses are used, the magnitude of the pulses must be kept high to ensure that the film is formed in a short span of time.

3.5.1.9 Gap between Tool and Auxiliary Electrode

The tool-auxiliary electrode gap (also known as interelectrode gap) plays a significant role in determining machining performance in ECSM. When non-conducting materials are machined, an auxiliary electrode is used which is kept at a spacing of a few millimetres from the tool. A small interelectrode gap leads to the formation of an unstable gas film over the tool and results in uneven sparking. As this gap increases, the thickness of the gas film decreases because the gap resistance increases. Stability of the film also increases because the current density over the tool surface decreases. Consequently, sparking becomes stable and uniform, resulting in an increase in the MRR and a superior machined surface quality [31].

3.5.2 PERFORMANCE PARAMETERS

3.5.2.1 Material Removal Rate

In ECSM, the MRR depends upon electrolyte type, level, and concentration; applied voltage, pulse frequency, and duty ratio. The MRR increases as the magnitude of the applied voltage increases due to the increase in spark intensity [31]. Researchers have observed that the MRR is higher when pulsed DC voltage is used rather than a constant DC voltage. The MRR also increases with an increase in the concentration of the electrolyte solution. This is because at higher concentrations, electrochemical processes get accelerated and gas bubbles are generated in a larger volume, resulting in a uniform gas film over the tool surface. Moreover, the MRR is higher when NaOH is used as an electrolyte in place of $NaNO_3$. The level of electrolyte in the machining chamber also affects the MRR in the process. With an increase in the level of electrolyte, the MRR decreases because more sparks get generated from the side wall of the tool instead of its tip [32].

3.5.2.2 Tool Wear Rate

Similarly to EDM, a key issue associated with the ECSM process is that of tool wear. The generated sparks melt and vaporize a volume of material from the tool as well, resulting in a rough and non-uniform surface. Due to this, the surface quality of the machined feature is adversely affected. Over a period of time, different methods have been proposed to minimize the tool wear in ECSM. By lowering the magnitude of the applied voltage and electrolyte conductivity, the intensity of sparks can be reduced and tool wear can be minimized. Researchers have also observed that tool wear rate is relatively higher when machining is performed with an aqueous $NaNO_3$ electrolyte as opposed to that obtained with NaOH [31]. Tool wear can also be minimized by imparting rotary motion to the tool. This prevents sparking from the same point all the time, on the tool surface [33].

3.5.2.3   Overcut in the Process

For any process to possess a high degree of machining accuracy, the overcut associated with it must be low. In ECSM, machining overcut is controlled predominantly by the magnitude of the applied voltage and the conductivity of the electrolyte solution used. When the applied potential and/or the electrolyte conductivity is high, gas bubbles are generated more vigorously, resulting in the formation of a thick gas layer over the tool surface. Consequently, this leads to a more frequent and intensified sparking from the flat surface as well as the curved surface of the tool. As a result of this, a larger volume of material gets melted and vaporized from the workpiece boundary and the feature size increases [34]. Therefore, both applied voltage and electrolyte concentration must be maintained at an optimal value such that neither machining accuracy nor machining efficiency is significantly compromised.

3.5.2.4   Recast Layer

The molten material expelled from the machining zone accumulates near the edge of the machined feature and solidifies. This layer of re-solidified material is known as the recast layer and it adversely affects the surface morphology of the feature. The thickness of the recast layer in ECSM is less than that in the EDM process. This is because most of the re-solidified material gets removed due to anodic dissolution [35].

3.5.2.5   Heat Affected Zone

Since the primary mode of material removal in ECSM is melting and vaporization due to the thermal energy of the sparks, a heat affected zone (HAZ) often gets created near to the edge of the machined feature. This HAZ is undesirable as it leads to the creation of micro-cracks and an irregular surface finish. Experimental observations by different researchers suggest that by decreasing the duty ratio and increasing the frequency of the voltage pulses, the dimensions of the HAZ can be diminished [36]. It is also proposed that by adding a surfactant in the electrolyte solution, the thickness of the gas film near the tool surface can be reduced. This minimizes the undesirable effect of stray corrosion and subsequently, the dimension of the heat affected zone. In addition, the use of a tool with an insulated curved surface can also minimize the HAZ significantly.

3.6   POWER SUPPLY SYSTEM IN ECSM

Electrical parameters such as applied voltage, pulse on-and-off time, frequency, discharge current, and electrode polarity plays a vital role in improving process

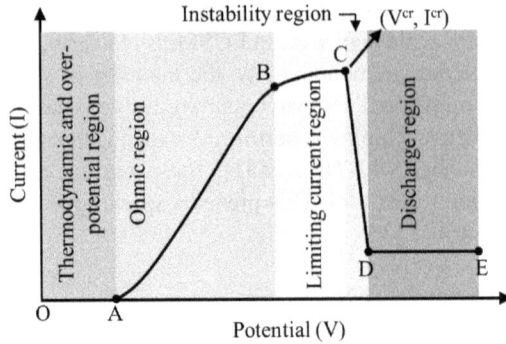

FIGURE 3.7   A typical plot of current versus voltage for ECSM process.

performance. Both constant and pulsating DC power sources can be used in the process. However, the use of a pulsating DC power source is often recommended as the released spark energy is more concentrated, which results in better machining accuracy. Also, the heat produce in this case is less and thus minimises the thermal damage to the workpiece. On increasing the magnitude of the applied voltage, the intensity of bubble formation also increases and consequently, sparks of high intensity form and increase the material removal rate of the process [36, 37]. A typical graph of current against voltage for the process is divided into five different regions as shown in Figure 3.7. They are as follows:

1   Thermodynamic and over-potential region: In this region, the current flowing in the circuit is negligible and as a result, the process of electrolysis does not initiate.
2   Ohmic region: Here, the current increases almost linearly with the applied potential.
3   Limiting current region: The circuit current in this region becomes almost constant and attains a peak value.
4   Instability region: In this region, the circuit current drops sharply, accompanied by gas film formation around the tool surface which is in contact with the electrolyte.
5   Discharge region: In this region, a current of very small magnitude flows in the circuit because the tool gets completely insulated by the gas film. Finally, in this region, breakdown of the ionized gas column takes place which is accompanied by a spark. Now, with the energy of this spark, machining can initiate.

3.7   TOOLS IN ECSM

The tool electrode plays a significant role in the ECSM process. When conductive materials are machined, tool and workpiece form the two electrodes of the system. However, when a non-conducting material is machined, the tool is connected to one terminal of the battery and an auxiliary electrode (made of conducting material) is used which is connected to the other terminal of the battery [35]. The surface area of tool is usually kept significantly smaller (about 100 times) than that of auxiliary electrode [38]. This is to ensure high current density in the tool. The tool is connected to the negative terminal of the battery. It has been observed that a tool with a pointed tip is more effective in concentrating the spark energy into a small area and thereby, enhancing the machining accuracy. In this section, different tool related factors that influence the machining performance are discussed.

3.7.1 SHAPE AND DIMENSION OF THE TOOL

The majority of research carried out in ECSM involves a tool with a cylindrical shape and diameter less than 1 mm. However, a tool with a spherical end and a drill tool have also been tested for their effectiveness in the process. The curved surface of the spherical tool enables the electrolyte to function optimally over its entire area. This helps in reducing the current density (concentration) which is required for uniform bubble formation and increasing the intensity of sparks [39, 40]. With a spherical tool, the shape accuracy of the machined feature is greater with the cylindrical tool. A drill tool, rotating about its axis, is found to be effective in enhancing the flushing of machining products out of the narrow gap and promotes the formation of a uniform film. Figure 3.8 shows the SEM image of holes machined with cylindrical and spherical tools.

In ECSM, tool dimensions are also important in determining the MRR. As the area of the active tool increases, the tool-electrolyte contact area also increases. This enhances the current flow between the two electrodes. However, in consequence, the

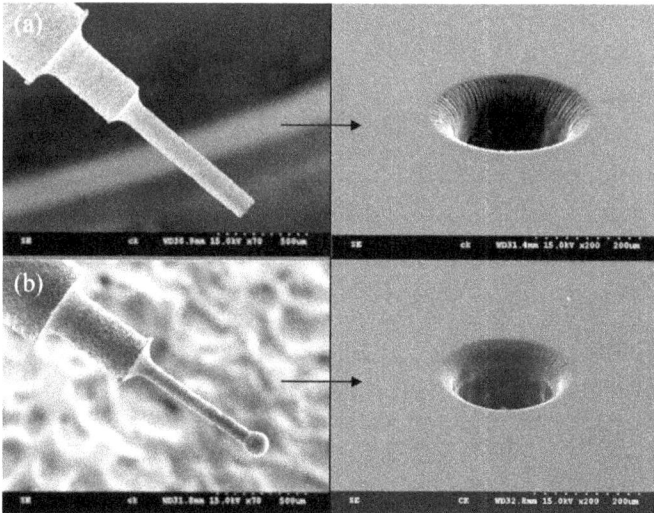

FIGURE 3.8 SEM image of (a) tool with a flat end and hole machined using it, and (b) tool with a spherical end and hole machined using it [41].

critical voltage needed for sparking to take place also increases. This is because now the breakdown of the gas film takes place over a larger area.

3.7.2 MATERIAL AND SURFACE MORPHOLOGY OF THE TOOL

Tool material is an important parameter in ECSM which has a significant impact on process performance. An ideal tool material must have: good electrical conductivity, good chemical stability, resistance towards corrosion, and good bending stiffness. Different tool materials tested by researchers in ECSM are: graphite, high speed steel, stainless steel 304, carbon alloy steel, copper, tungsten, and tungsten carbide. Researchers have shown that tool wear initiates at the highest applied voltage for tungsten carbide, followed by steels, and brass at the lowest voltage. This is mainly because the melting point of brass is lowest when compared to that of steels and tungsten carbide. It was also observed that the thermal expansion was highest in the case of steels and lowest in the case of tungsten/tungsten carbide.

Apart from the tool, different materials used for making auxiliary electrodes are: platinum, graphite, stainless steel 304, copper, and nickel. The material for the auxiliary electrode must possess good chemical stability and therefore, graphite and platinum are most widely used.

The thickness of the gas film which forms over the tool surface depends upon the wettability of the tool. The higher the wettability, the thicker will be the layer. The wettability of any surface depends upon its roughness [42]. Therefore, a tool with a rougher surface is more wettable and leads to the formation of gas bubbles of greater diameter. These bubbles coalesce and form a thick gas layer over the surface and enhance the MRR in the process [43]. Figure 3.9 shows a mechanism for material removal when a tool with surface textures is used in machining.

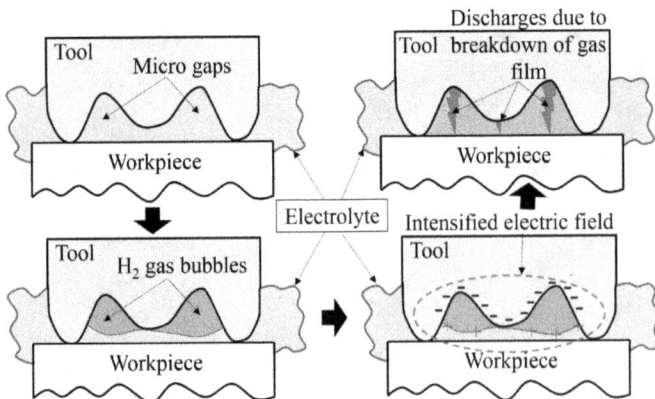

FIGURE 3.9 Schematic diagram showing the narrow gaps between tool and workpiece, entrapment of gas bubbles in these narrow gaps, intensification of electric field about this gap, and sparking in this gap.

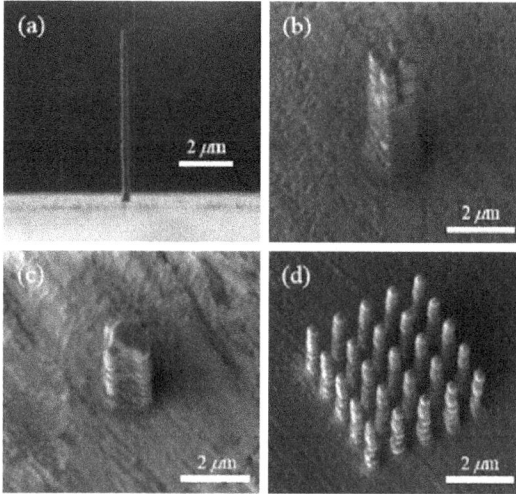

FIGURE 3.10 SEM images of nano-electrodes developed by FIB-CVD method: (a) pillar, (b) horn, (c) corner, (d) array of micro-tools [44].

3.7.3 ADVANCEMENT IN TOOL FABRICATION

Tools used in ECSM are generally made by wire electrical discharge grinding (WEDG), electrochemical machining, wire EDM, and micro grinding process. Amongst these, WEDG is the most widely used method since the chances of tool misalignment are low in this case. The tool material is made as the anode and is given a rotary motion about its axis. High aspect ratio macro as well as micro tools can be fabricated using this method. Recently, researchers have fabricated 3D micro tool electrodes by using the focus-ion-beam chemical vapor deposition (FIB-CVD) method. On the polished tip of a stainless-steel tool, a micro tool of diameter 10 μm and height 28 μm was developed using an ion beam of very low current (≈ 3977 pA). Nano-pillars of diameter 210 nm were also developed on a copper substrate as shown in Figure 3.10. However, this method of tool fabrication is very expensive and time consuming.

3.8 HYBRIDIZATIONS OF THE ECSM PROCESS

3.8.1 ELECTROCHEMICAL SPARK ABRASIVE DRILLING (ECSAD)

In ECSAD, a tool is coated with non-conductive abrasive particles such as silicon carbide, alumina, and diamond for carrying out the drilling operation. In this process, three different types of energies are used. They are chemical energy, thermal energy, and mechanical energy. A tool coated with abrasive particles rotates about its axis and is given a feed motion perpendicular to the surface of the workpiece. Chemical energy is used in chemical etching of the workpiece, thermal energy is used in the form of sparks to melt the workpiece material, and mechanical energy is used to remove the softened workpiece material by the ploughing action of abrasive particles. These abrasive particles also assist in maintaining a gap between the tool and workpiece (almost equal to the size of the protrusion of abrasive particles) which is necessary for the generation of gas bubbles near the tool to initiate sparking [13].

3.8.2 POWDER MIXED ELECTROCHEMICAL SPARK MACHINING (PM-ECSM)

In PM-ECSM, micro and nano-sized conducting and non-conducting particles are added in the electrolyte for machining. By mixing these particles, the spot size of sparks decreases and these sparks also become stable. Due to this, machining accuracy and the quality of the process improves. This also enhances the electrolyte conductivity, which subsequently lowers the critical voltage needed to initiate the sparking process [45].

3.8.3 ULTRASONIC ASSISTED ELECTROCHEMICAL SPARK MACHINING (UA-ECSM)

Ultrasonic-assisted electrochemical spark machining is another hybridization of ECSM which combines mechanical, thermal and chemical energy for improving machining performance. To machine high aspect ratio features, supply and circulation of electrolyte in narrow discharge gaps is essential. This is otherwise difficult to achieve using the standard ECSM process. Imparting ultrasonic vibrations to either tool, workpiece, or the electrolyte triggers the circulation of electrolyte especially in the sparking region. This aids in the flushing of sludge particles from the gap and promotes the formation of a thin gas film over the tool surface (necessary for the channelling of discharge energy) [46].

3.8.4 MAGNETIC FIELD ASSISTED ELECTROCHEMICAL SPARK MACHINING (MA-ECSM)

In MA-ECSM, magnetic field improves the circulation of electrolyte, uniformity of the gas film, depth of hole, MRR and machining quality. Figure 3.11 shows a schematic diagram of the setup used for the MA-ECSM process. Under the influence of a magnetic field, Lorentz force provides momentum to the ions present in the electrolyte and this phenomenon is known as magneto hydrodynamic convection [47].

In the presence of a magnetic field, gas bubbles in the electrolyte get detached readily from the tool surface, resulting in a thin and stable film near the tool. This is assistive in the channelling of spark energy in the machining area and improvement in the machining accuracy.

3.8.5 MAGNETIC FIELD ASSISTED TRAVELING WIRE ELECTROCHEMICAL SPARK MACHINING (MA-TWECSM)

To overcome the problem of sporadic movement of electrolyte in the machining area which often arises in traveling wire electrochemical spark machining, researchers have applied a magnetic field across it. This initiates magneto hydrodynamic convection in the electrolyte and stabilizes the flow. This subsequently leads to the stabilization of

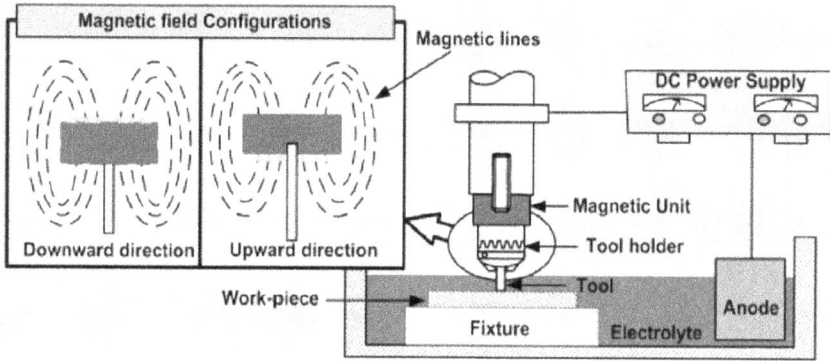

FIGURE 3.11 Schematic diagram showing the arrangement of magnetic field assisted electrochemical spark machining process [2].

sparks and an increment in the MRR. The heat affected zone and machining overcut also reduces in the presence of a magnetic field across the machining area [48].

3.9 VARIANTS OF THE ECSM PROCESS

3.9.1 ELECTROCHEMICAL SPARK DRILLING (ECSD)

The electrochemical spark machining process is analogous to a conventional drilling process. A single point tool, generally of the diameter of the required hole, is used and is given feed motion in the direction perpendicular to the workpiece surface. This process is capable of drilling both blind and through holes. Material is removed due to chemical etching and sparks which originate from both the flat and the cylindrical part of the tool. Rotation of the tool about its axis accelerates the flushing of electrolyte from the machining region. Apart from non-conductors, the process is also suitable for semi-conducting and conducting materials as well [49].

3.9.2 ELECTROCHEMICAL SPARK MILLING

The electrochemical spark milling process is analogous to the conventional milling process and is generally used to fabricate macro/micro channels, grooves, and 3D structures on non-conductive workpieces. A single point tool (generally with a flat end) is used to trace a pre-determined path on the workpiece surface. The process is most suited for machining 3D structures with a complicated shape. However, due to the bottleneck in electrolyte flushing at deep regions of the workpiece, the process might not be suitable for generating high aspect-ratio structures especially when the feature size is in micrometres.

3.9.3 WIRE ELECTROCHEMICAL SPARK MACHINING (WECSM)

WECSM is generally used to machine 2D contour profiles on flat as well as cylindrical non-conducting as well as conducting workpieces. Here, a conductive wire

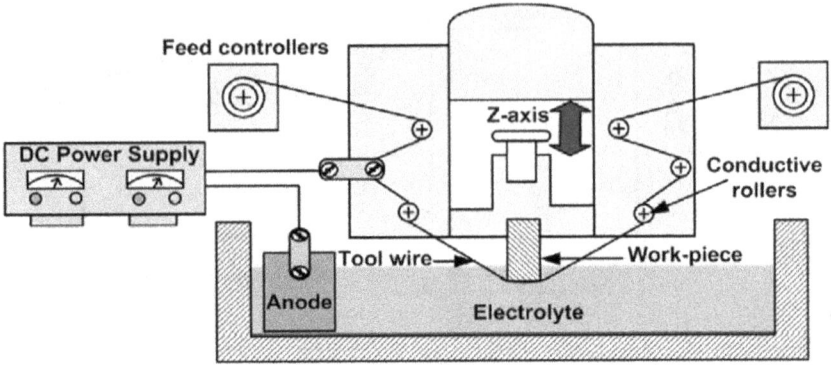

FIGURE 3.12 Schematic diagram showing the arrangement of wire electrochemical spark machining process [2].

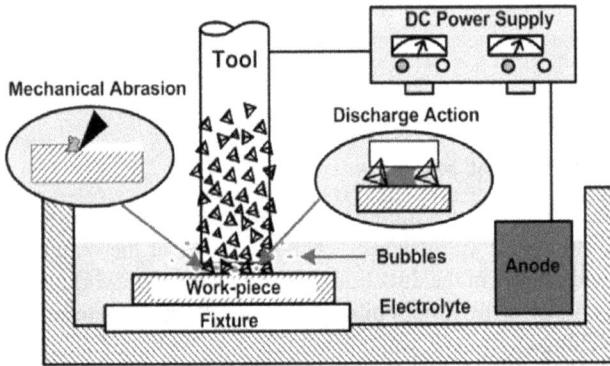

FIGURE 3.13 Schematic diagram showing the arrangement of electrochemical spark grinding process [2].

with diameter of a few hundreds of micrometres is made to trace a desired profile in a workpiece with a constant velocity as shown in Figure 3.12. This wire is also made to travel along its axis, while maintaining its tension, to flush out the debris from the narrow machining gap and stabilize the process. As a wire is used for machining, the need for a shaped tool and power supply with high voltage ratings is reduced. Care must be taken to maintain the correct tension in the wire so that tool radial vibrations won't hamper the uniformity of the machined feature [45].

3.9.4 ELECTROCHEMICAL SPARK GRINDING (ECSG)

ECSG works on the combined action of chemical etching, spark erosion, and mechanical grinding. A rotating wheel is used as a tool (negative polarity) in this process. This is coated with diamond or abrasive particles as shown in Figure 3.13. During sparking, material beneath the wheel gets softened due to sparking and subsequently gets removed due to grinding action. In this process, the material removal rate is about three times higher than that of a standard ECSM process and surface roughness is also superior [50, 51].

3.9.5 DIE SINKING-ELECTROCHEMICAL SPARK MACHINING (DS-ECSM)

In this process, the tool is a negative replica of a required feature shape and dimension on the workpiece. In this process, the tool does not rotate about is axis and is sunk into the workpiece to carve out cavities in them. The process is generally used for fabricating tiny and shallow dies on conductive as well as non-conductive workpieces. The material removal rate in DS-ECSM is superior to that of ECM and EDM. The process also offers better dimensional accuracy which is comparable to that achieved in an EDM process [52].

3.10 SOME DIRECTIONS FOR FUTURE RESEARCH

In this section, some of the possible areas where future research can be concentrated are highlighted. They are as follows:

1   Up until now, ECSM has been used for the machining of only non-conductive materials such as quartz and glass. There is a need to examine the machinability of other such non-conducting engineering materials such as ceramics, metal matrix composites, and other functionally graded materials.
2   Electrolyte solutions of alkalis such as NaOH and KOH have been used very extensively in ECSM for machining but this is not environment friendly. Neutral electrolytes such as $NaNO_3$, NaCl, and KCl can also be tested for their compatibility in machining as these electrolytes are relatively safer for underground deposition.
3   There is an evident deficit in an analytical model which can relate material removal rate and machining overcut in the process to different input parameters. Such a model will reduce the need for carrying out expensive trial experiments to identify optimum values for machining parameters.
4   Additionally, ultrasonic vibrations can be coupled with other configurations of ECSM to study their effectiveness over the existing configurations in terms of machining efficiency, accuracy, and quality.

3.11 CONCLUSIONS

Some of the noteworthy conclusions from the chapter are as follows:

1   ECSM process is a hybridization of the EDM and ECM processes and is capable of machining both conductive and non-conductive materials and at both macro and micro level.
2   Researchers around the globe have made attempts to improve the MRR, surface roughness, machining tolerance, and overcut generated in ECSM. Studies have also been conducted to study the effect of tool shape and surface

morphology, effect of electrolytic and power choices on various performance parameters.

3   Over time, different hybridizations and variants of ECSM have been developed. These include hybridizations such as: ECSAD, PM-ECSM, UA-ECSM, MA-ECSM, and MA-TWECS. Variants of ECSM include: ECSD, Electrochemical spark milling, WECSM, ECSG, and DS-ECSM.

4   When the ECSM process is assisted with ultrasonic vibrations, the flow of electrolyte in the narrow machining gap increases. This prevents arcing from taking place between the tool and workpiece which further results in improved machining efficiency and quality. The numbers of micro cracks, which usually get generated in the vicinity of the machined feature also decrease.

5   When abrasive particles made of silicon carbide are mixed with the electrolyte solution, micro cracks and the recast layer get removed from the machining region and this results in an improved surface quality.

REFERENCES

[1]   Holmberg, J.; Berglund, J.; Wretland, A.; Beno, T. Evaluation of Surface Integrity after High Energy Machining with EDM, Laser Beam Machining and Abrasive Water Jet Machining of Alloy 718. Int. J. Adv. Manuf. Technol. 2019, 100(5–8):1575–1591. DOI: 10.1007/s00170-018-2697-z.

[2]   Singh T.; Dvivedi A. Developments in Electrochemical Discharge Machining: A Review on Electrochemical Discharge Machining, Process Variants and Their Hybrid Methods. Int. J. Mach. Tools Manuf. 2016 Jun 1, 105, 1–3.

[3]   Kumar N.; Mandal N.; Das A. K. Micro-machining Through Electrochemical Discharge Processes: A Review. Mat. Manuf. Proc. 2020 Mar 11, 35(4), 363–404.

[4]   Karafuji, H.; Suda, K. Electrical Discharge Drilling of Glass. CIRP Ann. 1968, 16, 416–419.

[5]   Crichton, I. M.; McGeough, J. A. Studies of the Discharge Mechanisms in Electrochemical Arc Machining. J. Appl. Electrochem. 1985, 15(1), 113–119. DOI: 10.1007/BF00617748.

[6]   Basak, I.; Ghosh, A. Mechanism of Spark Generation during Electrochemical Discharge Machining: A Theoretical Model and Experimental Verification. J. Mater. Process. Technol. 1996, 62(1–3), 46–53. DOI: 10.1016/0924-0136(95)02202-3.

[7]   Jain, V. K.; Dixit, P. M.; Pandey, P. M. On the Analysis of the Electrochemical Spark Machining Process. Int. J. Mach. Tools Manuf. 1999, 39(1), 165–186. DOI: 10.1016/S0890-6955(98)00010-8.

[8]   Kulkarni, A.; Sharan, R.; Lal, G. K. An Experimental Study of Discharge Mechanism in Electrochemical Discharge Machining. Int. J. Mach. Tools Manuf. 2002, 42(10), 1121–1127. DOI: 10.1016/S0890-6955(02)00058-5.

[9]   Fascio, V.; Wüthrich, R.; Bleuler, H. Spark Assisted Chemical Engraving in the Light of Electrochemistry. Electrochim. Acta. 2004, 49(22–23), 3997–4003. DOI: 10.1016/j.electacta.2003.12.062.

[10] Wüthrich, R.; Bleuler, H. A Model for Electrode Effects Using Percolation Theory. Electrochim. Acta. 2004, 49(9–10), 1547–1554. DOI: 10.1016/j.electacta.2003.11.014.

[11] Ruszaj, A. Unconventional Processes of Ceramic and Composite Materials Shaping. Mechanik. 2017, 90(3), 39–44.

[12] Ghosh, A. Electrochemical Discharge Machining: Principle and Possibilities. Sadhana. 1997, 22(3), 435–447. DOI: 10.1007/BF02744482.

[13] Jain, V. K.; Choudhury, S. K.; Ramesh, K. M. On the Machining of Alumina and Glass. Int. J. Adv. Manuf. Technol. 2002, 42(11), 1269–1276.

[14] Gupta, P. K.; Dvivedi, A.; Kumar, P. Effect of Pulse Duration on Quality Characteristics of Blind Hole Drilled in Glass by ECDM. Mater. Manuf. Process. 2016, 31(13), 1740–1748. DOI: 10.1080/ 10426914.2015.1103857.

[15] Liu, J. W.; Yue, T. M.; Guo, Z. N. An Analysis of the Discharge Mechanism in Electrochemical Discharge Machining of Particulate Reinforced Metal Matrix Composites. Int. J. Mach. Tools Manuf. 2010, 50(1), 86–96. DOI: 10.1016/j.ijmachtools.2009.09.004.

[16] Sabahi, N.; Razfar, M. R. Investigating the Effect of Mixed Alkaline Electrolyte (NaOH+ KOH) on the Improvement of Machining Efficiency in 2D Electrochemical Discharge Machining (ECDM). Int. J. Adv. Manuf. Technol. 2018, 95(1–4), 643–657. DOI: 10.1007/s00170-017-1210-4.

[17] Jha, N. K.; Singh, T.; Dvivedi, A.; Rajesha, S. Experimental Investigations into Triplex Hybrid Process of GA-RDECDM during Subtractive Processing of MMC's. Mater. Manuf. Process. 2019, 34(3), 243–255. DOI: 10.1080/ 10426914.2018.1512126.

[18] Dong, S.; Wang, Z.; Wang, Y. High-Speed Electrochemical Discharge Drilling (HSECDD) for Micro-Holes on C17200 Beryllium Copper Alloy in Deionized Water. Int. J. Adv. Manuf. Technol. 2017, 88(1–4), 827–835. DOI: 10.1007/ s00170-016-8645-x.

[19] Lee, E. S.; Howard, D.; Liang, E.; Collins, S. D.; Smith, R. L. Removable Tubing Interconnects for Glass-Based Micro-Fluidic Systems Made Using ECDM. J. Micromech. Microeng. 2004, 14(4), 535. DOI: 10.1088/0960-1317/14/4/014.

[20] Hof, L. A.; Wüthrich, R. Industry 4.0–Towards Fabrication of Mass-Personalized Parts on Glass by Spark Assisted Chemical Engraving (SACE). Manuf. Lett. 2018, 15, 76–80. DOI: 10.1016/j. mfglet.2017.12.003.

[21] Fascio, V.; Langen, H. H.; Bleuler, H.; Comninellis, C. Investigations of the Spark Assisted Chemical Engraving. Electrochem. Commun. 2003, 5(3), 203–207. DOI: 10.1016/ S1388-2481(03)00018-3.

[22] Kumar Gupta, P.; Dvivedi, A.; Kumar, P. Effect of Electrolytes on Quality Characteristics of Glass during ECDM. Key Eng. Mater. 2015, 658, 141–145. DOI: 10.4028/www.scientific.net/ KEM.658.141.

[23] Jain, V. K.; Adhikary, S. On the Mechanism of Material Removal in Electrochemical Spark Machining of Quartz under Different Polarity Conditions. J. Mater. Process. Technol. 2008, 200(1–3), 460–470. DOI: 10.1016/j.jmatprotec.2007.08.071.

[24] West, J.; Jadhav, A. ECDM Methods for Fluidic Interfacing through Thin Glass Substrates and the Formation of Spherical Microcavities. J. Micromech. Microeng. 2007, 17(2), 403. DOI: 10.1088/0960-1317/17/2/028.

[25] Zhang, Z.; Huang, L.; Jiang, Y.; Liu, G.; Nie, X.; Lu, H.; Zhuang, H. A Study to Explore the Properties of Electrochemical Discharge Effect Based on Pulse Power Supply. Int. J. Adv. Manuf. Technol. 2016, 85(9–12), 2107–2114. DOI: 10.1007/ s00170-015-8302-9.

[26] Mehrabi, F.; Farahnakian, M.; Elhami, S.; Razfar, M. R. Application of Electrolyte Injection to the Electro-Chemical Discharge Machining (Ecdm) on the Optical Glass. J. Mater. Process. Technol. 2018, 255, 665–672. DOI: 10.1016/j. jmatprotec.2018.01.016.

[27] Laio, Y. S.; Wu, L. C.; Peng, W. Y. A Study to Improve Drilling Quality of Electrochemical Discharge Machining ECDM Process. Procedia CIRP. 2013, 6, 609–614. DOI: 10.1016/j. procir.2013.03.105.

[28] Sabahi, N.; Razfar, M. R.; Hajian, M. Experimental Investigation of Surfactant-Mixed Electrolyte into Electrochemical Discharge Machining (ECDM) Process. J. Mater. Process. Technol. 2017, 250, 190–202. DOI: 10.1016/j.jmatprotec.2017.07.017.

[29] Ranganayakulu, J.; Hiremath, S. S.; Parametric, P. L. Analysis and a Soft Computing Approach on Material Removal Rate in Electrochemical Discharge Machining. Int. J. Manuf. Technol. Manag. 2011, 24(1–4), 23–39. DOI: 10.1504/IJMTM.2011.046758.

[30] Cheng, C. P.; Wu, K. L.; Mai, C. C.; Yang, C. K.; Hsu, Y. S.; Yan, B. H. Study of Gas Film Quality in Electrochemical Discharge Machining. Int. J. Mach. Tools Manuf. 2010, 50(8), 689–697. DOI: 10.1016/j.ijmachtools.2010.04.012.

[31] Jawalkar, C. S.; Sharma, A. K.; Kumar, P. Investigations on Performance of ECDM Process Using NaOH and NaNO3 Electrolytes while Micro Machining Soda Lime Glass. Int. J. Manuf. Technol. Manag. 2014, 28(1–3), 80–93. DOI: 10.1504/ IJMTM.2014.064623.

[32] Gupta, P. K.; Dvivedi, A.; Kumar, P. Developments on Electrochemical Discharge Machining: A Review of Experimental Investigations on Tool Electrode Process Parameters. Proc. Inst. Mech. Eng. B. J. Eng. Manuf. 2015, 229(6), 910–920.

[33] Huang, S. F.; Liu, Y.; Li, J.; Hu, H. X.; Sun, L. Y. Electrochemical Discharge Machining Micro-Hole in Stainless Steel with Tool Electrode High-Speed Rotating. Mater. Manuf. Process. 2014, 29 (5), 634–637. DOI: 10.1080/10426914.2014.901523.

[34] Zheng, Z. P.; Su, H. C.; Huang, F. Y.; Yan, B. H. The Tool Geometrical Shape and Pulse-off Time of Pulse Voltage Effects in a Pyrex Glass Electrochemical Discharge Micro Drilling Process. J. Micromech. Microeng. 2007, 17(2), 265. DOI: 10.1088/ 0960-1317/17/2/012.

[35] Kang, X.; Tang, W. Micro-Drilling in Ceramic-Coated Ni- Superalloy by Electrochemical Discharge Machining. J. Mater. Process. Technol. 2018, 255, 656–664. DOI: 10.1016/j. jmatprotec.2018.01.014.

[36] Kim, D. J.; Ahn, Y.; Lee, S. H.; Kim, Y. K. Voltage Pulse Frequency and Duty Ratio Effects in an Electrochemical Discharge Microdrilling Process of Pyrex Glass. Int. J. Mach. Tools Manuf. 2006, 46(10), 1064–1067. DOI: 10.1016/ j.ijmachtools.2005.08.011.

[37] Singh, T.; Dvivedi, A. On Pressurized Feeding Approach for Effective Control on Working Gap in ECDM. Mater. Manuf. Process. 2018, 33(4), 462–473. DOI: 10.1080/ 10426914. 2017.1339319.

[38] Wüthrich, R.; Ziki, J. A. Micromachining Using Electrochemical Discharge Phenomenon; William Andrew: Oxford, 2009.

[39] Bhattacharyya, B.; Doloi, B. N.; Sorkhel, S. K. Experimental Investigations into Electrochemical Discharge Machining (ECDM) of Non-Conductive Ceramic Materials. J. Mater. Process. Technol. 1999, 95(1–3), 145–154. DOI: 10.1016/S0924-0136(99)00318-0.

[40] Behroozfar, A.; Razfar, M. R. Experimental Study of the Tool Wear during the Electrochemical Discharge Machining. Mater. Manuf. Process. 2016, 31(5), 574–580. DOI: 10.1080/ 10426914.2015.1004685.

[41] Yang, C. K.; Wu, K. L.; Hung, J. C.; Lee, S. M.; Lin, J. C.; Yan, B. H. Enhancement of ECDM Efficiency and Accuracy by Spherical Tool Electrode. Int. J. Mach. Tools Manuf. 2011, 51(6), 528–535.

[42] Yang, C. T.; Song, S. L.; Yan, B. H.; Huang, F. Y. Improving Machining Performance of Wire Electrochemical Discharge Machining by Adding Sic Abrasive to

Electrolyte. Int. J. Mach. Tools Manuf. 2006, 46, 2044–2050. DOI: 10.1016/j.ijmachtools.2006.01.006.

[43] Xu, Z.; Zhang, C. A Tube Electrode High-speed Electrochemical Discharge Drilling Method without Recast Layer. Procedia CIRP. 2018, 68, 778–782. DOI: 10.1016/j.procir.2017.12.154.

[44] Guo, D.; Wu, X.; Lei, J.; Xu, B.; Kometani, R.; Luo, F. Fabrication of Micro/Nanoelectrode Using Focused-Ion-Beam Chemical Vapor Deposition, and Its Application to Micro-ECDM. Procedia CIRP. 2016, 42, 733–736. DOI: 10.1016/j.procir.2016.02.310.

[45] Saranya, S.; Sankar, A. R. Effect of Tool Shape and Tool Feed Rate on the Machined Profile of a Quartz Substrate Using an Electrochemical Discharge Machining Process. In 2nd International Symposium on Physics and Technology of Sensors (ISPTS), Pune, India, 2015, 313–316.

[46] Tang, L.; Zhao, G. G. Discussing the Measure of Improving Pyrex Glass ECDM Removal Rate. Adv. Mater. Res. 2012, 411, 319–322. DOI: 10.4028/www.scientific.net/AMR.411.319.

[47] Cheng, C. P.; Wu, K. L.; Mai, C. C.; Hsu, Y. S.; Yan, B. H. Magnetic Field-Assisted Electrochemical Discharge Machining. J. Micromech. Microeng. 2010, 20(7), 075019. DOI: 10.1088/0960- 1317/20/7/075019.

[48] Rattan, N.; Mulik, R. S. Experimental Set up to Improve Machining Performance of Silicon Dioxide (Quartz) in Magnetic Field Assisted TW-ECSM Process. Silicon. 2018, 10(6), 2783–2791. DOI: 10.1007/s12633-018-9818-z.

[49] Tang, W.; Kang, X.; Zhao, W. Enhancement of Electrochemical Discharge Machining Accuracy and Surface Integrity Using Side insulated Tool Electrode with Diamond Coating. J. Micromech. Microeng. 2017, 27(6), 065013. DOI: 10.1088/1361 6439/aa6e94.

[50] Liu, J.; Lin, Z.; Guo, Z.; Jiang, S.; Yue, T.; Chen, X. A Study of the Materials Removal Mechanism of Grinding-Aided Electrochemical Discharge Machining of Metal Matrix Composites. Adv. Compos. Lett. 2018, 27(5), 096369351802700504. DOI: 10.1177/096369351802700504.

[51] Liu, J. W.; Yue, T. M.; Guo, Z. N. Grinding-Aided Electrochemical Discharge Machining of Particulate Reinforced Metal Matrix Composites. Int. J. Adv. Manuf. Technol. 2013, 68(9–12), 2349–2357. DOI: 10.1007/s00170-013-4846-8.

[52] Khairy, A. B. E.; McGeough, J. A. Die-Sinking by Electroerosion- Dissolution Machining. CIRP Annals. 1990, 39(1), 191–195. DOI: 10.1016/S0007-8506(07)61033-6.

4 Sequential EDM and ECM Process

R. M. Tayade and B. Doloi

Department of Production Engineering,
Jadavpur University, Kolkata, India

CONTENTS

4.1 INTRODUCTION

There is a need to explore different innovative methods and new manufacturing processes to achieve the desired improved accuracy of the items produced. At present various sequential machining combinations are being used effectively for the precision manufacturing of micro-parts. Sequential machining is defined as the plan of action whereby two or more machining processes are executed in a sequence on the same or on different machine tools [1]. EDM and ECM sequential machining have gained an edge over the other sequential machining processes due to their simplicity in setup design and machining ease. Advanced high strength engineering materials which have high hardness and cannot be cut easily by other machining processes can be machined by using an ECM-EDM sequential machining process. This sequential machining process has the capability to create complex micro-features on hard, brittle, ductile materials, as well as on workpieces with thin cross-sections. The application of non-traditional machining processes such as EDM and ECM in a sequential manner for machining sophisticated materials proves to be the best alternative. A wide range of non-traditional machining processes have been used in sequence, namely EDM and laser ablation, micro-milling and laser deburring, EDM and ultrasonic machining (USM), etc. Like in traditional machining, the different machining processes can be applied in a sequential manner in non-traditional machining also. Hence these processes can be broadly classified into two groups: traditional and non-traditional sequential machining processes [2]. Based on the purpose of the operation, the sequential machining processes have been classified into five groups, namely:

1. machining time-oriented
2. micro-tools making
3. surface quality improvement
4. energy efficiency
5. microstructure improvement [1].

Despite the numerous advantages of EDM–ECM sequential machining, comparatively less research work is reported in this sequential machining combination. One of the fundamental reasons must be the integration of both the processes at one station. The changing of machining chambers or the dielectric fluid and the management of the electrolyte for the machining process is quite complicated. Similarly, replacing the working fluid without altering the positions of the tool and of the workpiece is quite complicated. Despite these difficulties, some researchers have successfully integrated both machining processes at one station.

To improve the machining efficiency of sequential machining, Skoczypiec et al. [3] proposed a computer simulation software. The development of this software is considered to be the advanced step in the sequential machining process. In the EDM–ECM milling process, it is thought that the sequence of operation can be either electrochemical micro-milling (ECMM) first, then electro-discharge micro-milling (EDMM) or vice versa. Firstly, about 80% of the material is removed by the application of ECMM process, and then the remaining 20% of material can be removed by the EDMM finishing process. This sequential strategy is used to reduce machining time, whereas EDMM first and then ECMM is used when minimum changes in surface layer quality are required. Thus, by studying the characteristics of both the processes and considering the modeling of the first process as an initial condition for the second process, mathematical modeling has been performed. Simulation software thus developed allows the determination of sequencing strategy for both the approaches.

4.2   FUNDAMENTALS OF THE SEQUENTIAL EDM AND ECM PROCESS

In today's world of modern technology, there is a requirement for multifeatured components consisting of complicated contours, shapes such as channels, grooves, holes, projecting parts, walls, etc. Manufacturing of such elements by conventional machining methods usually takes a lot of time, because the manufacturing of different features using a single machine may not be possible. In such situations, the workpiece would have to be moved from one machine to another. Similarly, loading, setting, and unloading takes a lot of time and it hampers the dimensional accuracy of the part produced. In sequential machining, two or more machining techniques are implemented at one station. This approach lessens the time required for machining, and the reduced loading and unloading results in improved dimensional accuracy. In sequential machining, various operations are performed in series, one after the other on the workpiece, which is fixed in a specific holding device. The workpiece position will not be altered until all the predetermined machining operations have been executed. A single system setup is required to carry out both EDM and ECM processes on the workpiece. Some researchers have made an effort to design and develop the EDM–ECM system set up, so that the machining will be done without changing the tool and workpiece position. It is a well-known fact that every machining process has its own advantages and disadvantages. On the one hand, the sequential machining approach maximizes the benefits, and on the other hand, it minimizes weaknesses of the individual machining processes. The integration of two or more diverse machining processes for the development of multifunctional machine tools is a challenging task because, irrespective of distinct machining, kinematics, methodology, and machining environment, it is essential to maintain tool and workpiece position and the specific inter-electrode gap with respect to the type of process.

4.3 THE BASIC WORKING PRINCIPLE OF THE SEQUENTIAL EDM AND ECM PROCESS

Electrical discharge machining (EDM) and electrochemical machining (ECM) are two different processes. In both the process, the tool and the workpiece work in a different environment. The requirement of power supply ratings for both the processes are different. The voltage required to carry out EDM is 40 to 400V, whereas micro-EDM takes place in the range of 40 to 120V with a current value of less than 1A. The ECM takes place in the voltage range of 10 to 30V, and for micro-ECM, the voltage required is less than 10V, and the current requirement is less than 1A. In the EDM process, the workpiece is placed in the sparking zone where the inter-electrode gap is maintained in the range of 10 to 100 μm. The spark discharges produced between the tool and the workpiece melt and vaporize the material from both the electrodes. In ECM, the tool electrode and workpiece are both kept in a conductive electrolyte medium, and the material is removed smoothly, atom by atom, from the workpiece surface using electrochemical dissolution. Thus, in ECM, there is no heat affected zone, and no tool wear takes place. The performance of the ECM is superior when it is used in combination with the other machining processes [4]. Like ECM, EDM can machine any electrically conductive materials irrespective of their strength and hardness. Advanced materials such as carbides, high strength and high temperature resistant (HSHTR) materials, hardened steel, composite materials, etc., can be machined using EDM. The material removal from the workpiece surface takes place in the medium of non-conducting dielectric, whereas ECM is possible in the presence of conducting electrolyte solution. In EDM, during machining, high-intensity electric

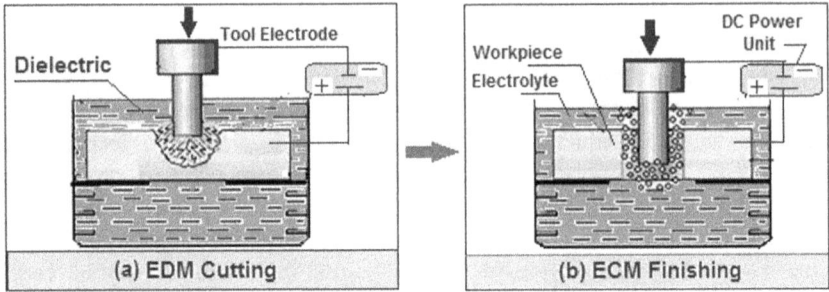

FIGURE 4.1   Working principle of sequential EDM and ECM process.

sparks are generated, which melt and vaporize the workpiece material, but in the ECM process, workpiece material is removed smoothly due to the electrochemical dissolution process. Despite the differences in basic working principles of both the processes, they have one common objective of machining of conducting materials. This is the decisive point that makes it possible to utilize EDM and ECM processes in a sequential manner. The sequence of operations of EDM and ECM is decided based on the end objective of machining. Sometimes, to maintain precision, the tool electrode of the required diameter is manufactured online, and then the same tool electrode is used for performing EDM and ECM processes. When the objective of the machining is to reduce the machining time and to improve the machining efficiency, or when a better surface finish is required on the workpiece surface, the sequence of machining preferred is EDM first and then ECM. Using EDM, the micro-features on the workpiece can be produced rapidly but as it is a thermal process, it inherently possesses certain drawbacks such as, generation of thermal stresses, deterioration of the machined surface due to the heat affected zone, induction of residual stresses, presence of the recast layer, micro-cracks and micro-pores in the HAZ. Tool wear takes place due to the spark generation during EDM machining, which deteriorates the shape and the profile of the machined part [5]. These adverse effects imparted by the EDM can be eliminated by subsequent application of the ECM process. When the sequential combination of EDM milling and ECM milling is used, then for achieving a high metal removal rate, the sequence of operation is ECM milling followed by EDM milling.

4.4   SELECTION OF PROCESS PARAMETERS FOR SEQUENTIAL EDM AND ECM PROCESSES

It has been observed that there is a significant influence of machining process parameters on machining performance. Hence to achieve precise part machining, the selection of appropriate process parameters is crucial. In the case of EDM, the important process parameters are open-circuit voltage, pulse frequency, polarity, pulse on-off time, peak current, and duty cycle, whereas, in the case of ECM, the influencing process parameters are inter-electrode gap, current density, tool feed rate, type and concentration of electrolyte, electrolyte flow rate and the reactions at the anode.

It has been mentioned that a particular sequence of machining processes has been decided by keeping certain objectives in mind, such as, to improve the surface quality or to improve the machining efficiency. Some researchers who have performed EDM and ECM processes in a sequential manner for the generation of micro-features are discussed in this chapter.

Masuzawa et al. [6] manufactured micro-nozzles by applying a sequential machining technique. Initially, the electrochemical deposition process was used to form a stainless-steel core, then a wire electro-discharge grinding (WEDG) process was used to generate the outer shape of the nozzle. The wire electrochemical grinding (WECG) process was finally applied to achieve a mirror finish on the nozzle surface.

4.5   IMPROVEMENT OF SURFACE CHARACTERISTICS BY SEQUENTIAL EDM AND ECM PROCESSES

Sequential Micro-EDM and micro-ECM combined milling processes are applied on SS-304 stainless steel plate of 400µm thickness for the generation of 3-D microstructures. Initially, the diameter of a tungsten rod tool electrode, used for machining, is reduced to 100µm by using an anti-copying block method. Machining is performed layer by layer by scanning movements of the tool electrode. The effects of various machining process parameters such as applied voltage, tool feed rate, and machining gap on the milled square cavity have been analyzed. The open-circuit voltage for EDM was 250V and the servo reference voltage was 180V. The machining voltage in ECM has been maintained in the range of 8 to 10V with separate power supply units used for machining.

In this EDM–ECM sequential machining, the EDM process is initially used for the rapid removal of material from the workpiece. The application of EDM affects the surface quality due to the generation of the recast layer. The ECM milling of a square profile is accomplished [7] using 8V, 9V, and 10V with varied machining gap of 10µm, 15µm, and 20µm. It was observed that at 8V and 9V, the recast layer from the cavity had been removed partially due to less anodic dissolution of the material. At 10V, the rate of electrochemical dissolution was sufficient to remove the recast layer completely. Similarly, machining at different machining gaps gives different results, as shown in Figure 4.2.

If the inter-electrode gap is reduced to below 10 µm, the bumps formed during EDM roughing cause sparking, which further deteriorates the surface quality. When the machining gap increased in the range of 15µm to 20µm, the current density decreases with an increased inter-electrode gap. Therefore, the micro-craters cannot be removed properly. At a 10µm machining gap, the micro-craters and micro-pores get removed completely, and a smooth surface finish is obtained. This may be due to the increased machining voltage. The energy required for anodic dissolution also increases, which efficiently removes irregularities present on the workpiece surface. Figure 4.3 represents the material removal rate with respect to the varying voltages of micro-ECM milling.

With the increase of voltage, the material removal rate increases. Since micro-EDM is used as a shaping process, the MRR is the highest. It is also observed that the tool feed rate is an important process parameter. During micro-ECM milling, if

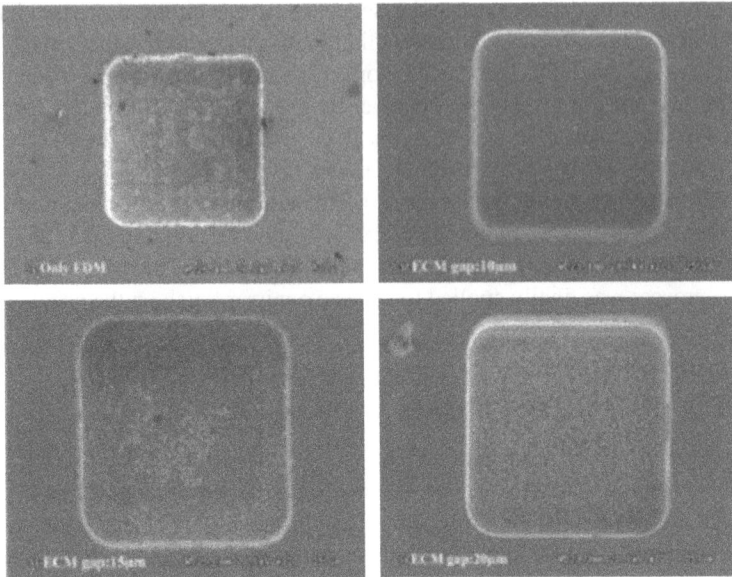

FIGURE 4.2   SEM photo of EDMed square cavity at different machining gap (machining voltage; 10V, tool feed rate; 10μm/s) [7]. With permission from Elsevier.

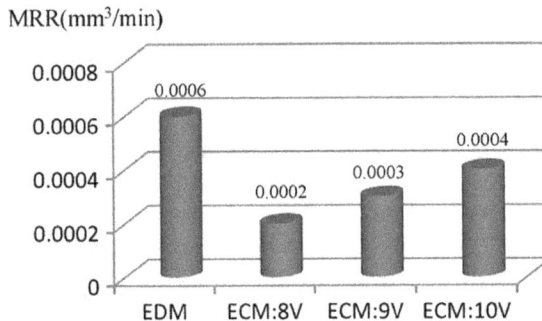

FIGURE 4.3   The material removal rate for micro-ECM finishing and micro-EDM roughing [7]. With permission from Elsevier.

the tool feed rate is less than 6 μm/s, the tool remains in the vicinity of the workpiece for a longer period of time, resulting in excessive anodic corrosion. If the tool feed rate is greater than 15 μm/s, the surface irregularities will not get removed. Whereas the tool feed rate of 10 μm/s appears to be optimum because all the surface defects get removed, and a smooth surface is obtained. EDM–ECM sequential milling allows machining of various complex 3-D microstructures.

Figure 4.4 shows a star-shaped cavity structured initially by using a micro-EDM process and then finally finished using the micro-ECM method. Micro-EDM micro-ECM sequential machining strategy for hole drilling in 1mm thick sheet of SS-304 grade stainless steel material was investigated [8]. Initially, the reduction of the

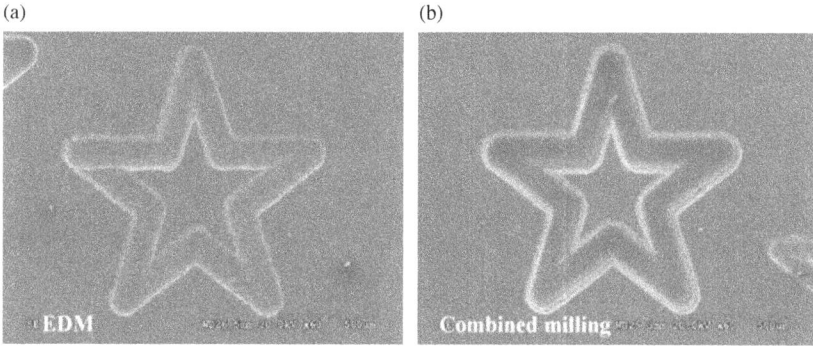

FIGURE 4.4 Star-shaped cavity milled using (a) Micro-EDM and (b) combined milling [7].

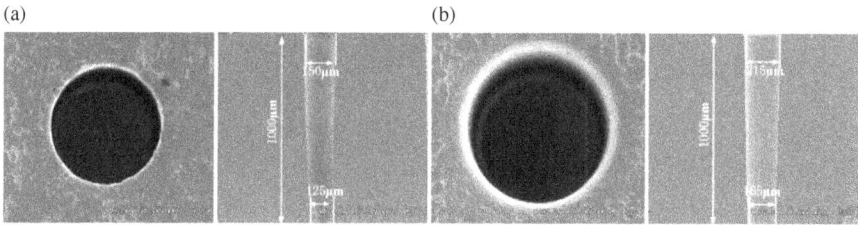

FIGURE 4.5 Hole drilled by (a) EDM and (b) sequential EDM and ECM with sectional views [8].

micro-tool diameter from 0.5 mm to 0.1 mm was carried out online by the Block Electric Discharge Grinding Process (BEDG). A through-hole drilling operation in 1 mm thick sheet has been performed using an EDM shaping process. The drilled hole surface contains an undesirable heat-affected zone and recast layer after a micro-ECM process executed in $NaClO_3$ electrolyte. The passivating nature of the electrolyte prohibits unrestrained corrosion during the process of electrochemical dissolution.

Furthermore, the important ECM process parameter, machining time, has been optimized to a great extent. These factors help in removing the recast layer and other surface defects from the drilled hole, resulting in an improved surface characteristic of the hole. Figure 4.5 represents two photographs of holes with their sectional views showing the progress of the taper, which is gradually decreasing from the top to the bottom surface.

The application of the micro-ECM process after micro-EDM shows considerable improvement in the surface characteristics of the hole. The smoothness of the internal surface of the hole as well as the top edge surface is appreciable in Figure 4.5.

T. Kurita [9] experimented with EDM–ECM sequential machining technique by using a formed tool electrode as well as by using the universal electrode for generating complex shapes in 61 HRC hardened steel material. Both the processes were carried out at one station in a single system setup, without changing tool and workpiece positions. In the first experiment, a complex shape was created in the workpiece using a formed tool electrode by applying EDM and ECM processes in sequence. The profile cutting is done using the EDM process, whereas glossy surface finish is

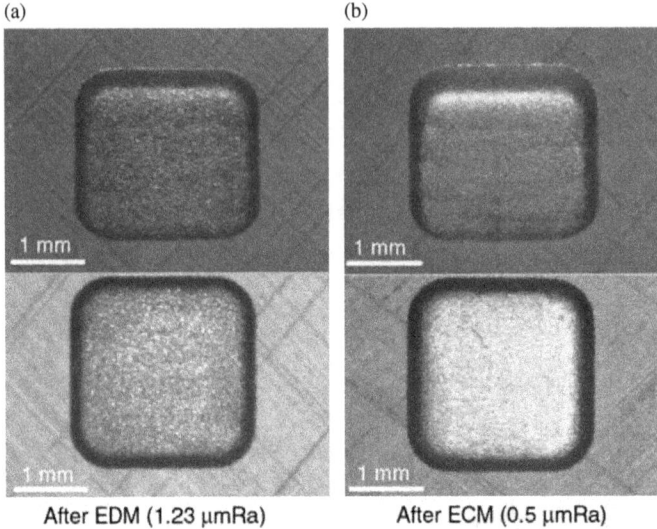

FIGURE 4.6   Photograph of the square cavity (a) after EDM and (b) after ECM lapping [9]. With permission from Elsevier.

imparted to the EDMed part by applying the ECM process. The second experiment was carried out by adopting scanning movements of a 3 mm diameter copper tool electrode to mill a cavity on a hardened steel workpiece. The results of EDM–ECM sequential machining processes are analyzed from the point of view of surface finish obtained on the workpiece. In the first case, EDM used for shaping the 3mm square cavity indicated surface roughness (Ra) of 1.23 μm at the bottom surface, whereas ECM applied later lowered the Ra value to 0.5 μm. During the ECM process, a passivating oxide layer formed on the workpiece surface resists the flow of electric current. This hampers the process of electrochemical dissolution of the workpiece material. The abrasive particles present in the electrolyte help remove this oxide layer, which enhances the rate of anodic dissolution of the material. Thus, material removal is achieved with high efficiency. EDM and ECM lapping is applied for the purpose of imparting a very smooth finish on the workpiece surface. Figure 4.6 represents actual photographs of the machined square cavity showing improvement in surface roughness value after performing the ECM process.

In the electrochemical lapping process, aluminum oxide (Al_2O_3) abrasive mixed with the water-electrolyte is used. ECM lapping produced a surface roughness (Ra) value less than 0.1 μm.

4.6   IMPROVEMENT OF GEOMETRIC ACCURACY USING SEQUENTIAL EDM AND ECM PROCESSES

Micro-ECM milling and micro-EDM milling processes are applied in a sequence for micro-groove generation. Micro-milling in both processes is carried out using a tungsten carbide electrode of 400 μm diameter [10]. Figure 4.7 shows that the shape

FIGURE 4.7 Photograph showing the groove milled using micro-ECM milling and micro-EC-ED sequential milling [10]. With permission from Elsevier.

of the micro-groove generated using micro-electrochemical milling is improved after employing the sequential machining approach. It has been mentioned that the irregular electric field distribution has an impact on the groove shape. The uncontrolled dissolution of the groove surface using universal electrodes during electrochemical milling is the major disadvantage of the process. The sequential EC-ED milling sequencing plays an important role in improving the geometric accuracy of the groove.

The machine element, a flexure hinge is generally used in high precision motion control instruments and in precision locators such as interferometers, gyroscopes, etc. Maintaining the dimensional accuracy of the flexure hinge during manufacturing is very difficult. The flexure hinges are generally made of low stiffness material and are manufactured using conventional machining processes such as boring and grinding. It is not possible every time to attain the required dimensional accuracy on hinge workpieces by using these traditional manufacturing processes. Xiaowei et al. [11] applied EDM and pulsed electrochemical machining (PECM) process sequentially to produce the flexure hinge. As EDM is a non-conventional machining process in which the tool and the workpiece do not contact each other during machining, no cutting forces are exerted by the tool on the workpiece. During EDM machining, orbital motion is imparted either to the tool or to the workpiece. The orbital motion improves the machining efficiency and reduced arcing improves the dimensional accuracy of the part produced. Similarly, due to orbital motion between tool and workpiece, the machined surface obtained is smooth in nature, and the neck area of the hinge can be machined precisely. With these advantages, EDM generates a heat affected zone, forms recast layers, and also induces residual stresses in the machined region. It is essential to remove the hardened recast layer as it affects the fatigue life of the flexure hinge. Hence to remove the irregularities generated due to EDM process, a microsecond-scale, pulsed electrochemical machining (MPECM) process has been applied on the EDMed part. Pulse on time (T_{on}) of 10 to 100 μs, inter-electrode gap of 0.1 to 0.15 mm, current density 15 to 20A/cm², imparts high dimensional accuracy as well as a mirror surface finish to the flexure hinge surface.

4.6.1 BLOCK-MICRO-EDM, MICRO-EDM, AND MICRO-ECM

In this approach, three different micromachining processes are integrated on a single machine tool in order to improve the machining precision, surface characteristics, and mechanical properties of the machined part. Initially, the tool electrode, which is required to be utilized for subsequent operations, is prepared by the online reverse copying block micro-EDM process; here, the larger tool diameter is reduced to the required smaller size to serve the purpose of micromachining. This online reverse copying block method is suitable to compensate for the tool wear and eliminates the possibility of repositioning. As the micro-EDM needs non-conductive dielectric fluid to generate electric spark discharges and micro-ECM requires conducting electrolyte medium for electrolysis to occur, two distinct working fluid circulation systems are required to be developed in the setup. Using micro-EDM, any difficult to cut conductive materials can be shaped irrespective of their mechanical properties, but it is predominantly a thermal process. The electrical discharge machined surface is generally rough in nature because of the presence of a multitude of discharge craters formed due to electric spark discharges. The machined surface also consists of white layers or recast layers and the heat affected zone. Extensive tool wear occurs during machining, hampering the machining accuracy. On the other hand, in micro-ECM, there is no tool wear, no re-solidified or recast layer formation, no cracks on the machined surface, no heat affected zone, and no residual stresses induced in the workpiece surface. The temperature in the machining zone is less than 100 °C, which offers stress-free machining. Irrespective of all these advantages, micro-ECM is not suitable for the precision machining of parts because of the low material removal rate (MRR). If the MRR is increased, frequent short circuits may occur, which degrades the surface quality. The phenomenon of stray current effect causes uncontrolled dissolution of anodic material in the tool facing area, which results in the formation of an unsatisfactory shape on the workpiece. The sequencing of micro-ECM and micro-EDM adds the advantages of both the processes by minimizing their disadvantages. Hence, sequencing of the processes makes it possible to make the surface characteristics better than that machining merely by micro-EDM. Additionally, the machining precision and shape accuracy obtained by micro-ECM is parallel to the level of micro-EDM. Figure 4.8 shows block-micro-EDM and micro-EDM plus micro-ECM sequential machining processes performed on a single machine tool without changing the tool workpiece positions.

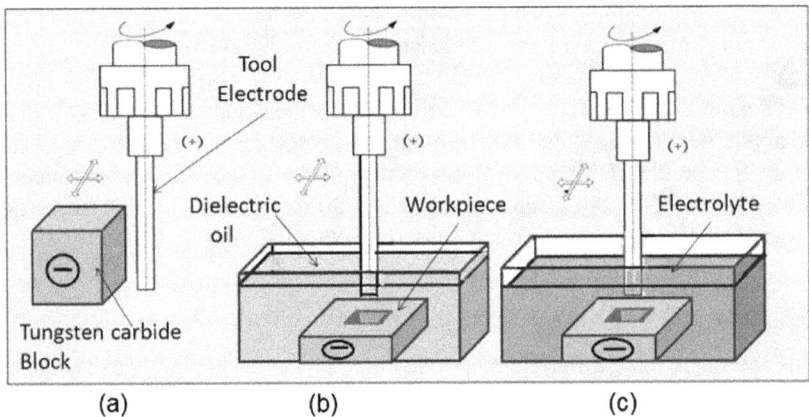

FIGURE 4.8 Schematic diagram of (a) block-micro-EDM, (b) micro-EDM, and (c) micro-ECM sequential machining process.

All the three processes, namely, online fabrication of micro-tool using tungsten carbide block, then micro-EDM milling and finally finishing the workpiece by micro-ECM process have been executed at one station. The system set up has been made in such a way that micro-EDM shaping and micro-ECM finishing can be performed sequentially, without changing the positions of the tool and the workpiece. The micro-ECM process assists in lessening the recast layers formed in the heat affected zone because of the pre-treatment of EDM cutting. Similarly, the surface roughness value Ra which was 0.143 μm, after applying micro-EDM, improved to 0.707 μm [12]. Hence, this sequential machining combination was satisfactory in improving the surface characteristics as well as generating the required cavity or contour on the component. Thus, the electrochemical and electro-discharge sequential micro-milling processes have been employed successfully for machining 3D metallic micro-structures with complicated shapes. Thus, sequencing methodology proves its usefulness in manufacturing parts with improved machining performance and mechanical properties.

4.7 DISCUSSION ON SEQUENTIAL WEDM AND WECM PROCESSES

It is a well-known fact that the sequential machining technique provides a structured approach towards the manufacturing of highly complex shapes on the workpiece. Exploring new sequential machining combinations is vital from the point of view of an increasing rate of metal removal from workpiece surface, improving dimensional accuracy and imparting excellent surface finish to the machined part. Sequential machining, on the one hand, reduces the time required for machining, and on the other hand, it improves the machining efficiency. Hence, by considering these benefits, it is worthwhile finding more and more sequential machining combinations. Two diverse non-traditional machining processes, namely, Wire Electro-discharge Machining (WEDM) and Wire Electrochemical Machining (WECM), have been applied in a sequential manner [13], for micro-groove generation. It is reiterated that, WEDM and WECM are different machining processes, which means that the working principles and the methods of metal removal of both the processes are totally different from each other. In WEDM, once the rated electric power is supplied to the circuit, the generation of a high electric field occurs due to the potential difference present between the wire electrode and the anode sheet. As the applied voltage exceeds the threshold value, spark discharge takes place, which removes workpiece metal by melting and vaporization. WEDM needs a dielectric medium, whereas WECM occurs in conducting electrolytes. The machining takes place due to the electrochemical dissolution of the anodic workpiece.

4.7.1 DEVELOPMENT OF A MICRO WIRE-ECM SYSTEM SETUP

The machining combination of WEDM and WECM is used to generate micro-grooves in a 400 μm thick sheet of titanium alloy (Ti6Al4V). The machining operations, such as micro-groove cutting and finishing the previously cut groove, is carried out on individual system setups. The micro-wire ECM system setup has

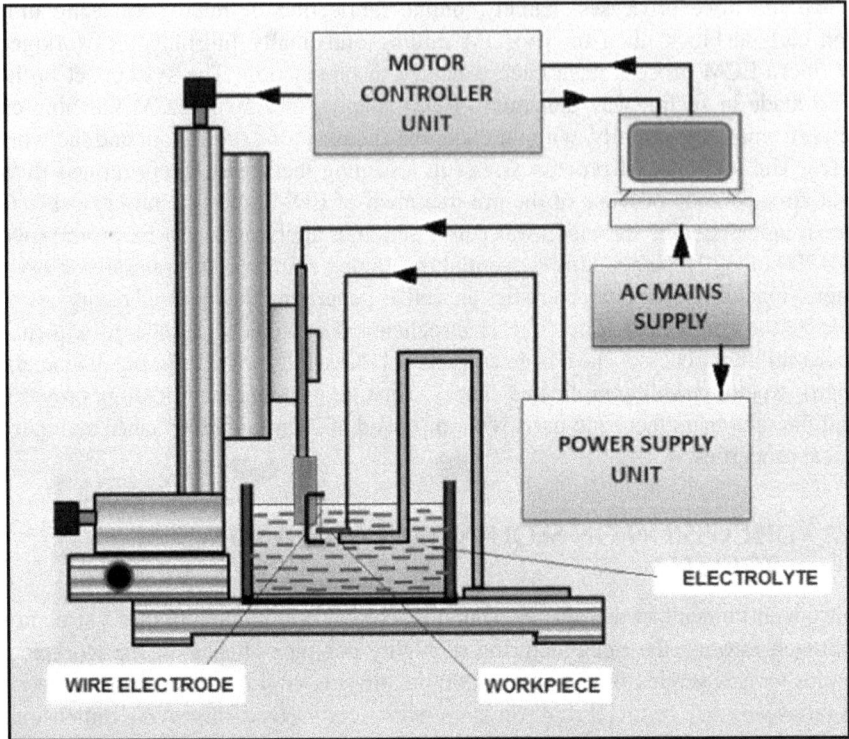

FIGURE 4.9 Schematic diagram of wire ECM system setup.

been developed by combining three different major units, the power supply unit, X-Y-Z translation stage, and machining unit. The three-phase, 440V A.C. supply line is connected to the stabilizer unit, which is further connected to the WECM power supply. A pulsed DC power supply is made available in the machining zone through a rectifier and function generator unit. X-Y-Z translation stage with resolution of 0.1 micron is used to execute the wire electrode's movements in the X, Y, and Z directions. This stage provides controlled movements to the wire electrode. The APT software provides commands to stepper motors, and these motors are installed at the end of each stage. The machining unit consists of a machining chamber containing electrolyte and the workpiece mounting unit. An arrangement is made to fix the template holding workpiece in the workpiece mounting unit. A schematic diagram of the wire ECM system setup which is developed for the experimentation is shown in Figure 4.9.

4.7.2 THE PROCEDURE FOR GROOVE GENERATION USING MICRO WEDM AND MICRO WECM

In the micro-WEDM and micro-WECM sequential machining process, the WEDM process is used to cut the groove effectively on the Titanium alloy sheet, and the

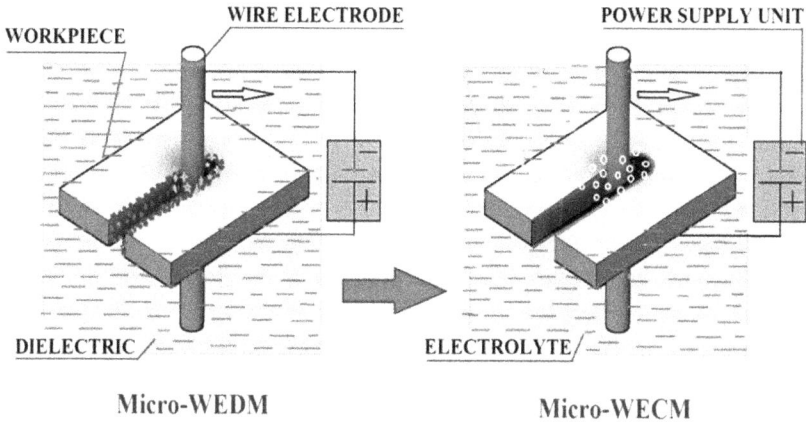

FIGURE 4.10   Microgroove generation using sequential micro-WEDM and micro-WECM machining [13].

WECM process is used to finish the machined groove surface. In Figure 4.10, schematic diagrams of the micro-WEDM occurring in dielectric fluid and the micro-WECM process occurring in electrolyte is shown. Initially, the groove cutting operation on the Ti6Al4V sheet is performed by means of a wire-EDM machine. The wire electrode of 250 mm diameter with material of commercial brass has been used to cut the groove in the Ti6Al4V sheet. The experiments conducted on a wire EDM machine, indicated the ideal values of pulse on time (T_{on}) as well as servo voltage (V) for the fabrication of a micro-groove. In this wire cut EDM machine, deionized water is used as a dielectric medium.

The process parameters such as wire-speed 6m/min and open-circuit voltage of 92V are maintained constant throughout the WEDM process. The design of experiments (DOE) with four levels has been prepared using Taguchi's method. The aim was to determine the optimum values of process parameters such as applied voltage and pulse on time, producing a micro-groove with minimum width. The cutting of the micro-groove using WEDM took only about 5 microseconds. Thus, with WEDM, the micro-groove of 800μm length cut very quickly. As we know, in WEDM, during machining, the metal on the workpiece surface melts and gets vaporized because of the generation of high intensity electric sparks. The running wire electrode and the metal sheet are both connected to the D.C. power supply. The release of very large thermal energy for a very short period of time in the form of sparks creates a heat affected zone (HAZ) in the machined region. In the HAZ, the presence of recast layers with numerous micro-craters and micro-cracks affect the workpiece surface characteristics. The white layers and the other anomalies present in the machined region alter the thermal properties of the material along with inducing residual stresses. The micro-grooves thus produced using WEDM with various parametric combinations showed that the servo voltage of 10V and pulse on time, T_{on} of 4μsec generated grooves of minimum width. Figure 4.11 represents the actual photographs

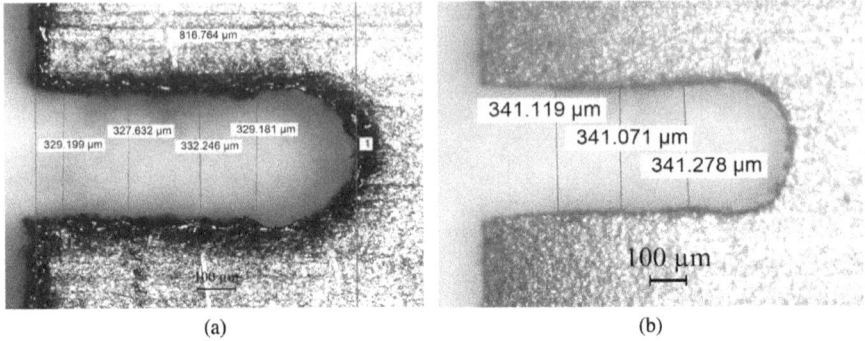

(a)   (b)

FIGURE 4.11 Photographs of microgrooves (a) after micro-WEDM and (b) after micro-WECM [13].

of the micro-groove cut by using the micro-WEDM process and later finished by using the micro-WECM process.

After groove cutting by micro-WEDM, finishing the groove using micro-WECM was necessary to remove the non-uniformity present on the groove edges which was formed due to the accumulation of the recast layers. WECM is a stress-free process used to generate grooves on any conductive material, irrespective of the mechanical properties. In WECM, a traveling wire and workpiece kept in the electrolyte are connected to negative and positive terminals, respectively. When the electric current starts, the flow of electrons through an electrolyte in the form of ions occurs from cathode to anode. Thus, with the start of electrolysis, electrochemical dissolution of the titanium sheet takes place. The sodium nitrate ($NaNO_3$) is a passive electrolyte that forms a protective oxide film on the surface. This film is helpful in maintaining precision, whereas metal removal rate has been augmented due to the application of non-passivating sodium chloride (NaCl) electrolyte [14]. Hence, a combination of NaCl and $NaNO_3$ is used as an electrolyte for machining Ti6Al4V effectively. In the micro-WECM process, there was a need to discover the most suitable process parameters, for the improvement of the surface finish by maintaining the minimum width of the groove. Hence, the design of experiments taking three factors with four levels using Taguchi's method has been applied during the micro-WECM experiments. The input process parameter/factors varied in the following way, machining voltage: (6V, 8V, 10V, and 12V), machining time: (30, 60, 90, and 120 sec), and duty ratio: (20%, 40%, 60%, and 80%), whilst maintaining the frequency of 100 kHz constant. Similarly, in wire ECM, an electrolyte combination of 1M of $NaNO_3$ and 1M of NaCl has been used for finishing the groove surface from the previous WEDM process by a 250μm wire diameter.

4.8 CONCLUSION

The novel sequential combination of micro-WEDM cutting and micro-WECM finishing has been applied successfully for the formation of micro-groove in difficult-to-cut titanium alloy (Ti6Al4V) sheet. In WEDM, the micro-groove with minimum width can be made by taking T_{on} as 4 μsec and servo voltage as 10V. The surface roughness (Ra) of 0.124 μm was noted at the groove surface after applying the micro-WEDM process.

When micro-WECM was performed for 30 seconds at 8V by keeping the duty ratio 60%, an excellent surface finish was obtained on the groove surface. The surface roughness (Ra) of 0.0734 μm was noted at the groove surface after applying the micro-WECM process. Thus, this sequential machining combination proved effective in the production of a micro-groove on titanium alloy (Ti6Al4V) sheet, in very little time. This sequential combination was found useful in improving the surface roughness, along with improving the machining efficiency.

REFERENCES

[1] S. Z. Chavoshi, S. Goel, P. Morantz, "Current trends and future of sequential micromachining processes on a single machine tool," Materials & Design, (2017), 127, 37–53.

[2] R. M. Tayade, B. Doloi, B. R. Sarkar, B. Bhattacharyya, "A State of the art on Sequential Micro-Machining," IOP Conference Series: Materials Science and Engineering, IOP Publishing, 653 (2019) 012026 doi:10.1088/1757-899X/653/1/012026.

[3] S. Skoczypiec. K. Furyk, A. Ruszaj, "Computer simulation and experimental study of a sequential electrochemical- electro-discharge machining process," Procedia CIRP, (2013), 6, 444–449.

[4] K. P. Rajurkar, d. Zhu, J. A. McGeough, J. Kozak, A. De Silva, "New developments in electrochemical machining," Keynote papers, https://doi.org/10.1016/S0007-8506(07)63235-1.

[5] H. P. Tsui, J. C. Hung, J. C. You and B. H. Yan, "Improvement of electrochemical microdrilling accuracy using helical tool," Materials and Manufacturing Processes, (2008) 23 (5), 499–505.

[6] T. Matsuzawa, C. L. Kuo, M. Fujino, "A combined electrical machining process for micronozzle fabrication," CIRP Annals, (1994), 43 (1), 189–192.

[7] Z. Zeng, Y. Wang, Z. Wang, D. Shan, X. He, "A study of micro-EDM and micro-ECM combined milling for 3D metallic micro-structures", Precision Engineering, (2012), 36 (212) 500–509.

[8] L. Xiaowei, J. Zhixin, Z. Jiaqi, L. Jinchun, "A combined electrical machining process for the production of a flexural hinge," Journal of Materials Processing Technology, (1997), 71, 373–376.

[9] T. Kurita, M. Hattori, "A study of EDM and ECM/ECM-lapping complex machining technology", International Journal of Machine Tools and Manufacture, (2006), 46(14), 1804–1810.

[10] S. Skoczypiec, A. Ruszaj, "A sequential electrochemical-electro-discharge process for micropart manufacturing", Precision Engineering, (2014), 38, 680–690.

[11] Li Xiaowei, JiaZhixin, Zhao Jiaqi, Liu Jinchun, "A combined machining process for the production of a flexure hinge," Journal of Material Processing Technology, 71 (1997) 373–376.

[12] H. Xiolong, W. Yukui, W. Zhenlong, Z. Zeng, "Micro-hole drilled by EDM-ECM combined processing," Key Engineering Materials Online, 562 (2013), 52–56.

[13] R. M. Tayade, B. Doloi, B. R. Sarkar, B. Bhattacharyya, Study of micro-groove generation on titanium alloy (ti6al4v) using novel WEDM and WECM sequential electro micro machining (SEMM) technique, Materials Today: Proceedings 22 (2020) 1638–1644.

[14] R. Leese and A. Ivanov, "Electrochemical micromachining: Review of factors affecting the process applicability in micro-manufacturing," Proceedings of the Institution of Mechanical Engineering Part B: Journal of Engineering Manufacture, 232(2) (2017) 195–207.

5 Recent Trends in Arc Machining Processes

Afzaal Ahmed[1], Jibin Boban[1] and M. Azizur Rahman[2]

[1] Department of Mechanical Engineering, Indian Institute of Technology Palakkad, Kerala, India

[2] Department of Mechanical and Production Engineering, AUST Dhaka, Bangladesh

CONTENTS

5.1 INTRODUCTION

Arc machining is a non-conventional manufacturing domain widely applied for faster cutting of materials leading to enhanced production rates in the industrial field. Irrespective of material hardness, the arcs generated in the machining process cuts or

DOI: 10.1201/9781003202301-5

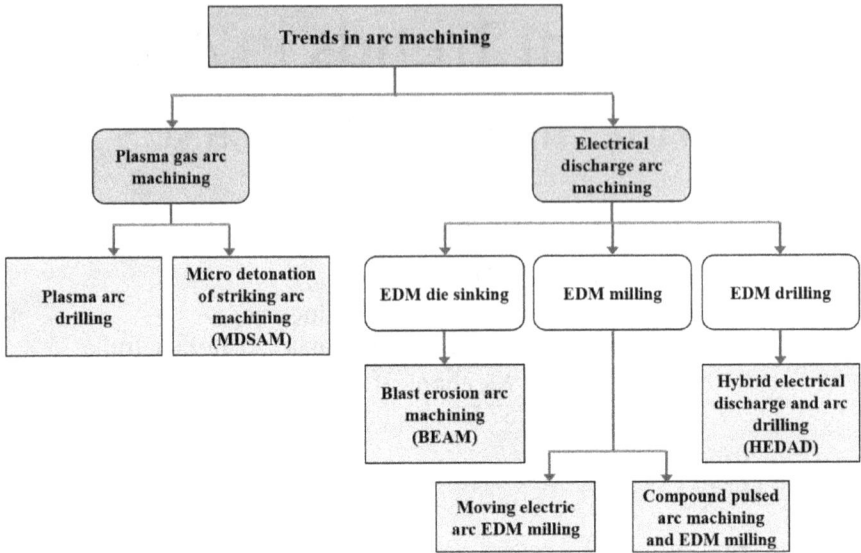

FIGURE 5.1   Broad classification of arc machining processes.

processes the material to the desired shape. However, surface quality and dimensional accuracy is compromised for any benefits achieved in production time. In this regard, extensive research is ongoing to overcome limitations in arc machining. Until now, multiple methods have been proposed by researchers in the field of arc machining. Along with the predominant plasma arc technology, recent trends in combining electrical discharge machining (EDM) with arc machining offer promising solutions that provide better machining performance in terms of material removal rate (MRR), tool wear ratio (TWR) and surface integrity. Nevertheless, many challenges are created by the high thermal energy and temperature involved in the aforementioned processes.

This chapter includes descriptions of recent trends in arc machining methods with special attention to the feasibility and setbacks of each method in detail. Figure 5.1 depicts the broad classification of arc machining processes.

5.2   THE PRINCIPLE OF ARC MACHINING

The basic principle of arc machining can be discussed with respect to the material removal mechanism involved in the process. The basic principle behind the working of arc machining methods is depicted in Figure 5.2. Arc machining involves melting of the material using an ionized plasma arc followed by subsequent flushing of molten material using a fluid at high pressure. The hydrodynamic action of the flushing fluid and the associated turbulence is responsible for breaking the arc plasma by which material removal takes place [1]. The removed molten particles will be larger in size and gradually stick to the nearby surface/wall leading to high surface roughness. However, a high MRR can be easily achieved using arc machining methods at the cost of surface quality. High material removal is attained on account of the high energy

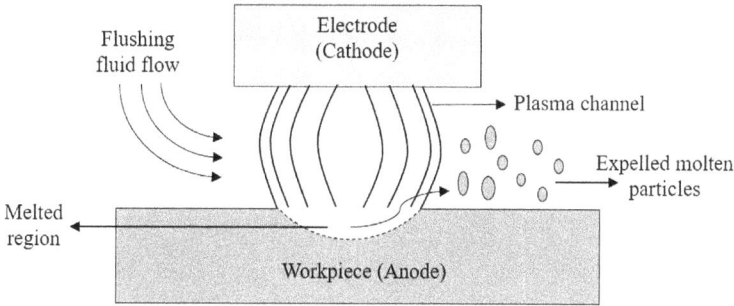

FIGURE 5.2 Principle of arc machining.

density in arc machining along with the high temperature of the arc plasma. The position of the discharge in arc machining is not shifted and remains the same owing to insufficient deionization. Accordingly, the same location is subjected to multiple discharges. Thus, arc machining can be regarded as a stable discharge process [2].

Although arc machining and EDM belong to an electro-thermal energy based material removal process, both differ in various respects. Unlike arc machining, EDM involves unstable discharges due to a randomly changing discharge position. Although surface quality is superior in EDM, arc machining dominates EDM with regard to a high MRR. The arc plasma temperature and associated energy in arc machining is relatively much higher than the temperature of the spark plasma in EDM. Moreover, the energy input is largely shared by tool electrode and workpiece in arc machining whereas the majority of the energy is dissipated in the machining gap and lost by heat conduction in EDM [1]. These two factors account for the higher material removal rate in arc machining compared to the EDM process.

5.3 RECENT TRENDS IN THE ARC MACHINING PROCESS

5.3.1 Plasma Gas Arc Machining

In plasma gas arc machining, the plasma is produced by means of heating the gas with an arc. The energy of the generated plasma is responsible for material removal from the workpiece. The recent trends in plasma gas arc machining includes plasma arc drilling and micro detonation striking arc machining (MDSAM) which are discussed in detail below.

5.3.1.1 The Drilling Method Using Plasma Arc Technology

Plasma arc technology is well established in the areas of welding, heat treatment and thermal cutting. With a view to incorporating plasma arc technology in drilling, Kusumoto and Sun [3] developed a drilling method using plasma arc technology. In this process, the workpiece undergoes melting on account of the high thermal energy generated by the high temperature plasma arc. Whenever the workpiece is exposed to a high velocity jet of hot ionized gas, it receives thermal energy from the plasma arc and undergoes melting. Subsequently, the molten metal is flushed and carried away

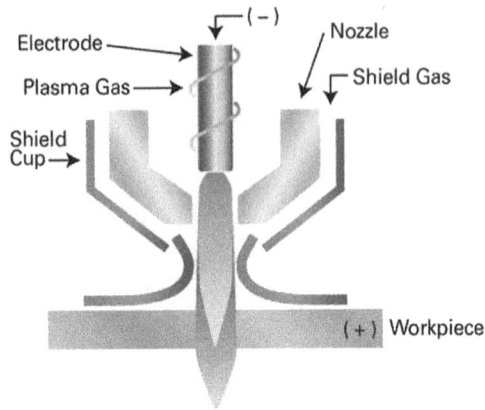

FIGURE 5.3   Schematic representation of plasma arc drilling [4].

from the workpiece in response to the high pressure flow of gas. The schematic of plasma arc drilling is shown in Figure 5.3.

Kusumoto and Sun [3] performed experiments to compare the influence on hole diameter, taper angle and volume of removed material with increasing processing time. Based on the observations, the diameters at the entry and exit of the hole over the workpiece showed an increasing trend with longer processing time. However, the holes produced by the plasma arc method exhibited good circularity. At the same time, the effect of taper angle is one of the prime concerns during any hole drilling process. It is in this context that plasma arc technology drilling is introduced as it can effectively produce zero tapered holes by proper control of the processing time [3].

However, plasma arc technology drilling faces challenges of localised deposition and dross formation owing to the insufficient pressure to completely blow away the molten material as well as the high viscosity of the molten metal. Further studies conducted by Kusumoto et al. [5] highlighted that dross formation occurs at both the entrance and exit sides of the hole. The amount of dross formation on the entrance side is found to be more than that on the exit side [5]. In spite of all these facts, another study by the same authors proved the excellent performance of plasma arc drilling in terms of MRR [6]. Nonetheless, the process can be employed for drilling only thinner samples due to inverted taper formation occurring in workpieces with increase in thickness.

5.3.1.2   Micro Detonation of Striking Arc Machining (MDSAM)

Micro Detonation of Striking Arc Machining (MDSAM) belongs to a thermal process that involves very high temperature. Whenever a pulsed voltage with high frequency is supplied between the anode and the cathode, the avalanche motion of electrons generates a spark discharge and the formation of an electric arc. The supply of high current and the consequent rapid gas ionization results in the generation of plasma. The plasma jet evolved is compressed and ejected through a narrow nozzle channel. The sudden volumetric expansion of the plasma jet generates micro detonation

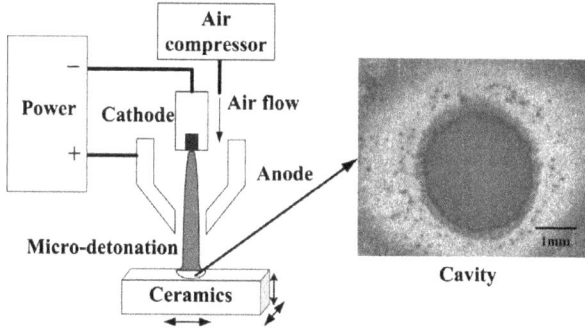

FIGURE 5.4 Schematic of MDSAM process [7].

followed by a strong shock wave which hits the workpiece at high temperature with high impact force thereby resulting in material removal [7]. The schematic of MDSAM process is depicted in Figure 5.4.

In MDSAM, the impact force is inversely proportional to the operating current whereas a direct relationship is prevalent with the nozzle diameter and the pressure of gas [7]. As a continuation to this study, the thermal influence of the micro detonation plasma jet was also explored in detail by the same researchers [8]. Followed by a single pulse of micro detonation at high temperature, the silicon nitride (Si_3N_4) ceramic material cools down rapidly resulting in the formation of recast layer over the surface of the cavity. The metallurgical characteristic of the workpiece revealed the presence of a micro crack network on the recast layer. Thus, the workpiece surface is severely affected after MDSAM and demands finish machining for achieving better surface finish [9].

Furthermore, a similar study on MDSAM was conducted on alumina ceramics as well [9]. The study was intended to evaluate how factors such as operating current, pressure of gas, pulse width, nozzle stand-off distance and diameter of nozzle have an effect on the MRR and the maximum temperature attained. The analysis was done using high-speed video monitoring in addition to a numerical simulation study and experiments. An increase in operating current, pulse width and pressure of gas was found to elevate the highest temperature, which in turn enhanced the MRR. Nevertheless, the temperature and MRR was found to decrease with the increase in the diameter and stand-off distance of the nozzle. The study also determined the optimal combination of the aforementioned parameters. The study also established that the material removal in MDSAM occurs as a result of the combined effect of sublimation, melting and phase transformation on account of high temperature (~13,540 K) as well as huge impact force caused by micro-detonation [9]. The large impact force together with the extremely high temperature of the micro detonation plasma jet is responsible for the spatter deposition on the alumina ceramic workpiece. Similar to a previous study on silicon nitrides, the machined surface for alumina is found to be very rough after MDSAM. This indicates that post processing is essential after MDSAM for improving the surface quality. Thus, it can be seen that MDSAM is applicable for machining only flat surfaces, which is a major setback of the process.

5.3.2   ELECTRIC DISCHARGE ARC MACHINING

EDM is an electro-thermal process predominantly employed in machining high strength temperature resistant alloys that are usually difficult-to-cut. Material removal in EDM occurs by means of melting and vaporization of the material caused by the momentary electrical sparks. It is well known that the low MRR in EDM struggles to meet the industry requirements of shorter lead time. Also, the demand for complex shapes and dimensionally accurate products led the researchers to explore the feasibility of other methods. As an end solution, electrical discharge arc machining methods have been developed, which remove material by the combined effect of arcing and sparking. Like sparking, the arcing resulting from prolonged discharges can also be utilized for machining materials. The compound EDM processes formed by incorporating arcing with EDM facilitates faster machining of hard and difficult-to-cut materials like nickel based superalloys [10]. In light of this, wide variants of electrical discharge arc machining have been developed over the years.

5.3.2.1   Blast Erosion Arc Machining (BEAM)

In EDM, electrical sparks are generated using electrical energy and the primary source behind material removal is the thermal energy of the generated spark [11]. For the same reason, EDM has a lower MRR in comparison to other material removal methods. Taking this into consideration, a more efficient method in terms of material removal, called Blast Erosion Arc Machining (BEAM) was put forward [12]. BEAM is a high speed material removal process which utilizes an intensive hydrodynamic force imparted by the flushing flow at high velocity to break the plasma channel in EDM. Unlike the momentary sparking phenomenon observed in the EDM process, a shockwave is created in BEAM on account of plasma channel collapse which in turn takes off molten material with an explosive action. The mechanism of material removal in BEAM is shown in Figure 5.5.

An intense flushing is the prime requirement of BEAM to bring out the substantial amount of heat produced by discharge that ultimately aids the system in sustaining a high magnitude of discharge current in comparison with EDM [12]. In order to ensure intensive flushing, a novel die-sinking electrode known as a bundled electrode (BE) is proposed [13]. A bundled electrode can be considered to be the discretization of a solid electrode (SE) into numerous tubular cell electrodes as shown in Figure 5.6. The special design of the BE with a coaxial hole-matrix structure facilitates strong tubular

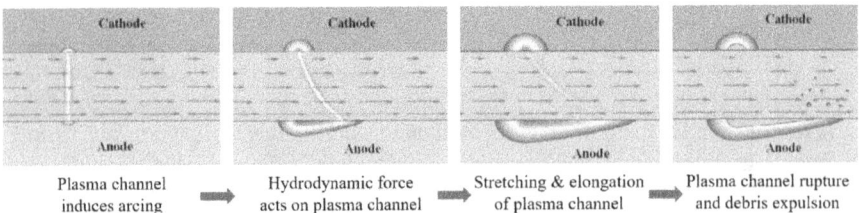

| Plasma channel induces arcing | Hydrodynamic force acts on plasma channel | Stretching & elongation of plasma channel | Plasma channel rupture and debris expulsion |

FIGURE 5.5   Material removal mechanism in BEAM process [12].

FIGURE 5.6   Geometry of bundled electrode (BE) [12].

flushing through multiple holes thereby improving the working conditions in the gap during machining [13].

A similar study on a die sinking process using a copper BE was conducted on Ti6Al4V alloy with oil based flushing [14]. The copper BE comprised 217 tubular cell electrodes with each having an internal diameter of 0.5 mm and an external diameter of 1.5 mm. Numerical simulations of BE multi-hole flushing was conducted based on the Eulerian–Lagrangian approach to evaluate the impact of flushing on gap discharge conditions. Similar to the previous study [13], the fluid flow velocity was improved in the gap leading to better MRR [14]. From the experimental results, the highest MRR achieved using a copper BE was ~150 mm^3/min which is nearly 5 times that of SE. Moreover, the tool wear rate (TWR) of BE was lower (< 7%) relative to the TWR of solid electrode (SE). The study was a milestone that widened the usage of bundled electrodes in BEAM processes for better flushing efficiency.

The MRR for the BEAM process is higher as compared to conventional EDM and other traditional cutting processes. This was further proved in a later study where BEAM of Inconel 718 alloy was performed using graphite BE and dielectric flushing of water at 1.5 MPa [12]. The system was connected to the power supply with positive workpiece polarity at an open voltage of 60V and 600A peak current. Whenever electro-arcing occurs in the BEAM process, both current and voltage undergo rapid and unpredictable fluctuations repeatedly which is a true indication of transient arcing phenomenon rather than a steady state arcing. As shown with red circles in the oscilloscope waveform in Figure 5.7, the current decreases in advance before the end of pulse's duration. This represents the breaking of the plasma column by the strong inner flushing offered by bundled electrodes. The MRR of Inconel 718 was much higher (~11,300 mm^3/min) than that of conventional EDM. The underlying reason can be attributed to the enhanced fluid flow velocity in the gap and improved debris

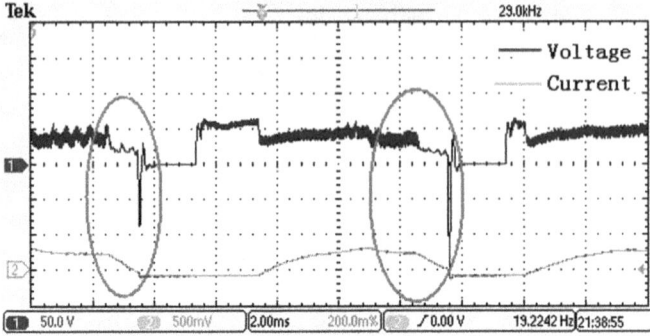

FIGURE 5.7   Voltage and current waveforms in BEAM process [12]. With permission.

removal ensured by the bundled electrode structure. In addition, the TWR obtained from experiments is lower than 1%. The low electrode wear ratio (EWR) and high material removal rate (MRR) offered by the BEAM process make it capable of machining super alloys together with difficult-to-cut materials.

Furthermore, the metallographic analysis done on the workpiece after the BEAM process disclosed that the HAZ thickness was small [12]. The formation of a small HAZ will result in shallower grain boundary cracks that can ultimately improve the mechanical properties of the workpiece. A small HAZ is formed owing to the use of multi-hole inner flushing that carries away heat from the discharge gap, thereby preventing the workpiece from overheating.

Other than the aforementioned studies, a study on machining performance of BEAM was attempted on Ni-based alloy GH4169 (similar to Inconel 718) [15]. The experiments were conducted under two different polarities namely positive (+ve) polarity (positive BEAM) and negative (–ve) polarity (negative BEAM). At a peak current of 500A, the –ve BEAM contributes to a higher MRR (up to ~14,000 mm³/min) and a lower TWR (down to ~1 %) while the +ve BEAM results in low MRR (3278 mm³/ min) and high TWR (18.3 %). On the other hand, SEM investigations revealed that +ve BEAM can ensure better surface integrity relative to –ve BEAM. The surface roughness (Ra) of ~31 μm on the +ve BEAM surface was lower compared to ~274 μm on the –ve BEAM surface. Thus, both +ve and –ve BEAM processes can be used in combination by employing –ve BEAM (high-current) for the roughing operation and subsequently +ve BEAM (low-current) for a relatively smoother surface [15].

Despite the fact that BEAM ensures a high MRR, it can be employed only for generating 2D surfaces and slots. The various BEAM experiments conducted on different materials considered only the milling process. This opens up the scope to explore its machining performance in ultra-fast drilling as well.

5.3.2.2   Moving Electric Arc High Speed EDM Milling

The material removal mechanism in EDM can be divided into four phases occurring in sequence. These phases include ignition delay (formation of the discharge channel), discharge channel expansion, discharge channel extinction followed by discharge

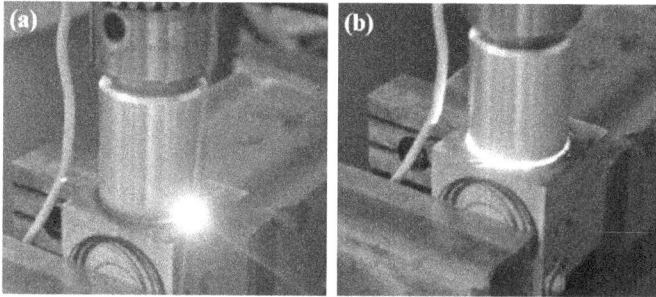

FIGURE 5.8 (a) Stationary arcs and (b) moving electric arcs [18]. With permission.

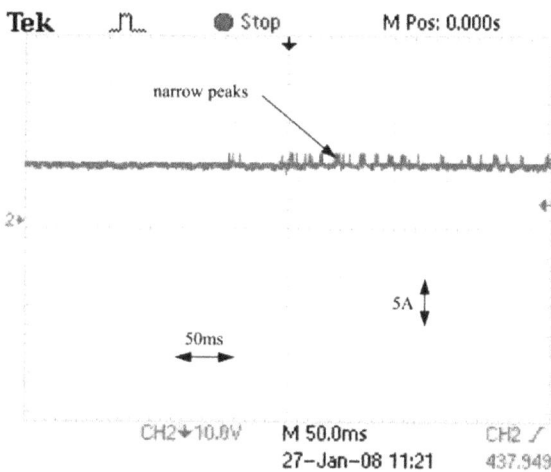

FIGURE 5.9 Waveform of moving electric arcs [18]. With permission.

interval [16]. Material erosion does not occur during the initial phase (ignition delay) and the final phase (discharge interval). Thus, reduction in ignition delay time and discharge interval can result in a high material removal rate (MRR). However, both of these phases cannot be eliminated as it makes EDM impossible [17]. Hence, if material removal can be achieved without interrupting the ignition delay and discharge interval, significant improvement can be achieved in EDM efficiency. This can be realized by performing EDM with burning electric arcs that must move continuously to prevent deterioration of surface finish. Unlike stationary arcs constricted to a comparatively small region, moving electric arcs are subjected to rapid movement within a large area. As a consequence, the energy is uniformly distributed along the trace as may be seen in Figure 5.8 (a,b) [18].

The stationary arc discharge is concentrated at a fixed position resulting in poor surface finish. In contrast, a moving electric arc maintains a uniform discharge gap in close proximity to and between the stationary workpiece and the rotating tool electrode. Figure 5.9 represents the waveform of the current corresponding to the moving

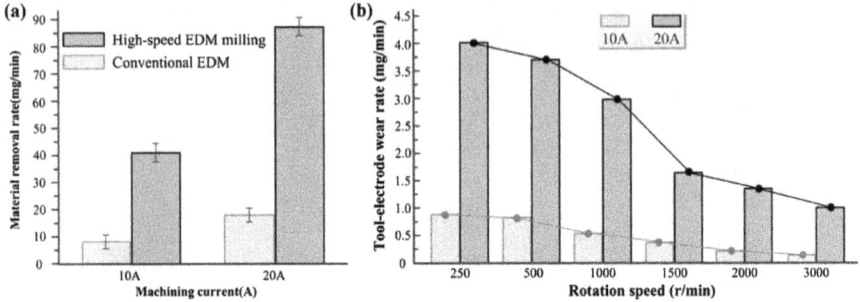

FIGURE 5.10 Machining performance of conventional EDM and high-speed EDM milling in terms of (a) MRR and (b) TWR [20]. With permission.

electric arc with a mean amplitude of ~6A. The waveform displays narrow peaks, which indicate a temporary short circuit caused by the presence of debris remaining in the gap owing to inefficient debris removal. Therefore, the discharge at any point along the moving arc path is impulsive and material removal occurs from the work piece without any concentration of discharge [18].

In arc machining, the ionization of plasma occurs much more effectively in comparison with EDM which in turn contributes to higher thermal energy to easily erode the material. Thus, a novel machining method was developed by combining moving electric arcs with high-speed EDM milling [19]. The proposed method is capable of ensuring continuous material removal without any discharge intervals. The intensive energy of the moving electric arc results in an extremely high temperature as well as high pressure within the plasma channel resulting in rapid melting and evaporation of the workpiece. In a study conducted by [20], moving electric arcs were incorporated with high speed EDM milling. A Ti6Al4V workpiece (anode) and copper tool electrode (cathode) were used for the study and were connected to a DC power supply. As the distance between the workpiece and the tool-electrode decreases, the intensity of electric-field increases. In high speed conditions, the plasma channel stretches toward the direction of tool electrode rotation. Consequently, the workpiece and the dielectric fluid undergo melting and vaporization respectively. As a result, the molten metal takes the form of debris once it gets expelled by the flushing dielectric from the melt surface. The MRR achieved in moving arc assisted high-speed EDM milling is nearly 25mg/min. Thus, the MRR attained is around five times more in comparison with conventional EDM. Moreover, the rate of electrode wear was ~0.5% on account of the formation of a protection layer during machining. A comparative analysis of the machining performance between conventional EDM and moving arc assisted high speed EDM milling can be visualized from Figure 5.10.

5.3.2.3 Compound Pulsed Arc Machining and EDM Milling

In due course, another compound EDM method was developed by combining high speed EDM milling with arc machining where pulsed arcs are produced with the help of a pulse generator. The method was proposed to facilitate the machining of difficult-to-cut materials like Ti6Al4V [21].

With years passing by, Ti alloy machining gained research thrust with the primary objective of reducing machining costs and ecological hazards. In light of this, a novel compound machining technique was developed by merging dry EDM with arc machining such that their parallel working would be capable of removing material from Ti6Al4V [22]. Usually, dry machining utilizes air as the dielectric fluid thereby eliminating the use of other cutting fluids.

As shown in Figure 5.11 (a), high-speed dry compound machining makes use of a power supply where an EDM module is merged with an arc machining module. The module of EDM has a maximum current of 30A and peak voltage of 310V whereas arc machining modules have a maximum current of 700A and peak voltage of 70V. The compound power supply in turn is connected to the rotating electrode and the workpiece. While arc machining supplies higher energy for rapidly melting the material, the EDM module ensures higher voltage in the discharge gap. In addition, the dielectric air flow occurs through an internal electrode at high speed towards the machining gap, leading to efficient debris disposal.

The compound machining process involves three important phases, namely the ignition, discharging and deionization phases. During the primary phase (ignition), the high voltage induced by the EDM module breaks down the dielectric air resulting in the formation of a plasma channel. Next, during the discharging phase, an intense energy is supplied to the plasma channel by the arc machining module for rapid melting of the material. The high-temperature plasma so generated is maintained with the module of EDM during the pulse interval stage. Next, the plasma channel is broken and subsequently debris formation and expulsion takes place from the workpiece in the final deionization phase owing to the high speed air flow. The current and voltage waveforms recorded during the process are shown in Figure 5.11 (b). High current and low voltage are observed during the discharge phase of the arc machining module and opposite effect takes place during the pulse interval stage.

A comparative analysis of machining performance offered by arc machining, EDM and dry compound machining of Ti6Al4V alloy from the study is shown in Figure 5.12. The maximum MRR is achieved in dry compound EDM (5862 mm^3/

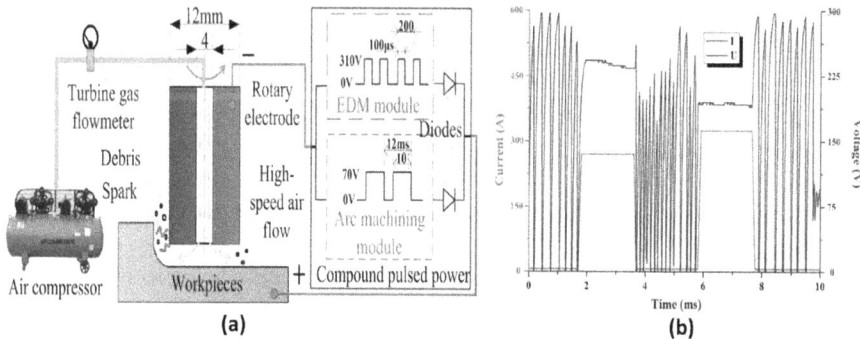

FIGURE 5.11 (a) Schematic diagram of high-speed dry compound machining and (b) waveforms of current and voltage [22]. With permission.

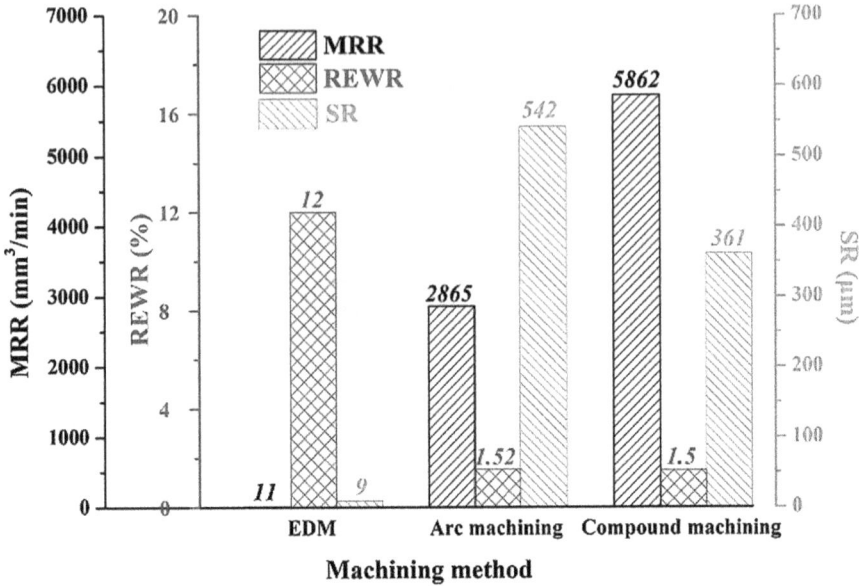

FIGURE 5.12 Machining performance of dry arc machining, dry EDM, and dry compound machining [22]. With permission.

min) and is 100 times higher in contrast to dry EDM. Nevertheless, it is worthy of note that the breakdown distance is shorter in arc machining than that in EDM, which ultimately hinders effective debris removal. Regarding surface quality, the roughness of the Ti6Al4V surface is worse in the case of arc machining compared to EDM ($SR_{EDM} < SR_{Compound\ machining} < SR_{Arc\ machining}$). It can be seen that compound dry machining combines the advantages of both arc machining and dry EDM in terms of machining performance.

5.3.2.4 Hybrid Electrical Discharge and Arc Drilling (HEDAD)

Over years, a wide variant of hybrid drilling methods have been developed by combining EDM drilling with other conventional or non-conventional drilling methods for achieving holes of better quality. Recent research on combining EDM drilling with conventional gun drilling was successful in obtaining deep holes with excellent surface integrity [23]. However, the possibility of achieving the aforementioned requirements by modifying the EDM circuit itself completely rules out the need for any additional methods. In light of this idea, a novel technique called Hybrid Electrical Discharge and Arc Drilling (HEDAD) is developed by incorporating arc machining with EDM drilling [24]–[25]. The working principle of the HEDAD process is demonstrated in Figure 5.13. The HEDAD process helps to eliminate the challenges arising from the faster removal of material from Nickel based superalloys.

Ultra-fast recovery diodes (2 Nos.) having a recovery time of 200 ns is used for connecting the two power supplies (DC and pulsed DC) (Figure 5.13 (a)). The diodes in the circuit are intended to prevent damage of both the power supplies by

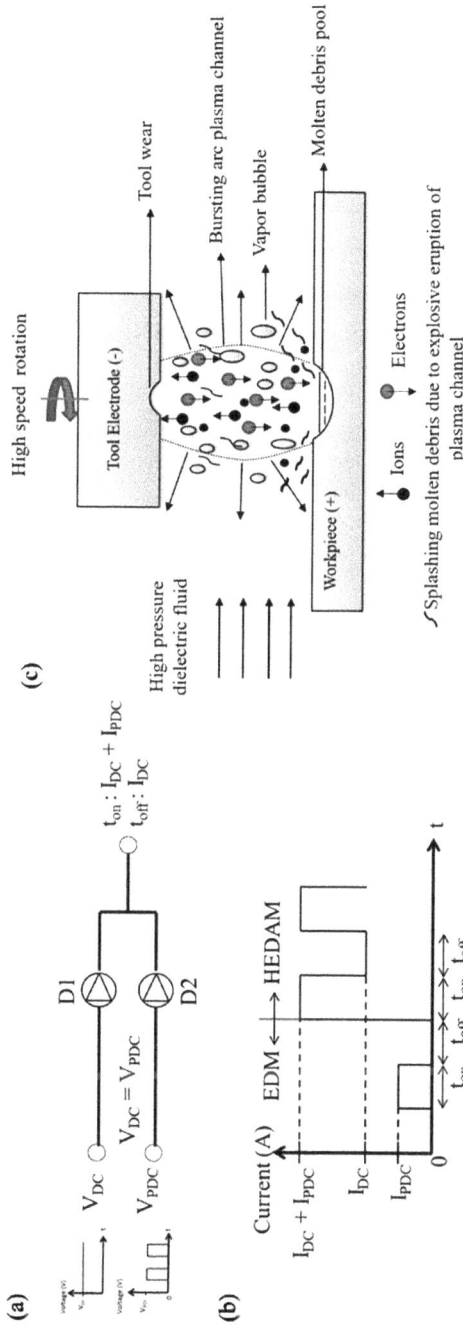

FIGURE 5.13 HEDAD (a) circuit schematic, (b) current waveform, and (c) working principle [24]–[25]. With permission.

blocking any cross flow of current. Thus, diodes ensure that no current flow occurs between pulsed power supply and DC supply. It is notable that an additional current is provided by the DC power supply in the system during machining. Consequently, the same additional current adds to the current from the pulsed DC power supply leading to electric arc formation [26]. However, no current flow occurs during pulse interval (T_{off}) as D2 gets reversed biased. Therefore, current is drawn only from DC supply through D1. The corresponding profile of current in HEDAM process is given in Figure 5.13 (b). Overall, EDM takes the role of a catalyst in HEDAD process by favoring the initial plasma channel formation. On account of the high pressure flushing and tool electrode rotation, the collapse of the plasma channel happens eventually followed by the vanishing of ionization, giving rise to beginning of the cycle further.

The mechanism of material removal in HEDAD can be interpreted from Figure 5.13 (c). At first, rapid thermal melting of the workpiece occurs due to the formation of discharge arcs within the machining gap. Subsequently, a crater is left over the workpiece surface and molten debris particles are generated. The combined effect of high speed tool electrode rotation as well as strong dielectric flushing break the plasma channel. The molten debris is thus expelled by the shockwave induced by plasma channel explosion. As shown in Figure 5.14, significant improvement in EWR and MRR can be observed in HEDAM drilling of Inconel 718. The results are plotted at conditions of 8 MPa dielectric flushing pressure, 50A DC current and 2000 rpm electrode rotation. The MRR achieved using HEDAM drilling is ~200 mm^3/min (at 50A), which is almost 12 times higher than that of normal drilling using EDM. At the

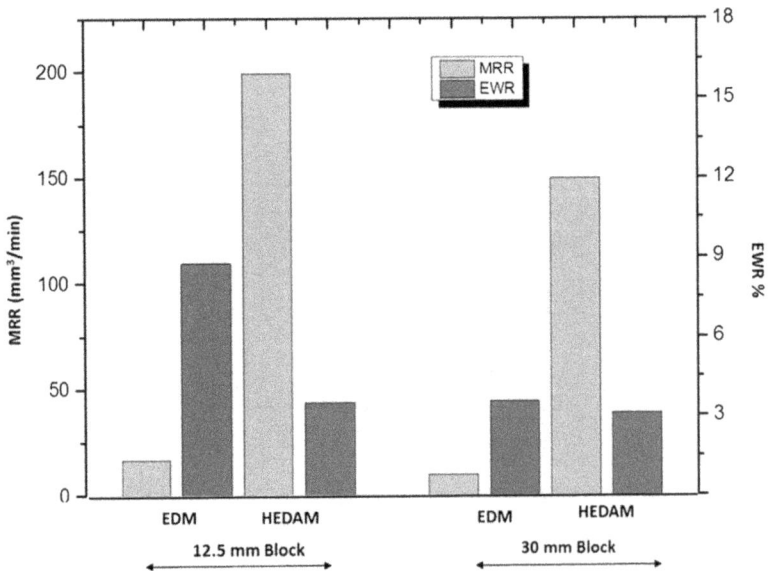

FIGURE 5.14 Machining performance of EDM and HEDAM processes for two distinct thickness of the Inconel 718 workpiece [26]. With permission.

same time, EWR is much lesser (~3.5%) compared to regular EDM. A remarkable observation is that an inverse relationship exists between the MRR and the workpiece thickness for both EDM and HEDAM. The factor responsible for this behavior is the ineffective flushing at the hole bottom with the increase in depth by which debris particles get trapped and accumulated. Consequently, the trapped debris causes frequent retractions of the tool electrode leading to secondary sparks that significantly reduces the cutting speed.

The flushing pressure of dielectric has also significant effect on EWR and MRR. It was reported that volume and size of debris particles generated in EDM depends on machining current [27]. Hence, extremely high dielectric pressure is needed for flushing out the debris particles from the machining zone. As higher flushing pressure aids the efficient removal of debris particles, the MRR also increases. Hence, it is important to use high flushing pressures for improved debris removal. However, as the flushing pressure increases, EWR exhibited a decreasing trend. This can be attributed to the inefficient debris flushing from the machining gap at low flushing pressures. This in turn reduces volumetric material removal by which EWR decreases. Furthermore, the conditions of high pressure flushing improve the cooling of electrode leading to lowered tool wear as well as minimal EWR [25].

5.4 FUTURE SCOPE OF THE ARC MACHINING PROCESS

Recent trends in arc based machining are reviewed in this chapter. Attempts have been made to identify the significant technological advancement made in recent years, including plasma arc generation, machine tool technology, flushing mechanisms, process control and optimization, and fundamental research in material removal mechanisms especially for 'difficult-to-cut' super alloys. Although increased industrial adoption of arc machining is observed, the future scope of research and development both in industry and academia is presented in the following subsections.

5.4.1 MACHINING DIFFERENT MATERIALS

Recent literature shows that most arc machining processes are carried out on steel-based alloys [1]. However, earlier research was conducted to explore the feasibility of micro-detonation plasma jet material removal on silicon nitride ceramics [7]. Therefore, future opportunities for machining different materials very much exist.

5.4.2 USING A DIFFERENT MEDIUM

Hydrodynamic pressure of the fluid is a universal controlling factor in arc machining [28]. However, majority of the researchers have employed water as a dielectric fluid [1]. Recently, it was observed that air is suitable for better finishing than air + tap water [29]. Moreover, machining with water in an oil nano-emulsion provides better results than that with water based dielectric [30]. Thus, performance analysis with different dielectric media will be an important area of future research.

5.4.3 ON-LINE MONITORING

As current industrial and manufacturing activities are accelerating towards Industry 4.0, the use of process data to optimize the arc machining process parameters is of significant importance [20]. Therefore, integration of real-time monitoring into the arc machining process will bring new prospects in future.

5.4.4 INTELLIGENT MACHINING

Compared to traditional EDM, arc machining time is significantly reduced because of faster electrode-tool feed in the machining area. Therefore, creating an automatic electrode-tool feed system is crucial for the arc machining process [31]. Moreover, research using an AI-based modelling and optimization technique for arc machining is lacking and thus, this area will be a future area for research.

5.4.5 COMPOUND/HYBRID MACHINING

The MRR is found to improve when vibration is employed in electro-arc machining [32]. Moreover, hybridization of arc machining proves very effective to enhance process performances [1]. For example, a novel hybrid machining method shows better performance when combined with arc machining (for example, BEAM) and grinding [33]. Additionally, merging the electric arcs under the field of composite energy increases discharge energy for higher material removal rate [34]. Hence, more attention is directed towards these trends.

REFERENCES

[1] Shrivastava, P. K., S. Pandey, et al. (2019). "Electrical arc machining: Process capabilities and current research trends." Proceedings of the Institution of Mechanical Engineers, Part C: Journal of Mechanical Engineering Science **233**(15): 519–5200.

[2] Shen, Y., Y. Liu, et al. (2016). "High-efficient dry hybrid machining of EDM and arc machining." Procedia CIRP **42**: 149–154.

[3] Kusumoto, K. and H. Sun (2010). "Development of drilling method using plasma arc technology." Quarterly Journal of the Japan Welding Society **28**(4): 415–420.

[4] Sharma, D. N. and J. R. Kumar (2020). "Optimization of dross formation rate in plasma arc cutting process by response surface method." Materials Today: Proceedings **32**: 354–357.

[5] Kusumoto, K., H. Sun, et al. (2011). "Observation of hole formation process in plasma arc drilling." Quarterly Journal of the Japan Welding Society **29**(3): 66s–70s.

[6] Sun, H. and K. Kusumoto (2010). "Drilling performance of plasma arc drilling method." Quarterly Journal of the Japan Welding Society **28**(4): 421–426.

[7] Zhang, B., J. Wang, et al. (2013). "Impact force for micro-detonation of striking arc machining of silicon nitrides using the Taguchi method." Journal of Alloys and Compounds **580**: 176–181.

[8] Zhang, B., J. Wang, et al. (2013). "Spectroscopic diagnostics of plasma jet in micro-detonation of striking arc machining of engineering ceramics." Nuclear Instruments and Methods in Physics Research Section B: Beam Interactions with Materials and Atoms **307**: 353–356.

[9] Zhang, B., X. Tian, et al. (2013). "Numerical and experimental investigations on micro-detonation of striking arc machining of alumina." Proceedings of the Institution of Mechanical Engineers, Part B: Journal of Engineering Manufacture 228(6): 918–930.

[10] Ahmed, A., M. Rahman, et al. (2020). Spark Erosion Based Hybrid Processes. Spark Erosion Machining: MEMS to Aerospace: 123. ISBN 9780367510107.

[11] Qu, J., A. J. Shih, et al. (2005). "Abrasive micro-blasting to improve surface integrity of electrical discharge machined WC–Co composite." Journal of Materials Processing Technology 166(3): 440–448.

[12] Zhao, W., L. Gu, et al. (2013). "A novel high efficiency electrical erosion process – blasting erosion arc machining." Procedia CIRP 6: 621–625.

[13] Li, L., L. Gu, et al. (2012). "Influence of flushing on performance of EDM with bunched electrode." The International Journal of Advanced Manufacturing Technology 58(1): 187–194.

[14] Gu, L., L. Li, et al. (2012). "Electrical discharge machining of Ti6Al4V with a bundled electrode." International Journal of Machine Tools and Manufacture 53(1): 100–106.

[15] Xu, H., L. Gu, et al. (2015). "Machining characteristics of nickel-based alloy with positive polarity blasting erosion arc machining." The International Journal of Advanced Manufacturing Technology 79(5): 937–947.

[16] Benedict, G. (1987). "Electrical discharge machining (EDM)." Nontraditional Manufacturing Processes, Marcel Dekker, Inc, New York & Basel: 211–213.

[17] Hayakawa, S. (1996). Analysis of time required to deionize an EDM gap during pulse interval. ICHMT Symposium on Molecular and Microscale Heat Transfer in Materials Processing and Other Applications.

[18] Han, F., Y. Wang, et al. (2009). "High-speed EDM milling with moving electric arcs." International Journal of Machine Tools and Manufacture 49(1): 20–24.

[19] Kou, Z. and F. Han (2018). "Machining mechanisms and characteristics of moving electric arcs in high-speed EDM milling." Procedia CIRP 68: 286–291.

[20] Kou, Z. and F. Han (2018). "On sustainable manufacturing titanium alloy by high-speed EDM milling with moving electric arcs while using water-based dielectric." Journal of Cleaner Production 189: 78–87.

[21] Wang, F., Y. Liu, et al. (2014). "Compound machining of titanium alloy by super high speed EDM milling and arc machining." Journal of Materials Processing Technology 214(3): 531–538.

[22] Shen, Y., Y. Liu, et al. (2015). "High-speed dry compound machining of Ti6Al4V." Journal of Materials Processing Technology 224: 200–207.

[23] Ahmed, A., M. T. Lew, et al. (2019). "A novel approach in high performance deep hole drilling of Inconel 718." Precision Engineering 56: 432–437.

[24] Ahmed, A., A. Fardin, et al. (2018). "A comparative study on the modelling of EDM and hybrid electrical discharge and arc machining considering latent heat and temperature-dependent properties of Inconel 718." The International Journal of Advanced Manufacturing Technology 94(5): 2729–2737.

[25] Ahmed, A., M. Tanjilul, et al. (2018). "On the design and application of hybrid electrical discharge and arc machining process for enhancing drilling performance in Inconel 718." The International Journal of Advanced Manufacturing Technology 99(5): 1825–1837.

[26] Ahmed, A., M. Tanjilul, et al. (2020). "Ultrafast drilling of Inconel 718 using hybrid EDM with different electrode materials." The International Journal of Advanced Manufacturing Technology 106(5): 2281–2294.

[27] Tanjilul, M., A. Ahmed, et al. (2018). "A study on EDM debris particle size and flushing mechanism for efficient debris removal in EDM-drilling of Inconel 718." Journal of Materials Processing Technology **255**: 263–274.

[28] Meshcheriakov, G., V. Nosulenko, et al. (1988). "Physical and technological control of arc dimensional machining." CIRP Annals **37**(1): 209–212.

[29] Chen, X., J. Zhou, et al. (2019). "A study on machining characteristics of nickel-based alloy with short electric arc milling." The International Journal of Advanced Manufacturing Technology **105**(7): 2935–2945.

[30] Dong, H., Y. Liu, et al. (2019). "High-speed compound sinking machining of Inconel 718 using water in oil nanoemulsion." Journal of Materials Processing Technology **274**: 116271.

[31] Savelenko, G., Y. L. Yermolaev, et al. (2016). "Optimization of arc ignition process for machines of arc dimensional machining." Technology Audit and Production Reserves **6**: 44–51.

[32] Zhu, G., M. Zhang, et al. (2019). "High-speed vibration-assisted electro-arc machining." The International Journal of Advanced Manufacturing Technology **101**(9): 3121–3129.

[33] Gu, L., G. He, et al. (2020). "High performance hybrid machining of γ-TiAl with blasting erosion arc machining and grinding." CIRP Annals **69**(1): 161–164.

[34] Zhang, J. and F. Han (2021). "Rotating short arc EDM milling method under composite energy field." Journal of Manufacturing Processes **64**: 805–815.

6 Electric Discharge Hybrid-Turning Processes

Jees George¹, R. Manu² and Jose Mathew²
¹Amal Jyothi College of Engineering, Kanjirappally, Kerala, India
² National Institute of Technology Calicut, Kerala, India

CONTENTS

6.1 INTRODUCTION

The recent advancements in the aviation industry, sophisticated biomedical instruments, defense industry, and micro-electro mechanical systems have paved the way for an increase in demand for the manufacturing of microelectrodes and micro-shafts made from difficult-to-machine materials. Micro-shafts have a wide application in micro-fluidic systems, micro pumps, data storage systems, micro engines, and turbines, and the like. In recent times, the industry is driving towards miniaturization and manufacturing components with advanced alloys and materials. Traditional machining processes have been found to be problematic while machining such advanced, extremely hard

DOI: 10.1201/9781003202301-6

materials. However, the problem of turning advanced difficult-to-machine materials can be solved by combining the features of various machining processes and thus constructing a new technique to manufacture the desired products with a higher production rate and economy. Electrical discharge machining (EDM), being a non-traditional machining method, has the benefit of non-contact machining regardless of strength and hardness, and is a low cost process, which makes it one of the mainstream technologies to manufacture advanced difficult-to-machine materials [1,2]. EDM uses quick, repeated spark discharges with the aid of a pulsating direct current power supply applied between the workpiece and electrode, which are usually kept under a dielectric medium [3]. At the onset of every discharge, an extremely high temperature is formed due to the formation of plasma, which causes local melting or evaporation of the molten material from the workpiece. This plasma channel formed gets collapsed, depending upon the plasma flushing efficiency (%PFE), leading to intense suction and extreme bulk boiling of a portion of the molten metal and flushing it away from the molten crater [4].

Wire electrical discharge machining (WEDM) uses the electro-thermal concept of the EDM process, by forming a plasma channel for the electric current between the wire electrode (tool) and the workpiece, which causes melting and evaporation of the workpiece material and hence enables the machining of intricate and complex profiles. Wire electrical discharge turning (WEDT) is a relatively new emerging area in the machining of cylindrical components by the addition of a rotational unit to the existing 5-axis WEDM machine. The WEDT process is applicable for producing both macro-level and micro-level cylindrical geometries of intricate shape and size.

WEDT is an exclusive alteration of the WEDM process which facilitates the generation of components in cylindrical form regardless of its initial shape. Qu et al. [3] pioneered the concept of Cylindrical Wire Electrical Discharge Turning (CWEDT) for producing precise cylindrical shapes. Figure 6.1 illustrates the schematic diagram of the WEDT process which shows the concept of generating cylindrical geometry.

The wire is passed through the rotating workpiece attached to the WEDT setup at a desired feed rate and depth of cut. The thin wire diameter (0.01-0.3 mm) enables the machining of very small internal radius [4]. Furthermore, due to its non-contact nature, it eliminates cutting force and this facilitates the generation of good quality cylindrical components, as defined by the desired product specifications, namely, material used, dimensional accuracy and surface properties.

6.2 TURNING PROCESSES

Turning processes to generate cylindrical components can be categorized as illustrated in Figure 6.2.

Turning is a type of manufacturing process; a metal removal process, for the generation of cylindrical components by removing the undesired material. Additionally, turning can also be utilized to produce parts having various features, for example, grooves, tapers, holes, different diameter steps and contoured surfaces. Turning on a lathe is one of the oldest techniques for manufacturing cylindrical products that is

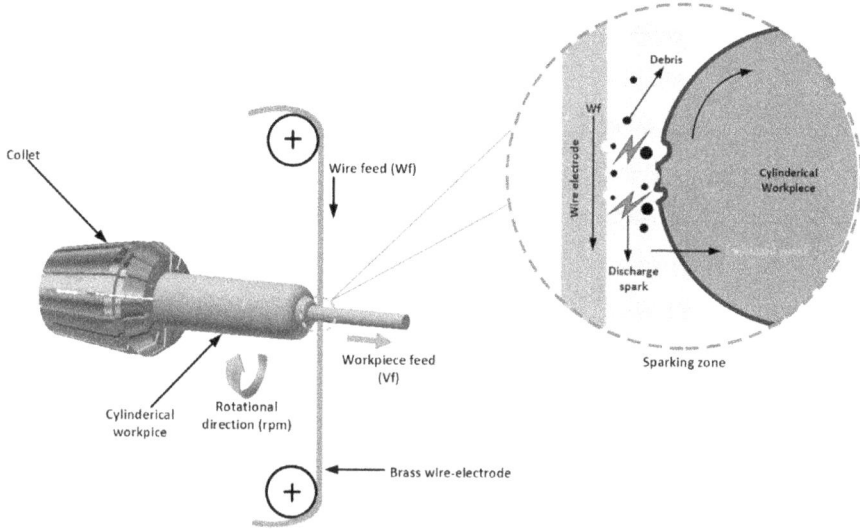

FIGURE 6.1 Schematic diagram of WEDT process.

still relevant today. The process consists of a rotational unit which holds the workpiece (the headstock assembly) and a cutting tool (made of high-speed steel, carbide, carbon steel, cobalt high-speed steel, and the like). Another important variant of the turning operation is the turning machine which can be controlled with the help of a computer, referred to as a computer numerically controlled (CNC) lathe. The workpiece rotation and movement of the cutting tool of a CNC lathe in the desired path is based on the commands pre-programmed into the computer, and this offers high precision. Diamond turning is yet another ultraprecision conventional turning process for the generation of intricate surfaces and ultra-fine microstructures with the aid of geometrically defined diamond cutters. Diamond turning is mainly utilized to generate ultra-high precision components, which have exceptional levels of form precision and surface finish for advanced industrial applications. In the early days, diamond turning was primarily used to manufacture optical components like lenses and reflectors, which are usually machined directly from stock material. However, now the process has been found to have wide applications in various industries, such as the aerospace, electronics, biomedical and defense and semiconductor industries [5]. Nevertheless, one of the major drawbacks of diamond turning is its inability to turn ferrous materials due to the excessive diamond tool wear while turning them [6]. Micro turning is one of the most popular conventional turning technologies to produce miniaturized cylindrical parts and components. It consists of a solid tool to remove material with a different cutting mechanism compared to the macro turning process. In the macro turning process, material is mainly removed by shearing bulk material along the average shear plane with high defect density and this forms a larger chip cross section. On the other hand, in case of micro turning, the material is removed across the grains, where the size effect becomes a dominant factor that leads (significantly) to a high specific cutting energy. One of main disadvantages of

```
                           ┌─────────────────────┐
                           │  Turning Processes  │
                           └──────────┬──────────┘
        ┌──────────────┬─────────────┼──────────────┬──────────────┐
        ▼              ▼                             ▼              ▼
┌───────────────┐ ┌───────────────┐         ┌───────────────┐ ┌───────────────┐
│ Conventional  │ │     Non-      │         │    Hybrid     │ │   Material    │
│               │ │ Conventional  │         │               │ │   Addition    │
└───────────────┘ └───────────────┘         └───────────────┘ └───────────────┘
```

Conventional	Non-Conventional	Hybrid	Material Addition
Lathe	WJT	Laser-assisted	Rapid Prototyping (Eg. Stereolithography, Selective Ground Curing, Selctive Laser Sintering, Selective Powder Binding, Fused Deposition, Laminated Object Manufacturing, etc.)
CNC Lathe	WEDG	Ultrasonic-assisted	
Diamond Turning	TF-WEDG	WEDG and ECM	
Micro-Turning	Twin-wire EDM	Micro Tuning and Micro-EDM	
	Self-drilled holes	LIGA and micro-EDM	
	BEDG		
	EDG-TBE		
	REDM		
	Laser Micro Turning		
	WEDT		

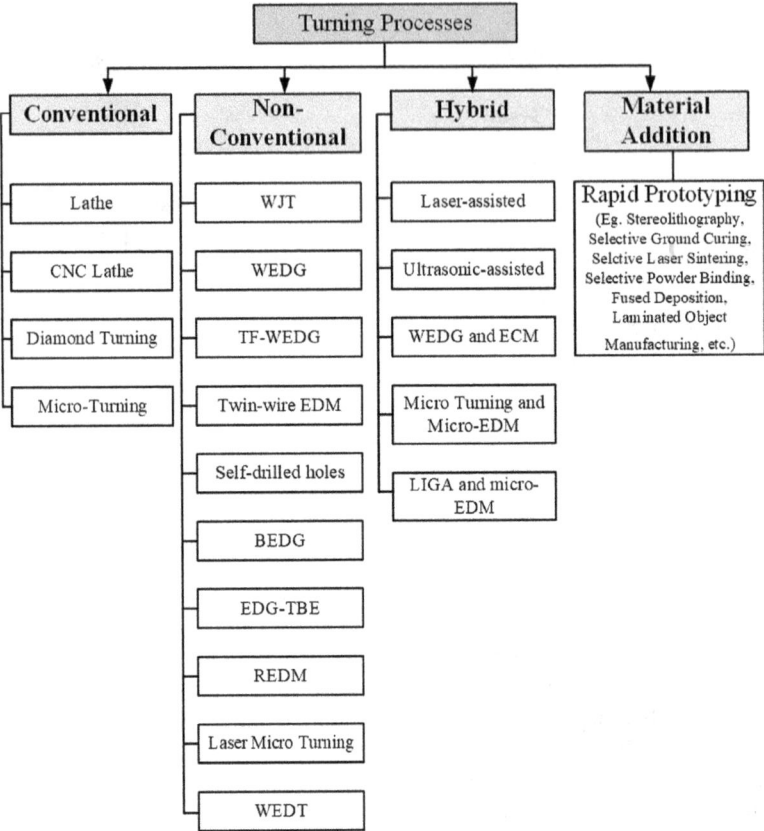

FIGURE 6.2 Classification of turning processes.

using conventional micro turning is its inability to turn difficult-to-machine material beyond 100 µm as forces exerted by the tool leads to the deflection of the micro-electrode and it often breaks during the turning process.

Various types of electrical discharge turning processes are discussed as follows:

6.2.1 WIRE ELECTRICAL DISCHARGE GRINDING (WEDG)

Wire electrical discharge grinding (WEDG) is a microfabrication technique that makes use of the electrical discharges produced in the dielectric medium to remove metal from electrically conductive wire blanks and generate micro-rods [7]. Masuzava et al. [8] introduced WEDG for the first time to fix various problems associated with the online production of microelectrodes and it became possible to manufacture microelectrodes with high machining efficiency. A micro-shaft of diameter 2.8 µm could be fabricated with the WEDG method. In comparison to the traditional grinding process that depends on forces caused by tools with higher hardness or abrasive materials for the removal of soft workpiece material, the WEDG technique uses sparks generated

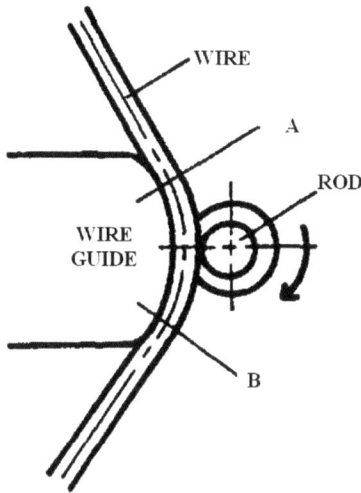

FIGURE 6.3   WEDG method [10]. With permission from Elsevier.

between tool and electrode or thermal energy to remove material from focus areas of the component and hence to create the desired features and shapes. Some of the disadvantages of using WEDG are its complex setup, poor surface roughness and high investment in contrast to the WEDT process [9]. Here, the cylindrical part is rotated at a rate of 10 to 3000 rpm depending on the desired finish and depending on the feed. A thin wire is used as the electrode and thus can be considered as an analogy of the WEDM process, solving the problem of the electrode wear which occurs in a conventional grinding process. In comparison to the WEDM process, the wire vibration and the deviation of the wire caused by the original curl of the machining force is negligible at the point where the wire and the wire guide are in contact. Figure 6.3 shows the setup that ensures the precise position of the wire edge which helps in deciding the accuracy of the final finished micro shaft. The side view of the WEDG setup is shown in Figure 6.4. As depicted in the figure, the discharge area is restricted to the front edge of the wire since the rod is being fed downwards. WEDG makes use of the NC system for controlling the movement of the axes.

6.2.2   TANGENTIAL FEED (TF)-WEDG

TF-WEDG is a technique to enhance the machining accuracy of the existing WEDG. In TF-WEDG proposed by Zhang et al. [10], the electrode is fed in the direction of the tangent of the wire-guide arc. Negative polarity machining is employed which leads to a significant improvement in the material-removal resolution of the micro-shaft diameter. A comparison to the WEDG method is schematically depicted in Figure 6.5.

With the use of TF-WEDG, the error in diameter of the machined electrode is reduced and it is easy to control. Additionally, TF-WEDG can produce a large number of micro electrodes continuously with consistent diameter by setting the

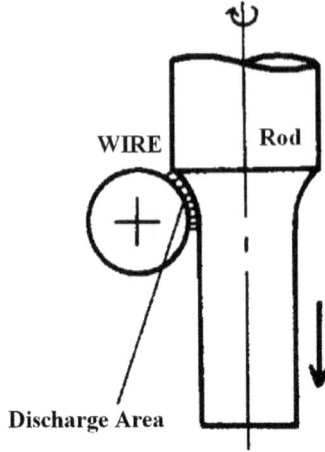

FIGURE 6.4 Discharge area in WEDG [10]. With permission from Elsevier.

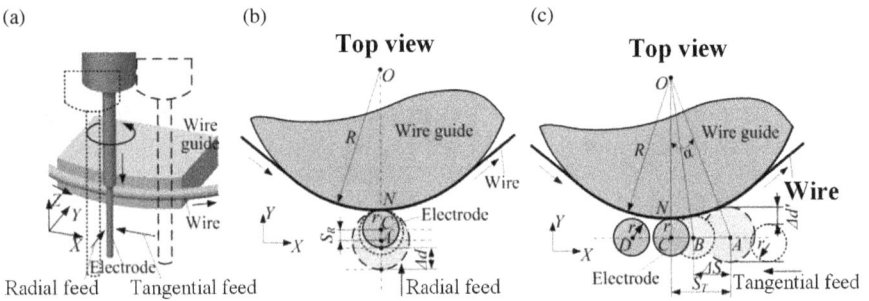

FIGURE 6.5 (a) WEDG setup, (b) radial feed WEDG, and (c) TF-WEDG [10]. With permission from Elsevier.

Y-axis location and feeding to the location-D as shown in Figure 6.5. One major drawback of the setup is that it can machine only cylindrical parts and is very complex to build. In this method, during the machining of micro electrodes with high accuracy, the sensitive direction avoids a position error. In this technique, a significant improvement in the removal resolution of the micro shaft is achieved and controlling the error in the diameter of the machined micro shaft is relatively simple and thus it reduces the positioning error.

6.2.3 TWIN-WIRE EDM

Twin-wire EDM (TW-EDM) is a method to generate micro tools that incorporates TrC and RC pulse generators, which enables the rough machining and the finish machining to take place side by side on one machine [11]. This simultaneous process makes it unique and allows efficient fabrication of micro tools in comparison to conventional WEDG. This novel technique utilizes only a single direction servo-circuit by keeping high discharge energy (DE) for rough machining, again in contrast to the twin-WEDG methodology. Twin-wire EDM can generate micro electrodes as low as 5 μm in diameter. A diagram showing the twin-wire EDM system is depicted in Figure 6.5

The twin-wire EDM machining process consists of two individual and isolated wires, which move on v-shaped guides, as shown in Figure 6.6. The two specific copper wire electrodes deliver two individual pulse generators on a single machine. A transistor pulse generator circuit is applied to the upper wire for rough machining while the lower wire utilizes a RC-type pulse generator circuit for finish machining. The transistor pulse generator consisting of a large capacitor delivers enough discharge energy to attain a high material removal rate for rough machining. Furthermore, the RC-type pulse generator, consists of a capacitor with a lower capacitance value to achieve reduced DE for a precise finish machining process. The reduced capacitance and lower DE of a RC-type pulse generator helps in generating micro shafts, which makes it a crucial discharge circuit. The TW-EDM has the capability of fabricating micro electrodes having a diameter as thin as 5 μm.

The cylindrical workpiece is initially fed through the wire located above, to roughly modify and reshape the rod, which uses a TrC electro-pulse generator as shown in Figure 6.6. Once the length of the machined rod surpasses the distance between two electro wires, the twin-wire EDM system acts as a finishing machining procedure through the RC electro pulse generator simultaneously. Traditional EDM systems require a T_{on} and T_{off} and duty factor using TrC pulse-generators to achieve high-precision machining. However, in the case of the TW-EDM setup, the precision of the micro shafts and the finishing process is determined without considering the transistor pulse generator. Furthermore, through the ability of high precision position control during the finishing process, this method can generate straight, dimensionally accurate, thin micro shafts. Hence, it can be concluded that micro shafts using electrically conductive materials can be efficiently fabricated using TW EDM technology. Additionally, in the TW EDM system, a one direction servo circuit only, is enough for the finishing process by creating large material removal during rough machining. Thus, this process has a significant advantage over WEDG technology by saving machining time (typically, a reduction of two thirds of the total time).

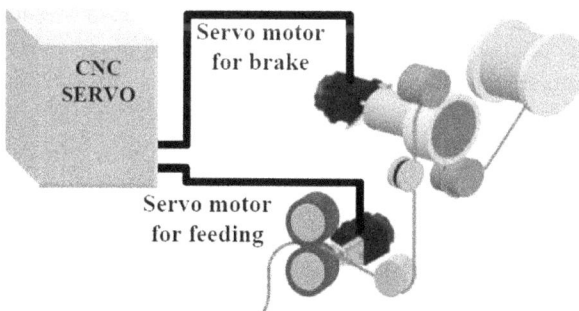

FIGURE 6.6   Twin-wire EDM setup [12]. With permission from Elsevier.

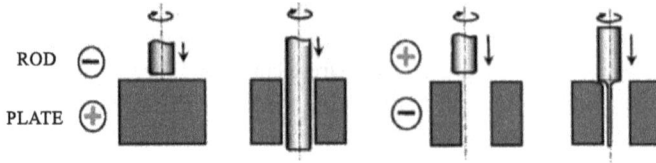

FIGURE 6.7   Self-drilled holes [13]. With permission from Elsevier.

6.2.4 SELF-DRILLED HOLES

The self-drilled holes process is a technique where the cylindrical workpiece is given negative polarity and fed with a rotating motion into a plate electrode with positive polarity to generate holes. The workpiece is moved back to the original position and then the axis of the workpiece is offset from the center of the hole by a desired amount. Afterwards, polarity is reversed and again the workpiece is delivered into the plate as shown in Figure 6.7. A precise cylindrical part can be fabricated as the holes are generated by the workpiece itself. The method eliminates the requirements around the original positioning of the tool electrode with respect to the plate electrode, which makes the process short and simple. A micro-shaft of diameter 4 µm could be produced using this method. One of the disadvantages of using the self-drilled holes process, is its low machining accuracy and taper.

A straight micro shaft is achieved if wear doesn't occur in the outlet hole while significant wear takes place in the inlet hole [14]. A rod electrode with the desired diameter can be achieved if the gap between the cylindrical electrode and machining hole are measured before conducting the machining process, as depicted in Figure 6.7. In the case of conventional methods, the micro-shaft diameter must be measured after machining and thus fix the initial point of the machining, taking previously measure results into account. The benefit of using this technique is that the prerequisite alteration of the initial point is not needed and it offers better machining effectiveness. Figure 6.8 shows the offsetting concept during the self-drilled holes method.

In comparison to reverse EDM and WEDG, this technique is capable of machining reference holes with the shaft electrode without the requirement to position the tool electrode about the rod electrode, and hence good forming precision can be obtained. Moreover, the machining process is also simpler and significantly faster. Additionally, a steady electrical discharge machining is obtained because of the larger area of electrical discharge in comparison to the WEDG process and thus, complex configurations are generated. Figure 6.9 shows the experiments conducted to showcase the principles of self-drilled holes.

6.2.5 BLOCK ELECTRODE DISCHARGE GRINDING (BEDG)

BEDG is a method where a specific rectangular conductive block is taken as a machining electrode and an electrode rod is considered as the workpiece in the EDM method as illustrated in Figure 6.10. The electrode that is required to be turned is delivered against the conductive block [15,16].

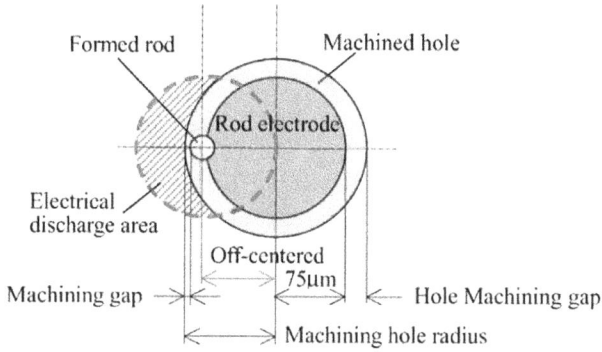

Formed rod radius = Machining hole radius − Off-centered distance − Machining gap

FIGURE 6.8   Off-centring concept chart [13]. With permission from Elsevier.

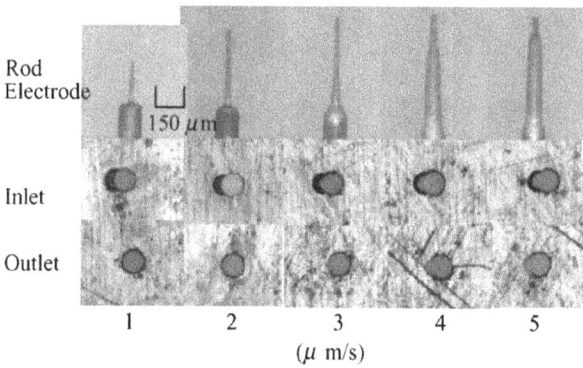

FIGURE 6.9   Experiments conducted to show the concept of self-drilled holes [13]. With permission from Elsevier.

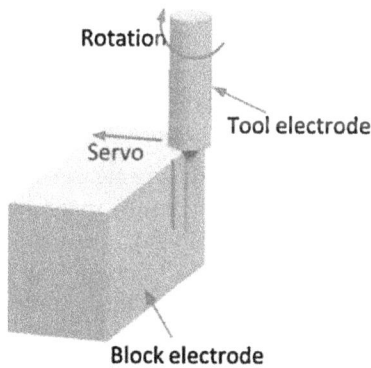

FIGURE 6.10   Schematic of BEDG setup [17]. With permission from Elsevier.

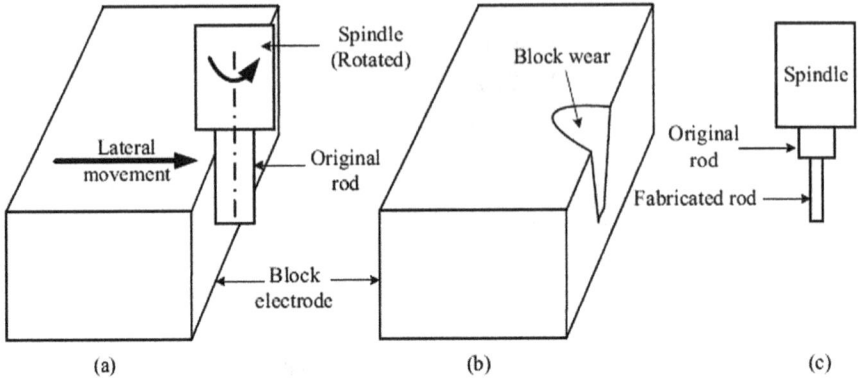

FIGURE 6.11   (a) BEDG process, (b) block wear in BEDG process, and (c) shaft generated [14]. With permission from Elsevier.

Application of a controlled electric spark initiates the machining process and the flushing of dielectric present between the spark gap helps in removing the debris particles eroded from both the block and the rod. Unlike WEDG, the EDM block electrode technique requires only a low expenditure and the setup is simpler. Furthermore, the method enables the generation of the electrode on the machine, which significantly diminishes the electrode installation error and cost. The wear on the block and the generated shaft is depicted in Figure 6.11.

The BEDG method uses a conductive block as the tool electrode and the shaft as a workpiece. It is recognized as an effective procedure for generating micro shafts because of its low cost and fast installation. Due to the application of voltage between the electrode and block, a sporadic spark is generated between the inter electrodes. The spark generated leads to the increase in the temperature at the surface of the shaft electrode and the block together up to a state where the temperature goes beyond the boiling or melting point of the metal [14]. Subsequently, a portion of the material is flushed away from both electrodes and flushed away due to flushing. Figure 6.11 shows the concept of BEDG before and after feeding.

6.2.6 ELECTRICAL DISCHARGE GRINDING USING TWO BLOCK ELECTRODE (EDG-TBE)

EDG-TBE is a unique method to produce micro electrodes where the workpiece with a given diameter is fed via a narrow slit and the spark gap. To achieve the desired diameter, the area of the slit is fixed accurately after the power supply factor is set. This is because the spark gap is estimated from the power supply factor [17]. The advantage of using two block electrodes is that this method does not need any online measuring device. A highly efficient micro-shaft can be manufactured by eliminating the influence of electrode wear. Furthermore, the facility for online measurement by this method reduces the complexity of the process. Figure 6.12 illustrates the

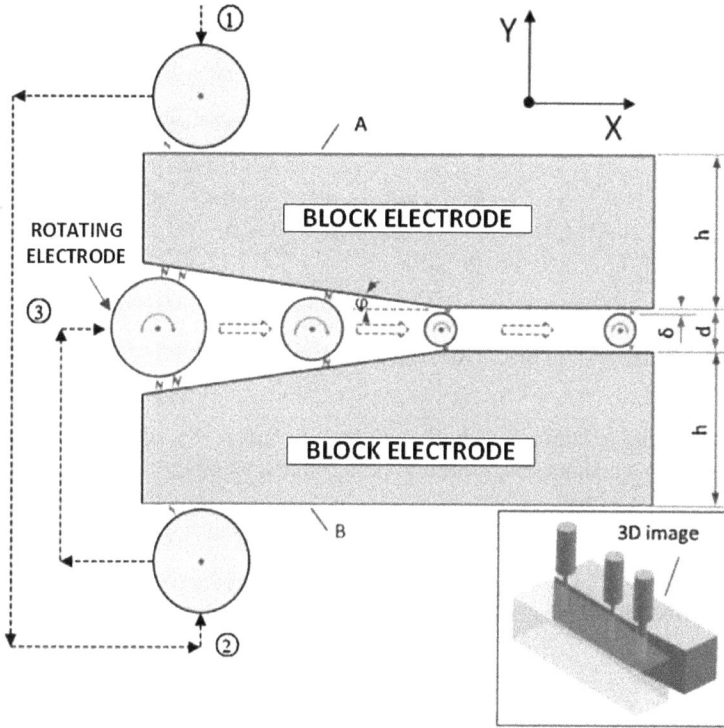

FIGURE 6.12 EDG-TBE process [17]. With permission from Elsevier.

EDG-TBE method. As discussed in previous section, BEDG can be more efficient in fabricating micro shafts in comparison to WEDG as a large space of block electrode is utilized while grinding micro shafts. However, in this method, a change in the dimension of block-electrodes occur and hence, the fabricated tool's diameter can't be predicted in general. A measuring device is required for both BEDG and WEDG to achieve micro shafts of desired diameter. The online measuring device measures the diameter of the shafts once the micro shaft is turned close to the targeted diameter [18]. Unlike BEDG and WEDG, an online measuring device is not needed in case of EDG-TBE.

In this method, two-block electrodes with the same configurations are set in the working table having a gap of d as illustrated in Figure 6.12. The block electrode consists of a plane and a plane with an inclination angle of φ as depicted in Figure 6.13 (b).

Once the location of plane A and plane B is obtained along the Y axis by edge find discharge, the location of two blocks' center lines in the Y-axis direction could be estimated. A rotational unit consisting of a rotating cylindrical shaft is passed in the center line direction. The grinding of the cylindrical rod is initiated due to the spark when the space between the cylindrical rod and block electrode is less than

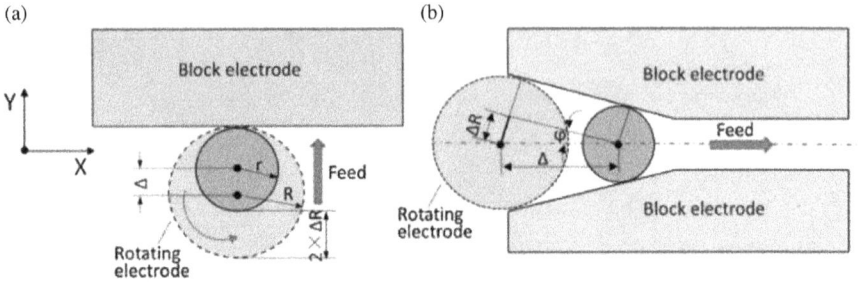

FIGURE 6.13 Schematic diagram showing removal resolution considering two machining methods: (a) BEDG and (b) EDG TBE [17].

the discharge gap δ. During machining, the rod diameter is gradually decreased. The diameter of the machined electrode can be estimated using the following equation:

$$d_a = d - 2 \times \delta \qquad (1)$$

One of the drawbacks of using EDG-TBE is its high demand for positional accuracy.

6.2.7 REVERSE ELECTRICAL DISCHARGE MACHINING (REDM)

REDM is a machining technique for producing micro-electrodes where a positive voltage is connected to the workpiece material and the workpiece is delivered down into holes already present in a metal plate and thus creating the desired micro shaft. Here, the area that corresponds to the holes remains unaffected, which results in the fabrication of micro-electrodes at every hole. Figure 6.14 shows a comparison between a normal EDM and REDM. There are a few drawbacks while machining with REDM, namely that high aspect ratio shafts can't be produced using this method, it has a complex setup requirement, low machining accuracy and taper formation is found. As the electrodes cannot be rotated, it is difficult to flush out the debris from the machining area thereby causing short circuits in the case of high aspect ratio micro shafts.

Micro EDM can be used to generate holes and structures. However, it can also be utilized to generate micro electrodes. For example, a few holes are drilled on a metal plate with the aid of the EDM process or other machining process. Some researchers have also worked on non-conventional hybrid turning processes to improve machining accuracy and efficiency. Afterwards, polarity is inverted, and the electrode shaft is delivered downwards into the holes previously generated in the metal plate. However, since the polarity is reversed, the regions corresponding to the holes are not affected, instead the electrode rod is machined, and thus micro electrodes are fabricated. REDM can also be used to fabricate multiple structures and micro shafts simultaneously.

The following subsections discuss some of the non-conventional hybrid processes.

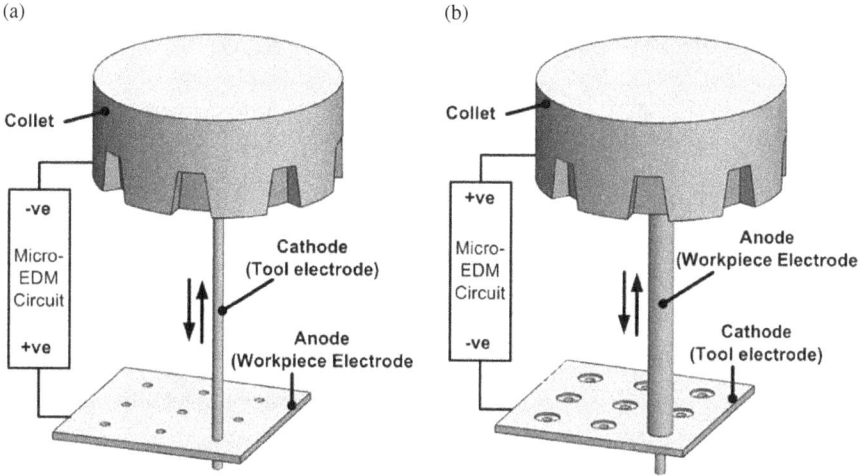

FIGURE 6.14 (a) μ-EDM process for drilling microhole and (b) R-EDM for fabricating microtool electrode [19]. With permission from Taylor & Francis.

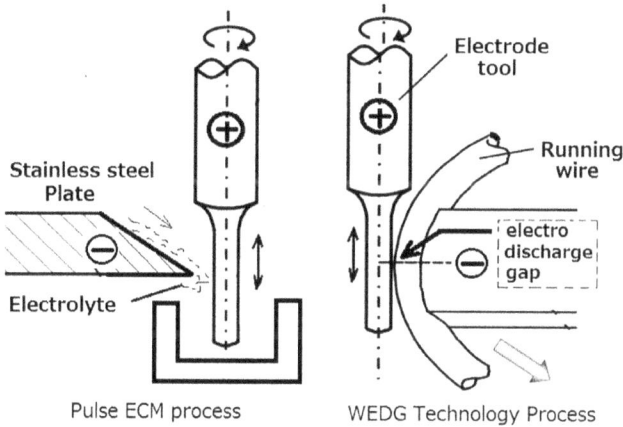

FIGURE 6.15 Schematic of WEDG and ECM process [21].

6.2.8 WEDG AND ELECTRO CHEMICAL MACHINING (ECM)

This is a combination of two techniques, namely WEDG and ECM for the improvement of ultra-fine micro-shafts machining efficiency [20]. A hybrid machine incorporating both WEDG and ECM technologies is constructed for the fabrication of micro pins. The technology could generate micro electrodes of diameter as low as 0.3 μm with good surface finish. Figure 6.15 illustrates the schematic of WEDG and the ECM process. The shaft is initially shaped using WEDG technology and diameter is reduced with the finish machining process. Once these micro shafts are fabricated, it is shifted to the pulse ECM position where chemical erosions take place on the electro-isolated tank.

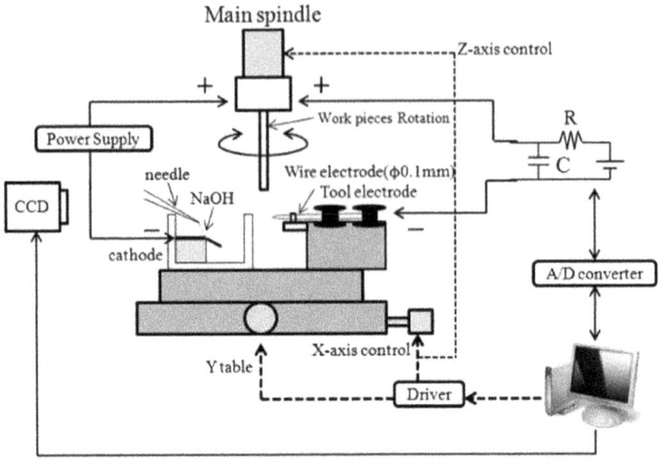

FIGURE 6.16 WEDG and pulse ECM hybrid process principle [21]. With permission from Elsevier.

Figure 6.16 shows the experimental setup for the hybrid process. Both WEDG and ECM units are fixed on the same worktable X. For the hybrid process to run simultaneously, two electric powers, namely, 80V (high voltage) and 12V (low voltage) are applied. The WEDG unit requires high voltage whereas lower voltage is used for the ECM unit. A linear scale and PC based controller are used on the worktable for high precision positioning. The WEDG unit makes use of RC electro discharge for generating pulses whereas a power transistor is utilized in the case of the ECM unit. One of the drawbacks of using this hybrid process is that it cannot produce dimensionally accurate micro pins consistently.

6.2.9 MICRO TURNING AND MICRO EDM

In this combination of two machining processes, after fabrication using the micro turning process, the electrode is then machined with the aid of a micro-EDM process. With the aid of this technique, the clamping error can be eliminated and the deviation of workpiece rod is significantly reduced [21]. Most of the operations discussed in the previous sections take up a lot of time to manufacture micro-electrodes. However, using this hybrid process, faster production of micro pins is possible due to the incorporation of the micro-turning process. Although micro-turning causes deflection in the micro shafts beyond 100 μm, the problem can be solved using the EDM process. Figure 6.17 shows the concept of micro turning and the micro EDM hybrid process.

In this type of experimental setup, EDM is conducted using a micro-shaft. Initially, a micro-shaft of the desired length is turned with the aid of the micro-turning method. The hybrid technique facilitates avoiding clamping error and minimizes the deviation error of the shaft, subsequently the accuracy of micro electrode fabrication can

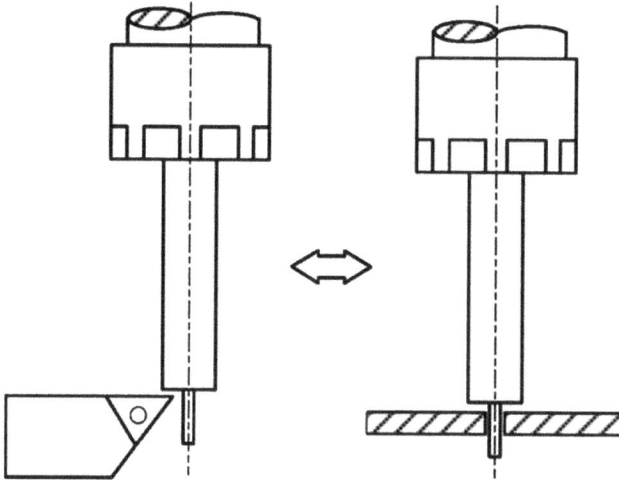

FIGURE 6.17 Hybrid micro turning and micro EDM process [21]. With permission from Elsevier.

be enhanced. Considering different electrode diameter, micro-turning can effectively lower the duration of electrode fabrication in comparison to other electrode generation techniques. Additionally, the hybrid technique can also be utilized to generate cylindrical shafts having non-rotational parts like a flat bar or a key slot with the aid of the EDM method and using micro-turning subsequently.

6.2.10 LIGA AND MICRO EDM

In this process, LIGA (Lithographie Galvanoformung Abformung) and micro-EDM are combined to produce extremely fine finished micro components. Here, fabrication of electrodes of negative configuration using the LIGA method where electroplating of the metal plate is done for the application in the micro EDM [22]. In this technique, any large conductor can be used as the workpiece, which is to be converted into a micro structure. This method solves the problem of lower shape accuracy of the EDM surface because of the electrode positioning error or the dimensional inaccuracy of the electrodes. It has the advantage of producing ultra-fine high precision micro structures developed from different types of bulk materials, that are extremely difficult of fabricate using other conventional methods. The LIGA process was originally proposed by Takhata et al. [23] for fabricating electroplated electrodes for micro-EDM. The combination of these two processes is illustrated in Figure 6.18. Steps 1–5 explain the LIGA process whereas, steps 6–7 use the micro EDM method to obtain the high aspect ratio micro shafts with patterns.

This hybrid methodology helps to obtain ultra-fine patterned highly accurate microelectrodes composed of different types of bulk-materials that are impossible to fabricate with other traditional microfabrication methods. Figure 6.19 shows the setup used for the hybrid LIGA and micro EDM technique.

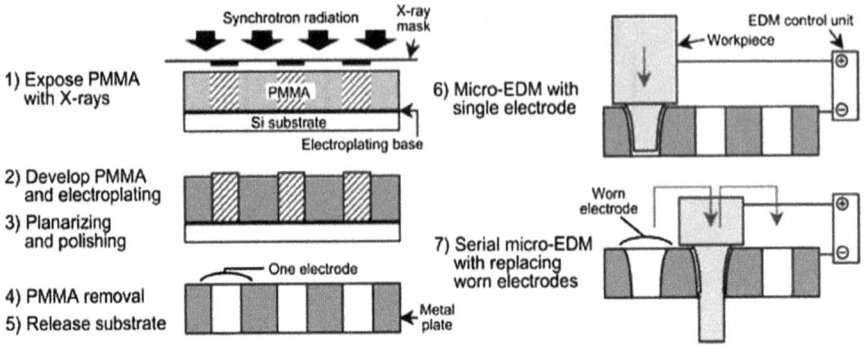

FIGURE 6.18 Steps of LIGA and micro-EDM hybrid method [23]. With permission from Springer.

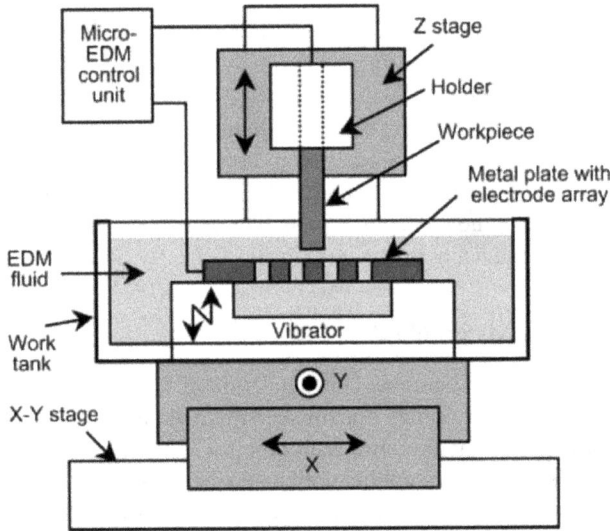

FIGURE 6.19 Setup for hybrid LIGA and micro-EDM process [23]. With permission from Springer.

A silicon substrate is used to develop the electrodes which are the negative kinds of microstructures having extremely-fine structures. A number of ultra-fine rods are made with ease in an electro-plated metal surface. As the negative types of electrodes are self-supporting, the metal plate consisting of electrodes is taken out from the substrate as shown in Step 5. The position of the micro-rods in the metal plate has high precision as the photolithographic methods are used for the fabrication of micro-shafts. The workpiece is linked to the positive terminal and a patterned micro-structure is generated by delivering the workpiece into one of the micro-rods with discharge as depicted in step 6. At this stage, the shape of the electrode is damaged because of its wear. To get a homogenous cross section throughout the workpiece's length, the electrode replacement is done just before the deformation appears along the thickness. The deformed electrode is replaced with a fresh one by locating its X-Y position. This process can be achieved very precisely due to the precise position of the electrodes as depicted in Step 7.

6.2.11 WIRE ELECTRICAL DISCHARGE TURNING (WEDT)

WEDT is a non-traditional turning process to generate cylindrical parts by adding a rotary attachment to the WEDM process. WEDT can be used to turn cylindrical components having a wide range of diameters unlike other non-conventional turning processes. A diagram depicting the WEDT process is illustrated in Figure 6.20.

As shown in Figure 6.20, the wire is passed through the rotating workpiece attached to the WEDT setup at a desired feed rate and depth of cut. The thin wire diameter (0.01–0.3 mm) enables the machining of a very short internal radius [6]. Here, material removal occurs due to repeated sparks forming craters. The energy present in the wire and electrode gap, known as discharge energy, is responsible for the crater formation leading to material removal. Controlling this energy helps in effective machining. Many researchers have proposed predictive models based on thermal modelling to estimate the crater morphology. There are various controlling parameters as in the WEDM process, namely, T_{off}, T_{on}, wire tension, flushing pressure, and servo voltage. Furthermore, as the process doesn't have any contact, it eliminates the cutting force and this facilitates the generation of good quality cylindrical components, closely matching the desired product specifications, that is material used, dimensional accuracy and surface properties. Furthermore, the WEDT method could be utilized in the manufacture of asymmetrical cylindrical components, turning of face surfaces on prismatic rods and in generation of sporadic curved surfaces. Researchers have conducted various experimental studies based on surface roughness, roundness and cylindricity of the cylindrical parts and attempted to optimize the input parameters. Figure 6.21 demonstrates the capability of the WEDT process to fabricate various microstructures.

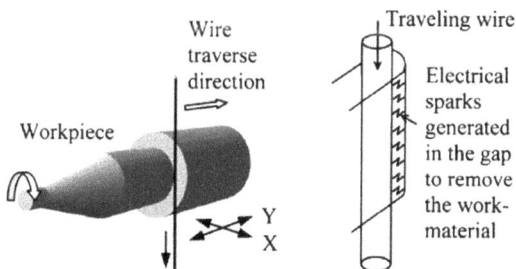

FIGURE 6.20   Schematic diagram of WEDT process [23]. With permission from Elsevier.

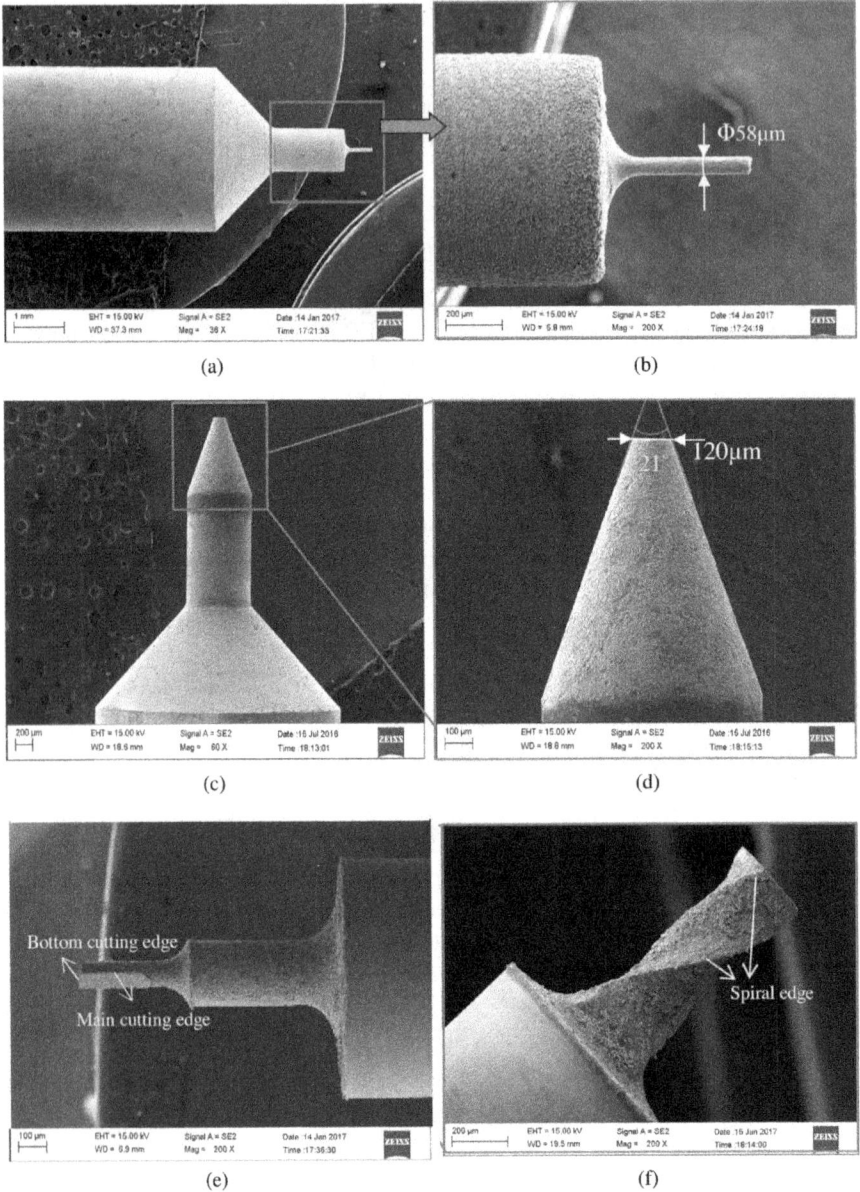

FIGURE 6.21 Different microelectrodes machined by LS-WEDT observed under SEM (a) Micro shaft, (b) enlarged view, (c) conical micro-electrode, (d) enlarged view, (e) D-shaped micro-cutting tool, and (f) micro-cutting tool with 3 spiral structures machined by LS-WEDT [24]. With permission from Elsevier.

6.3 COMPARISON

The metal removal method of both WEDT and WEDG are identical and thus WEDT holds significant benefits over WEDG such as the absence of electrode wear in WEDT because of the continuous movement of the wire electrode in one direction. This enables the production of non-tapered microelectrodes unlike BEDG and self-drilled hole techniques. Additionally, WEDT could perform both rough and finishing machining steps without the requirement of changing to fresh tool electrodes during the finishing step compared to BEDG. Furthermore, in WEDT, the measurement of microelectrode diameter can be done accurately by estimating the feed amounts rather than fixing online measuring equipment, which significantly reduces the cost and process complexity in comparison to TF-WEDG and the hybrid-processes. But there are some important difference between the WEDT and WEDG techniques. Primarily, the WEDG process needs a special and costly setup in comparison to the WEDT process. WEDT machining can be carried out on a conventional WEDM machine by attaching a rotational unit, which significantly reduces costs as well as expanding the WEDM application field. Furthermore, the workpiece feed mechanism for WEDT is horizontal while in WEDG the feed mechanism is in the longitudinal direction [25]. The dimensional precision of the turned parts is influenced significantly by the wire vibration and feed direction. Due to the gravitational pull while the longitudinal feed is taking place in WEDG, the workpiece is susceptible to deflection in comparison to the WEDT process [26]. In WEDT, the movement is not constrained by the idler pulley unlike WEDG which increases its flexibility and thus widens its application. Additionally, WEDT takes all the benefits of NC program technology which facilitates the fabrication of complex cylindrical structures effectively and flexibly. Hence, WEDT is a new and effective machining technology to turn a wide range of diameters including micro-shafts and micro-cutting tools with different microstructures.

6.4 CONCLUSION

A general idea about different types of non-conventional turning processes is introduced in this chapter. Turning with WEDM has been found to be cost-effective and is a precision machining alternative for difficult-to-machine metals. The option of turning intricate geometry with difficult-to-machine materials make WEDT one of the most trending and highly researched non-traditional machining processes. WEDT is now a new and novel research domain for turning micro-shafts of high hardness. Surface roughness, roundness, cylindricity, recast layer, and so forth, are the factors to be considered while conducting studies on different non-traditional machining processes. Researchers have conducted various experimental and theoretical studies for all the non-conventional turning processes to enhance the machining performance to achieve turned products with great dimensional precision.

REFERENCES

[1] Chen, S. L., Hsieh, S. F., Lin, H. C., Lin, M. H., and Huang, J. S., 2008, "Electrical Discharge Machining of a NiAlFe Ternary Shape Memory Alloy," J. Alloys Compd., **464**(1–2), pp. 446–451.

[2] Muthuramalingam, T., and Mohan, B., 2014, "Performance Analysis of Iso Current Pulse Generator on Machining Characteristics in EDM Process," Arch. Civ. Mech. Eng., **14**(3), pp. 383–390.

[3] Qu, J., Shih, A. J., and Scattergood, R. O., 2002, "Development of the Cylindrical Wire Electrical Discharge Machining Process, Part 1: Concept, Design, and Material Removal Rate," J. Manuf. Sci. Eng., **124**(3), pp. 702–708.

[4] Ho, K. H., Newman, S. T., Rahimifard, S., and Allen, R. D., 2004, "State of the Art in Wire Electrical Discharge Machining (WEDM)," Int. J. Mach. Tools Manuf., **44**(12–13), pp. 1247–1259.

[5] Zhang, X. Q., Woon, K. S., and Rahman, M., 2014, "Diamond Turning," *Comprehensive Materials Processing*, Elsevier, pp. 201–220.

[6] Gong, M., Li, P., Li, Y., and Tong, W., "Diamond Turning on Nitriding Diffusion Layer of Pure Iron," Integr. Med. Res., (14), pp. 2–6.

[7] Morgan, C. J., 2004, "Micro Electro-Discharge Machining: Techniques and Procedures for Micro Fabrication," University of Kentucky.

[8] Masuzawa, T., Fujino, M., Kobayashi, K., Suzuki, T., and Kinoshita, N., 1985, "Wire Electro-Discharge Grinding for Micro-Machining," CIRP Ann. – Manuf. Technol., **34**(1), pp. 431–434.

[9] Sun, Y., Gong, Y., Liu, Y., Cai, M., Ma, X., and Li, P., 2018, "Experimental Investigation on Effects of Machining Parameters on the Performance of Ti-6Al-4V Micro Rotary Parts Fabricated by LS-WEDT," Arch. Civ. Mech. Eng., **18**(2), pp. 385–400.

[10] Zhang, L., Tong, H., and Li, Y., 2015, "Precision Machining of Micro Tool Electrodes in Micro EDM for Drilling Array Micro Holes," Precis. Eng., **39**, pp. 100–106.

[11] Sheu, D. Y., 2008, "High-Speed Micro Electrode Tool Fabrication by a Twin-Wire EDM System," J. Micromechanics Microengineering, **18**(10), pp. 105–111.

[12] Takayama, Y., Makino, Y., Niu, Y., and Uchida, H., 2016, "The Latest Technology of Wire-Cut EDM," Procedia CIRP, **42**(Isem Xviii), pp. 623–626.

[13] Yamazaki, M., Suzuki, T., Mori, N., and Kunieda, M., 2004, "EDM of Micro-Rods by Self-Drilled Holes," J. Mater. Process. Technol., **149**(1–3), pp. 134–138.

[14] Kar, S., and Patowari, P. K., 2019, "Effect of Non-Electrical Parameters in Fabrication of Micro Rod Using BEDG," Mater. Manuf. Process., **34**(11), pp. 1262–1273.

[15] Ravi, N., and Huang, H., 2002, "Fabrication of Symmetrical Section Microfeatures Using the Electro-Discharge Machining Block Electrode Method," J. Micromechanics Microengineering, **12**(6), pp. 905–910.

[16] Lim, H. S., Wong, Y. S., Rahman, M., and Edwin Lee, M. K., 2003, "A Study on the Machining of High-Aspect Ratio Micro-Structures Using Micro-EDM," J. Mater. Process. Technol., **140**(1–3), pp. 318–325.

[17] Qingfeng, Y., Xingqiao, W., Ping, W., Zhiqiang, Q., Lin, Z., and Yongbin, Z., 2016, "Fabrication of Micro Rod Electrode by Electrical Discharge Grinding Using Two Block Electrodes," J. Mater. Process. Technol., **234**, pp. 143–149.

[18] Ravi, N., and Chuan, S. X., 2002, "The Effects of Electro-Discharge Machining Block Electrode Method for Microelectrode Machining," J. Micromechanics Microengineering, **12**(5), pp. 532–540.

[19] Singh, A. K., Patowari, P. K., and Deshpande, N. V., 2016, "Experimental Analysis of Reverse Micro-EDM for Machining Microtool," Mater. Manuf. Process., **31**(4), pp. 530–540.

[20] Wang, J., and Sheu, D. Y., 2016, "Developing a Process Chain with WEDG Technology and Pulse ECM to Fabricate Ultra Micro Pins," Procedia CIRP, **42**(18), pp. 815–818.

[21] Asad, A. B. M. A., Masaki, T., Rahman, M., Lim, H. S., and Wong, Y. S., 2007, "Tool-Based Micro-Machining," J. Mater. Process. Technol., **192–193**, pp. 204–211.

[22] Takahata, K., Shibaike, N., and Guckel, H., 2000, "High-Aspect-Ratio WC-Co Microstructure Produced by the Combination of LIGA and Micro-EDM," Microsyst. Technol., **6**(5), pp. 175–178.

[23] Haddad, M. J., and Fadaei Tehrani, A., 2008, "Material Removal Rate (MRR) Study in the Cylindrical Wire Electrical Discharge Turning (CWEDT) Process," J. Mater. Process. Technol., **199**(1), pp. 369–378.

[24] Sun, Y., and Gong, Y., 2018, "Experimental Study on Fabricating Spirals Microelectrode and Micro-Cutting Tools by Low Speed Wire Electrical Discharge Turning," J. Mater. Process. Technol., **258**(2), pp. 271–285.

[25] Lin, Y. C., Chen, Y. F., Lin, C. T., and Tzeng, H. J., 2008, "Electrical Discharge Machining (EDM) Characteristics Associated with Electrical Discharge Energy on Machining of Cemented Tungsten Carbide," Mater. Manuf. Process., **23**(4), pp. 391–399.

[26] Giridharan, A., and Samuel, G. L., 2016, "Analysis on the Effect of Discharge Energy on Machining Characteristics of Wire Electrical Discharge Turning Process," Proc. Inst. Mech. Eng. Part B J. Eng. Manuf., **230**(11), pp. 2064–2081.

7 Electric Discharge Grinding (EDG)

Dinesh Setti
Department of Mechanical Engineering,
Indian Institute of Technology Palakkad, Kerala, India

CONTENTS

7.1   INTRODUCTION

Electrical discharge grinding (EDG) is an unconventional machining process for electrically conductive hard and brittle materials. EDG is a variant of the electrical discharge machining (EDM) process, in which a rotating electrode is used instead of a stationary electrode. The wheel used in the EDG process rotates about its axis without having any bonded abrasive particles. Due to the similarities in the configuration with conventional grinding (rotating wheel and moving workpiece) and the material removal by the electrical discharge, this process is called electrical discharge grinding [1]. Similar to the EDM process, spark erosion is the material removal method in the EDG process. The wheel-speed is transmitted to the dielectric fluid between the workpiece and wheel, resulting in the effective removal of the molten metal and higher flushing efficiency. Thus, obtaining a higher material removal rate in the EDG process with a better surface finish is possible.

7.2   THE EDG PRINCIPLE

In the EDG process, the wheel is fed to the workpiece by a servo-control system. The workpiece acts as an anode (positive terminal), and the wheel acts as a cathode (negative terminal). The material is removed from the work surface due to the multiple sparks between the two electrodes in a dielectric medium. Every spark removes a small amount of material from the workpiece surface, forming a crater at the discharge site by melting and vaporization [2].

DOI: 10.1201/9781003202301-7

127

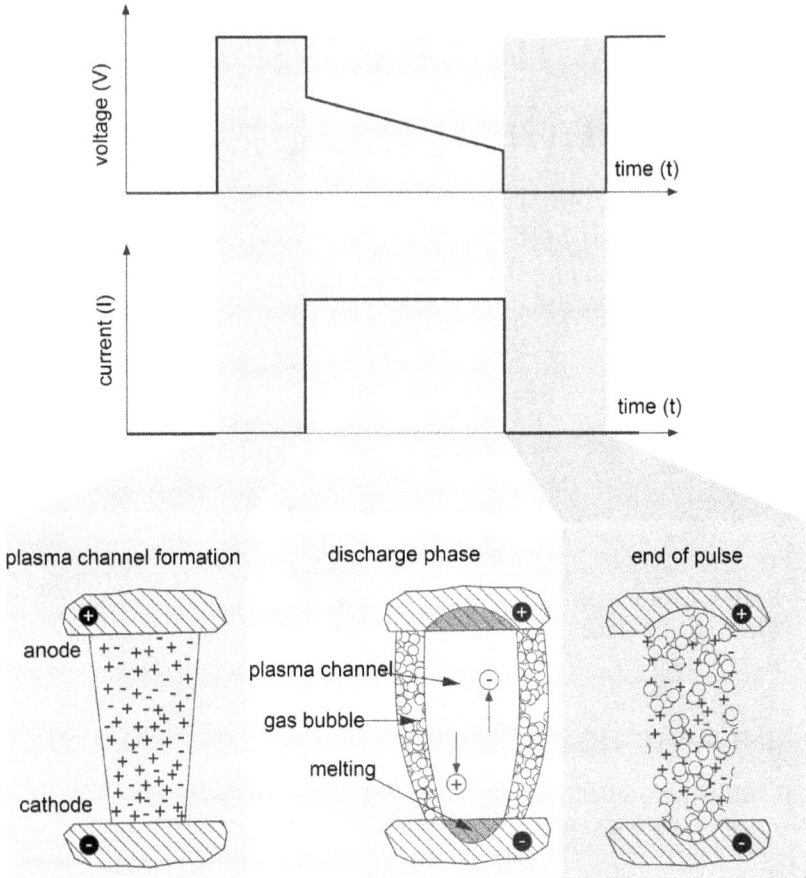

FIGURE 7.1 Schematic of the EDG working principle.

A pulse generator is used to produce high-frequency pulses of DC power. During the pulse-on period, the ionization and striking of ions and electrons occur at the respective electrodes and result in a spark generation in the interelectrode gap. Due to the spark, a high temperature (8000-12000° C) generated at the interface causes the material to melt at both the electrodes. At the same time, the DC pulse power supply is deactivated (pulse-off) to deionize the dielectric fluid, thus the spark discharges [3]. In addition, dielectric fluid also evaporates through forming dielectric gases. The heated dielectric gas bubbles collapse, form a cavity at the electrodes and remove the debris. Due to the flushing, the molten material moves from the gap and creates a crater on the work surface. The schematic of the EDG working principle is shown in Figure 7.1.

The parameters which affect the EDM process will affect the EDG process in a similar way. A summary of these is shown in Figure 7.2 in a cause and effect diagram. They are set voltage, peak current, pulse-on-off duration, pulse frequency, capacitance, polarity, type of dielectric, wheel rotational speed, and so forth. As the wheel

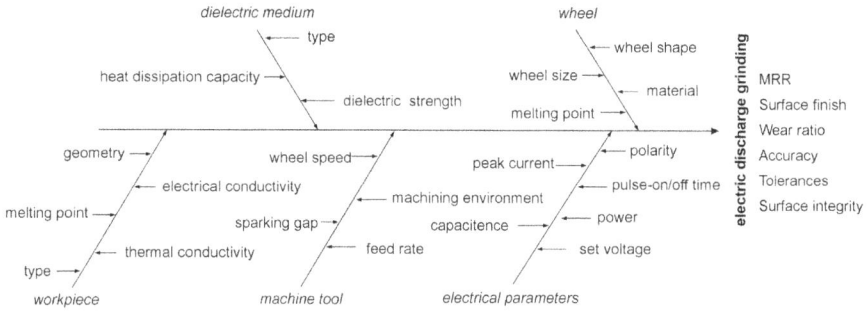

FIGURE 7.2 Cause and effect diagram of EDG process.

TABLE 7.1
Typical process parameters and their values [4,5]

Parameter	value/material
voltage	30-400 V
current	30-100 A
frequency	5-200 kHz
Wheel speed	30-180 m/min
sparking gap	12.5-75 μm
wheel material	graphite, brass, mild steel, copper
wheel diameter	100-300 mm
wheel width	0.25-150 mm
dielectric fluid	paraffin oil, transformer oil, kerosene, de-ionized water

peripheral speed transfers to the dielectric fluid, the effect of flushing pressure and method of flushing are insignificant in the EDG process. The typical values of the parameters involved in the EDG process are given in Table 7.1.

Porous and low-grade graphite is a typical material for EDG wheels. Other materials mentioned in Table 7.1 can also be used for some applications. Because of the material composition, EDG wheels can be easily formed in the desired profile.

7.3   THE EDG SYSTEM

Firstly, at the beginning of the process cycle, a very high voltage is applied between the wheel and the workpiece to control the gap between them via a servo drive. When the applied voltage is high, the wheel rapidly moves towards the workpiece until the first spark occurs. After the initial spark, the voltage drops to the normal range (40-60V), and the servo system senses the distance between the wheel and the work material and simultaneously feds the wheel into the workpiece [6]. The higher voltage between the wheel and the workpiece results in a larger spark gap. Higher spark gaps are desirable for flushing with the dielectric fluid in the case of plunge grinding. A higher electric

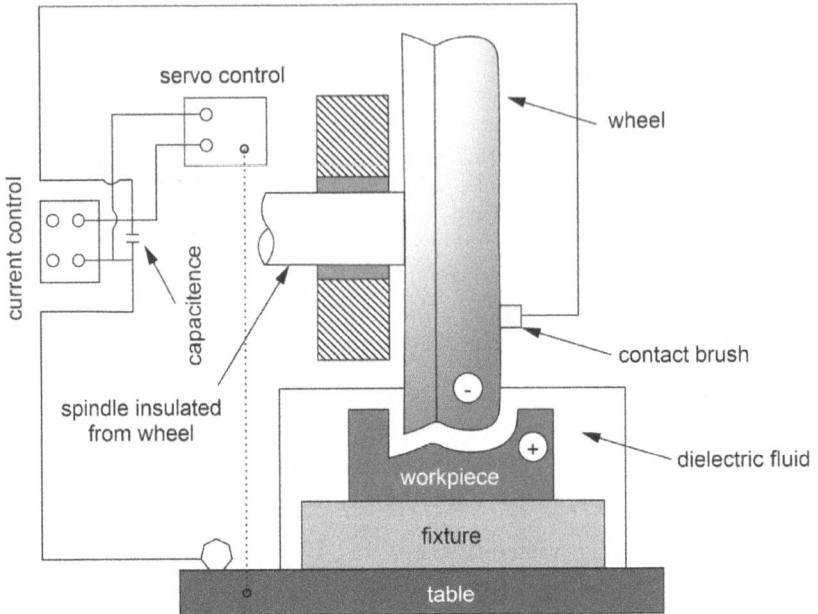

FIGURE 7.3 Schematic of EDG setup.

current and capacitance result in higher material removal rates, however this forms a rougher and more damaged surface. A higher pulse frequency is generally used to produce a smooth surface. The schematic EDG setup is shown in Figure 7.3.

7.4 PROCESS CHARACTERISTICS

The EDG process is used to produce the parts that require greater accuracy. A tolerance of ±0.005 mm is typical. In some applications, a tolerance of ±0.001 mm can be achieved. The material removal rates are low, ranging from 160 to 2500 mm³/ hour. Surface finish primarily depends on the metal removal rate. Because of the effective flushing action of the dielectric fluid, a surface finish (Ra) of up to 0.25 μm is possible. However, a surface finish (Ra) in the range of 1.6-3.2 μm is typical. With the increased pulse frequency, it is possible to achieve a Ra of 0.4-0.8 μm. The wear ratio (volume of material removed from the workpiece per unit volume of wheel wear) for the EDG process can range from 100:1 to 0.1:1 with an average of 3:1. It mainly depends on current density, dielectric fluid, wheel and work material, and the sharpness of corners. In profile grinding, there is a probability of maximum wear occurring at the high points and sharp edges of the profile, as shown in Figure 7.4. In this case, more frequent redressing is required than for uniform wear.

Low wheel speeds may result in non-uniform wheel wear, leading to a concentricity error. Due to the centrifugal effect, the dielectric cannot reach the sparking zone and this results in discharge instabilities at high speeds.

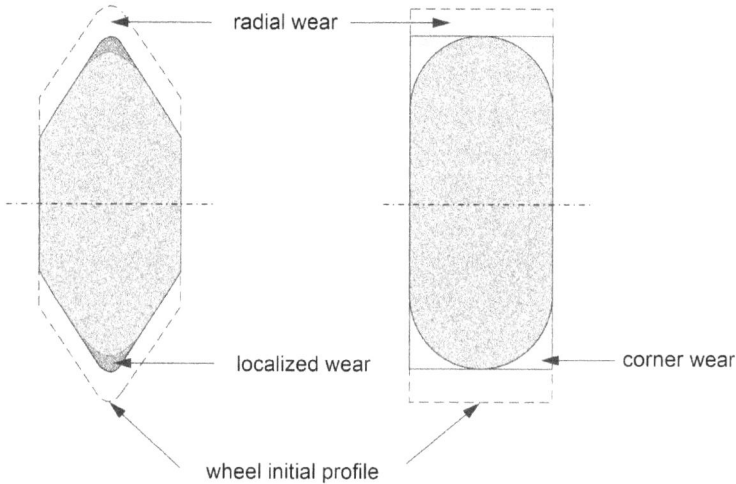

FIGURE 7.4 Illustration of uniform wear and corner wear of the wheels with different profiles.

Higher material removal rate conditions such as higher amperage, longer pulse duration, and the positive polarity of the wheel lead to increased wheel wear. In the case of ferrous materials, the carbon deposition on the wheel surface protects it from wear. In other instances, vaporized debris may splash onto the wheel and stick to it thus increasing the weight of the wheel [7].

The attained high temperatures result in a thin layer (0.0025 to 0.035 mm) on the work surface during the process [7]. For this reason, the EDG process is most widely used for machining and profiling carbide tools rather than for finishing. For high-stress applications, the surface layer formed during the EDG process may have to be removed.

Overall, process response aspects such as material removal rate, wear ratio, accuracy, and surface finish are better in the EDG process than in the EDM process. This is mainly due to the effective flushing and to the rotating electrode.

7.5 TYPES OF EDG PROCESS

Depending on the nature of the wheel, the EDG process is classified into two types. The first one has a wheel without abrasive particles, as mentioned earlier, while the other has a wheel with abrasive particles.

In the EDG process with abrasive particles, the metal wheel is replaced with a metal-bonded abrasive wheel. This process is called electro-discharge abrasive grinding (EDAG). In this process, the abrasives on the wheel surface have the primary purpose of enhancing material removal by abrasive action to achieve a better surface finish with lower grinding forces [8]. Further, the EDAG process benefits from wheel conditioning, such as self-dressing, unlike the conventional grinding process with metal bonded wheels. In the EDG process, the electrical discharge

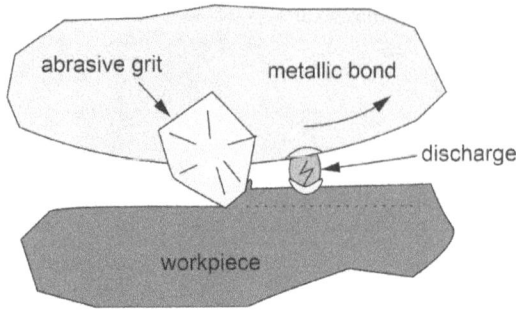

FIGURE 7.5   Principle of EDAG process.

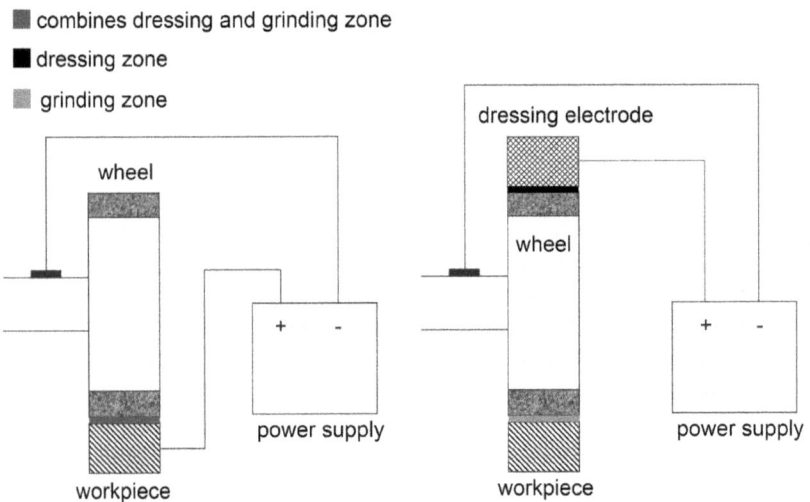

FIGURE 7.6   Basic configurations of EDAG process.

occurring at the wheel site wears away the metal bond and exposes fresh abrasive grit, thus eliminating frequent dressing. This process is also known as electric discharge diamond grinding (EDDG) [9]. A schematic diagram of the EDAG is shown in Figure 7.5.

Considering the locations of the grinding and discharge dressing zones, the EDAG process can be classified into two types, as shown in Figure 7.6. In the first configuration (Figure 7.6, left), the workpiece itself acts as the dressing electrode, thus it is under the simultaneous action of spark discharges and grinding. The main advantage of this configuration is thermal softening of the work material resulting from the sparks reducing the grinding forces. However, it is difficult to implement the independent optimized process parameters of grinding and dressing. Continuous in-process dressing might result in excessive wheel wear.

In the second configuration (Figure 7.6, right), a separate electrode is used to accomplish the wheel dressing outside of the grinding zone. This configuration does not have any influence on the work- material and facilitates the maintenance of the wheel topography alone.

At the desired time intervals, dressing can be done independently, irrespective of the grinding conditions and this improves the wheel life. In this configuration, the selection of the electrode material is vital. At temperatures of 900°C and above, in the presence of iron, considerable graphitization of diamond occurs and results in excessive wear. Out of many choices, copper and gold are the only metals that cause no thermal erosion to the diamond. Hence, a copper electrode with positive polarity is preferred as the dressing electrode [10].

EDAG is a mechanical grinding-assisted controlled electrical erosion process. Compared to EDM and EDG, the mechanical grinding action in EDAG helps in attaining a higher material removal rate. The material removal rate in EDAG is typically five times higher than EDM and twice as high as EDG for the given set of electrical parameters. This EDAG process can be controlled either in an abrasion dominant state or electrical erosion dominant state. There is less effect from electrical erosion in the abrasion dominant condition in producing a surface with a minimized heat-affected layer. In the electrical erosion dominant state, there is a relatively less effect from abrasion to reduce the process forces. Spark discharges thermally soften the work, hence the softened material is easily removed by the abrasion action of the grit and, consequently this decreases the forces. Therefore, a well-balanced state between the abrasion and erosion is desired to produce components with minimized HAZ and subsurface damage [11].

In the EDAG process, only the grains on the wheel surface with protrusion height greater than the interelectrode gap interact with the work. However, to remove material from the work surface, abrasive grit must satisfy the conditions like those in conventional grinding, such as that the uncut chip thickness must be greater than the minimum chip thickness. Also, the rake angle of the grit should be less than the critical value to facilitate chip formation [12]. All the factors that affect the EDG process can also affect the EDAG process. However, when the abrasive grit is added to the metallic wheel, the size of the abrasive grit, its type, the bond material, and concentration also affects the EDAG performance. It is also noted that a higher applied voltage increases the erosion rate and thermally softens the material, thus decreasing the process forces. However, increased erosion causes higher wheel wear [11]. Since the EDAG process is complemented by the abrasion action, the specific energy of the EDAG process is much lower than EDG despite higher material removal rates.

Figure 7.7(a) shows the influence of wheel speed on material removal rate at different currents. The initial increase in material removal rate with wheel speed diminishes at higher wheel speeds at lower current. At a higher current, the material removal rate increases with wheel speed [12].

The flow of current through the grinding zone accelerates spark discharges and increases material removal rate. At lower currents, grinding action dominates the electric discharge dressing, resulting in wheel glazing. The wear-flat formation on the abrasive grain reduces the material removal due to reduced uncut chip thickness or increased ploughing and rubbing actions.

Figure 7.7b shows the variation of the average radial wheel wear rate with the current. Enhanced in-process dressing with increased current helps prevent the loss

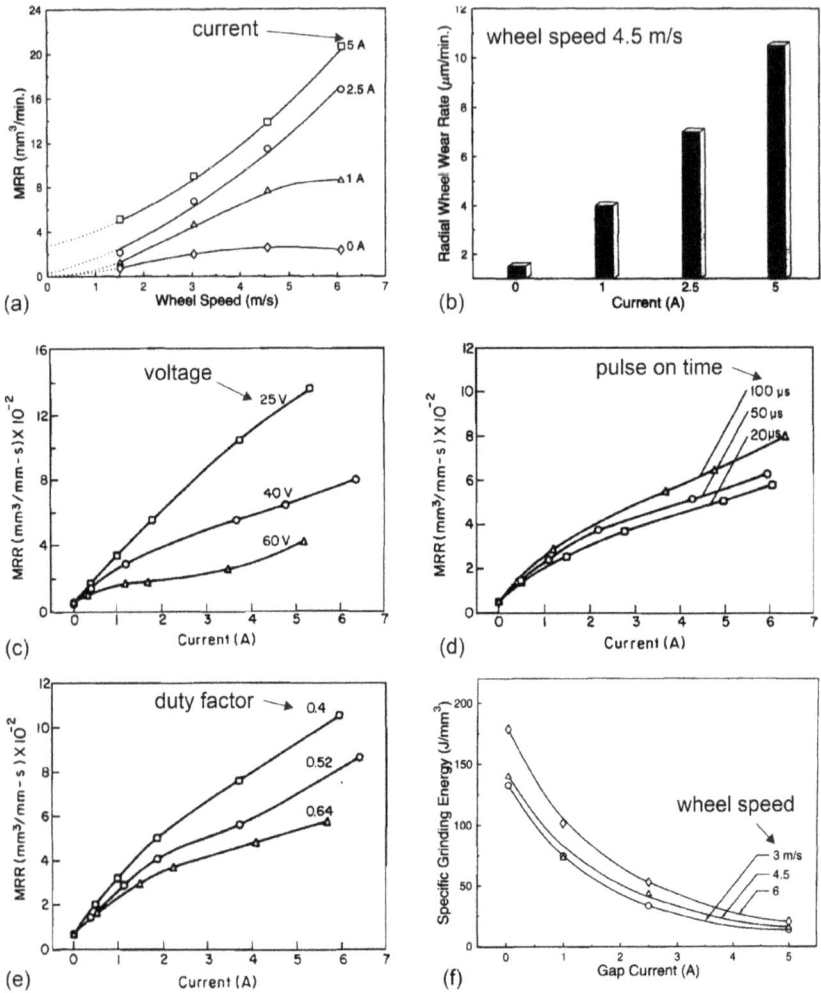

workpiece: HSS (20%W, 4%Cr, 5%V), hardness:1200HV, 5.6 mm thickness
wheel: diamond wheel, 5.7 mm thickness, G80/100 C75, bronze bond
wheel dressing conditions: gap voltage 60V, gap current 10A, pulse on time 100 μs
wheel speed 60m/min, duration 2 min, duty factor 0.5, dielectric-kerosene
grinding conditions: gap voltage 40V, pulse on time 100 μs, duty factor 0.5

FIGURE 7.7 Effect of process conditions on responses in EDAG [10,12]. With permission from Elsevier, Taylor & Francis.

of the sharpness of the grit, renews the wheel topography, and improves grinding performance. The degree of interaction between the wheel and the work, governed by the inter-electrode gap also decides the material removal rate. Increased wheel speed leads to an effective flushing action and reduces debris accumulation in the grinding zone, thus improving material removal [12].

Figure 7.7c indicates the variation in the material removal rate with the current, at different voltages. The data corresponding to zero current is related to grinding without the electrical discharge dressing. The material removal at zero current is predominantly by the abrasive action of the diamond grit and hence wears quickly with the ferrous materials due to graphitization. However, with the other current values, the material removal rate increases with the increase in current due to the in-process dressing and continuous renewal of the wheel topography. As mentioned earlier, higher applied voltage increases the gap between the wheel and work material (inter-electrode gap). Due to the increased gap, the uncut chip thickness value of the grit decreases and this results in reducing the material removal rate [10]. Figure 7.7d shows the effect of current at different pulse-on times on material removal rate. With the increased pulse-on time, the material removal rate increases because sufficient time is available for the heat conduction into a large amount of the work material and thermal softening. A similar effect can be found with a reduction in duty factor (Figure 7.7e) [10].

Specific grinding energy is the energy required to remove a unit volume of the material. This specific energy value indicates the mechanism of material removal and the nature of abrasive-work interaction. Generally, the specific energy value for the grinding process will be much higher due to large negative rake angles associated with the abrasive grit. Figure 7.7f shows the variation of the specific grinding energy in the EDAG process with wheel speed and current. The specific energy decreases exponentially with the current. It is the increased thermal softening of the work material with the increased current that causes effective material removal [12].

Further, considering the wheel and workpiece relative motion, EDAG is classified into three types in the same way as conventional grinding is classified. These types are electric discharge abrasive cut-off grinding, electric discharge abrasive face grinding, and electric discharge abrasive surface grinding. A summary of these is given in Table 7.2.

7.6 APPLICATIONS

The EDG process finds its primary application in making carbide form tools. It is used to machine carbide in shapes such as lamination dies and crushing rolls. It is also used to produce gear racks and closely spaced thin slots in hard materials, to grind brittle or fragile parts, and to create complex shapes due to its effectiveness in preventing the chipping of the brittle workpieces. Although cast iron is an electrically conducting hard material, it is not processed by the EDG because of non-conductive impurities like sand and slag, which cannot be removed by the spark discharge and consequently damage the wheel [3]. The EDAG process finds application in machining super-hard materials, engineering ceramics, sintered carbides, and metal composites [5].

TABLE 7.2
Summary of EDAG processes

Process	wheel rotation axis	feed direction	application
electric discharge abrasive cut-off grinding	horizontal axis	perpendicular direction to the machine table	to cut workpiece into pieces or to produce grooves
electric discharge abrasive face grinding	vertical axis	perpendicular direction to the machine table	To machine side surfaces of the cylindrical components
electric discharge abrasive surface grinding	horizontal axis	perpendicular direction to the machine table	to machine the flat surfaces

REFERENCES

[1] K.M. Shu, G.C. Tu, Study of electrical discharge grinding using metal matrix composite electrodes, International Journal of Machine Tools and Manufacture 43 (2003) 845–854. https://doi.org/10.1016/S0890-6955(03)00048-8.

[2] G. Li, M.Z. Rahim, S. Ding, S. Sun, J. Mo, Experimental study on quality of PCD tools machined by different electric discharge grinding processes, Cogent Engineering 3 (2016) 1228234. https://doi.org/10.1080/23311916.2016. 1228234.

[3] J. R. Davis, ASM Handbook: Machining, ASM International 16 (1989) 565–567. https://doi.org/10.31399/asm.hb.v16.a0002165.

[4] V.K. Jain, Advanced machining processes, Twentieth repr, Allied Publishers, New Delhi, 2013.

[5] H. El-Hofy, Advanced machining processes: Nontraditional and hybrid machining processes, McGraw-Hill, New York, 2005.

[6] E.C.Jameson,Electricaldischargemachining,SocietyofManufacturingEngineers,2001. https://books.google.co.in/books/about/Electrical_Discharge_Machining.html?id= FJbdIKGmfSgC&redir_esc=y.

[7] B. Bhattacharyya, B. Doloi, Machining processes utilizing thermal energy, in: Modern Machining Technology, Elsevier, 2020, pp. 161–363.

[8] H.R. Shih, K.M. Shu, A study of electrical discharge grinding using a rotary disk electrode, International Journal of Advanced Manufacturing Technology 38 (2008) 59–67. https://doi.org/10.1007/s00170-007-1068-y.

[9] X. Rao, F. Zhang, C. Li, Y. Li, Experimental investigation on electrical discharge diamond grinding of RB-SiC ceramics, International Journal of Advanced Manufacturing Technology 94 (2018) 2751–2762. https://doi.org/10.1007/s00170-017-1102-7.

[10] S.K. Choudhury, V.K. Jain, M. Gupta, Electrical discharge diamond grinding of high speed steel, Machining Science and Technology 3 (1999) 91–105. https://doi.org/ 10.1080/10940349908945685.

[11] M.K. Satyarthi, P.M. Pandey, Modeling of material removal rate in electric discharge grinding process, International Journal of Machine Tools and Manufacture 74 (2013) 65–73. https://doi.org/10.1016/j.ijmachtools.2013.07.008.

[12] P. Koshy, V.K. Jain, G.K. Lal, Mechanism of material removal in electrical discharge diamond grinding, International Journal of Machine Tools and Manufacture 36 (1996) 1173–1185. https://doi.org/10.1016/0890-6955(95)00103-4.

8 Electric Discharge Assisted Milling

Vipindas K.[1], Basil Kuriachen[2] and Jose Mathew[2]
[1]Department of Mechanical Engineering,Indian Institute of Information Technology, Design and Manufacturing Kurnool, Kurnool, India
[2]Advanced Manufacturing Centre, Department of Mechanical Engineering, National Institute of Technology Calicut, Kerala, India

CONTENTS

8.1 INTRODUCTION

The machining of difficult to cut materials like titanium and nickel-based alloys is always a challenge for the manufacturing industry. Due to poor thermal conductivity the machining temperature is very high while machining Ti and Ni based alloys that would result in rapid tool wear, dimensional inaccuracy and poor surface integrity of the machined surface [1]. Researchers have tried to improve the machinability of Ti alloys by improvising the tool material, optimizing the cutting parameters [2] and introducing new coatings to the tool material [3]. Even though these techniques can improve the machinability of the Ti and Ni based alloys to an extent, poor quality of the machined surface and tool wear still possess a challenge. Moreover, these techniques are not able to produce much improvement in reducing the tool wear. Another technique used to improve the machinability of Ti and Ni based alloys has been to adopt sequential machining processes including milling, burnishing and electrochemical polishing [4]. This technique has been very effective in getting good surface finish on the machined surface, but the manufacturing cost and time are very high.

Non-conventional machining technologies such as electrical discharge machining (EDM) and laser-based machining are viable solution, but these processes are costly and less productive. EDM is considered to be one of the preferred machining techniques especially when machining hard to cut materials. The strength and hardness of the materials puts no restriction on the EDM process. Additionally, there is no mechanical force involved in EDM and burr formation is nil [5]. Over time, many variants of the EDM process have been introduced to improve the material removal rate, machining efficiency and surface quality of the product. Wang et al. [6] combined EDM with arc machining to improve the material removal rate (MRR). Gao and Liu [7] used ultrasonically aided micro-EDM and found that machining efficiency and surface quality showed significant improvement.

Another commonly used technique to improve machinability is employing assisted machining. The most widely used assisted machining processes for Ti and Ni based alloys are laser-assisted machining (LAM), electrical discharge-assisted grinding (EDAG), and electrochemical-assisted grinding (ECAG). In LAM machining material properties such as hardness, yield stress and other characteristics change due to the impact of the high energy laser beam on the workpiece [8]. While turning Ti-6Al-4V with laser assistance, surface roughness reduced and tool life increased by 2–3 times [9]. Shu and Tu [10] used an EDAG process that helped in softening the workpiece surface before removal of the material.

Even though non-traditional and assisted machining have demonstrated significant improvements in the machining efficiency and surface quality of the machined product there are still some shortcomings. The LAM and EDM processes generate severe surface damage due to the heat affected zone, white layer formation and residual stresses on the machined surface. Additionally, in the EDAG process, abrasive grains easily fall off from the grinding wheel, which ultimately results in a defective product. Considering these drawbacks, manufacturing industries still prefer conventional tool-based machining to process such hard to machine materials.

Latest trends in the hybrid machining of Ti and Ni based alloys are combinations of EDM and end milling commonly known as electrical discharge assisted milling (EDAM). Many researchers have worked on this machining method and observed considerable improvement in machining performance in terms of tool life, machining efficiency [11–12] and less burr formation [13]. This chapter reviews the work of various researchers.

8.2 ELECTRICAL DISCHARGE ASSISTED MILLING (EDAM)

8.2.1 EDAM TOOL DESIGN

EDAM tooling generally consists of a milling tool insert and an electrode insert as shown in Figure 8.1.

8.2.2 EDM ELECTRODE DESIGN

Electrode material commonly used is copper. In Figure 8.1 the light grey (copper electrode) color insert indicates the electrode for electric discharge and the dark grey (Carbide insert) color insert indicates the milling insert. Some of the critical parameters of the EDAM tool design shown in the figure are outer diameter of the electrode (D_1), electrode height (H_1), outer diameter of the milling tool (D_2) and milling cutter height (H_2). During the EDM phase of the machining, there should not

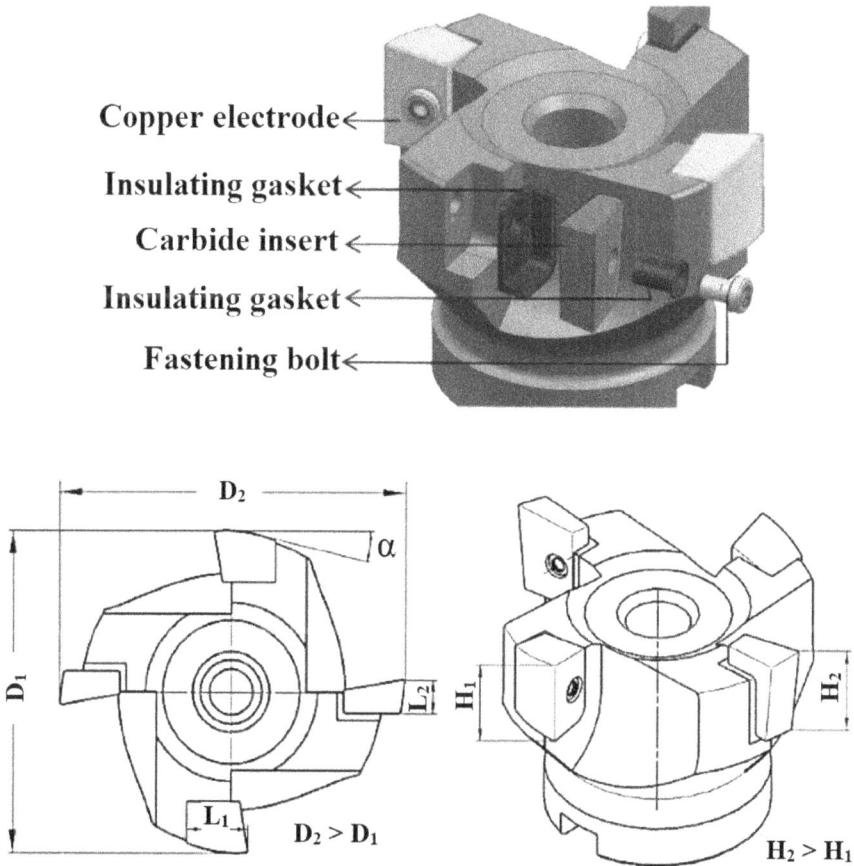

FIGURE 8.1   EDAM tool structure [14]. With permission from Elsevier.

be any direct contact between the tool and workpiece in order to maintain the necessary gap for producing the spark between the tool and workpiece. Hence, the outer diameter of the EDM electrode (D_1) should be kept smaller compared to the outer diameter of the milling insert (D_2) so that the required discharge gap (L_{gap}) can be maintained. The relationship between L_{gap}, D_1 and D_2 can be calculated as [14]

$$L_{gap} = \frac{1}{2}(D_2 - D_1)$$ (8.1)

Similar to the diameter, the height of the electrode (H_1) should also be kept smaller than the height of the milling insert (H_2). The discharge gap is generally governed by various factors such as discharge energy, electrode and workpiece material, and discharge environment.

8.2.3 MILLING INSET DESIGN

Special care should be taken to insulate conductive inserts like high speed steel or cemented carbide, because there is a possibility that the EDM spark will erode these inserts and ultimately result in the faster tool wear and breakage. Even the fastening bolt used to fix the milling insert should be made of non-conductive material as shown in Figure 8.1. However, insulation is not required for non-conductive inserts like ceramics or cBN.

8.3 THE PRINCIPLES OF EDAM

EDAM operation is achieved through a special cutting tool design as discussed in the previous section (Figure 8.1) and can generate an electric discharge and remove material from the workpiece through a conventional milling operation. Figure 8.2 shows the basic machining principle of the EDAM process. During the EDM phase of the EDAM operation, material to be removed from the workpiece will be changed into an easily machined layer consisting of recast and heat affected layers with the help of electrical discharge. When the milling insert contacts the workpiece, it will remove this transformed layer and a small amount of the bulk/matrix material. Hence, the depth of cut per tooth (D) can be expressed as [14]

$$D = D_c + D_r + D_h + D_m$$ (8.2)

where D_c is the crater depth, D_r the thickness of the recast layer, D_h is the thickness of the heat affected layer and D_m is the matrix thickness. The thickness of the easily machined layer can be controlled by adjusting the discharge energy and time. In the EDAM process, both milling and EDM are performed alternatively and repeatedly in order to reduce the machining force and for precise removal of the material from the workpiece.

Once electrical discharge occurs, the temperature and material properties of the modified layer change over time. The change in time or activation time (t_{change}) can be calculated as

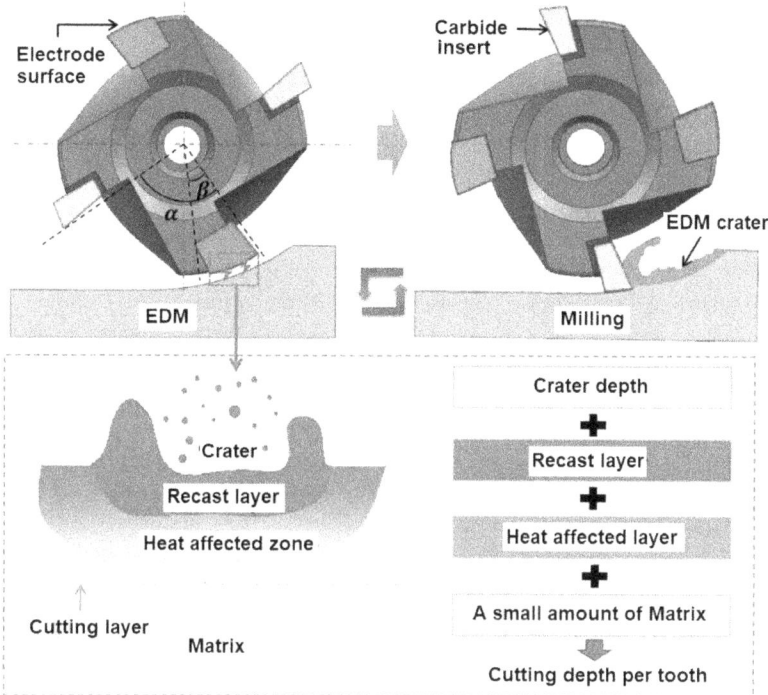

FIGURE 8.2   Working principle of EDAM [14]. With permission from Elsevier.

$$t_{change} = \frac{60\alpha}{2\pi N} \tag{8.3}$$

which is mainly influenced by spindle speed N. The discharge time, $t_{discharge}$ of each electrode can be calculated as

$$t_{discharge} = \frac{60\beta}{2\pi N} \tag{8.4}$$

where α is the angle between the milling cutter insert and discharge electrode and β is the angle between electrodes. Discharge time can be adjusted by changing the size of the electrode surface.

8.3.1   CHARACTERISTICS OF THE MODIFIED LAYER

It is important understand the characteristics of the modified layer in the EDAM process as it determines how the EDM process assists in improving the machinability of difficult to cut materials. Li et al. [14] carried out an experimental study in this are to analyze the characteristics of the modified layer. For this purpose, they performed

EDM experiments using an EDAM tool (two copper electrodes) but without milling inserts with different discharge energies. Discharge energy was varied by adjusting the capacitance value of the RC circuit (1000, 10,000 and 100,000 pF). In order to understand the influence of the discharge time on the material structure, two discharge operating times, namely, 2 and 10 s, were selected. Figure 8.3 shows the SEM and 3D morphology of the single crater after 2 s of electrode rotation with different capacitance.

From Figure 8.3 it can be seen that when capacitance was varied from 1000 to 10,000 pF, the corresponding change in crater depth is not very significant. However, with 100,000 pF capacitance, the depth of the crater increased significantly. In order to study the structural morphology of the EDM surface, the thickness of the recast layer and heat affected zone were analyzed after discharge times of 2 and 10 s. Figure 8.4 shows the cross-sectional view of the surface that has undergone the EDM with 100,000 pF capacitance.

From Figure 8.4 it can be seen that the recast layer thickness is in the range of 10–15 μm irrespective of the discharge time. However, the thickness of the heat affected zone was found to be influenced by the discharge time. It is clear from Figure 8.4 that when the discharge time is 10 s, a heat affected layer is visible; with 2 s discharge time, there is no visible heat affected layer.

The hardness of the modified layer is another important characteristic, which would have significant influence on the machinability of the workpiece material. Li

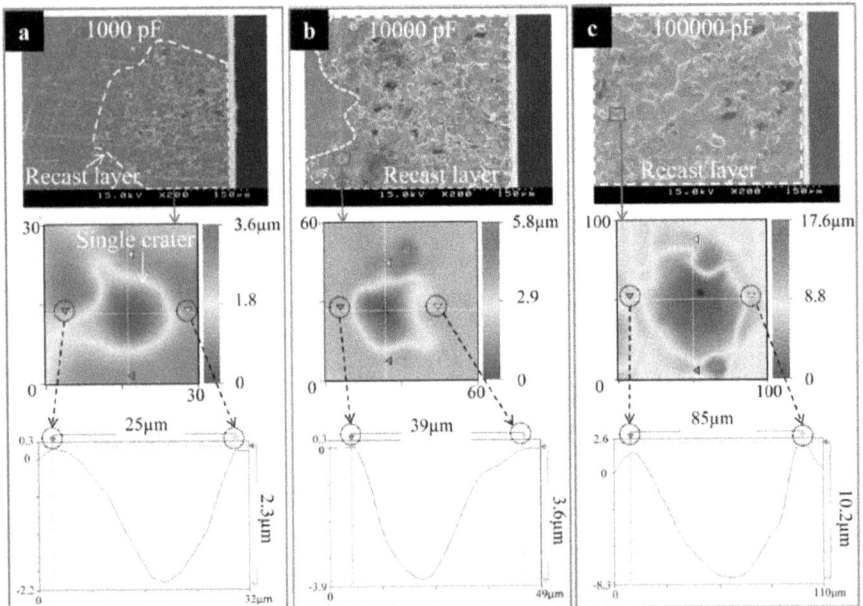

FIGURE 8.3 SEM image and 3D morphology of a single crater on the work surface after 2 s of discharge [14].

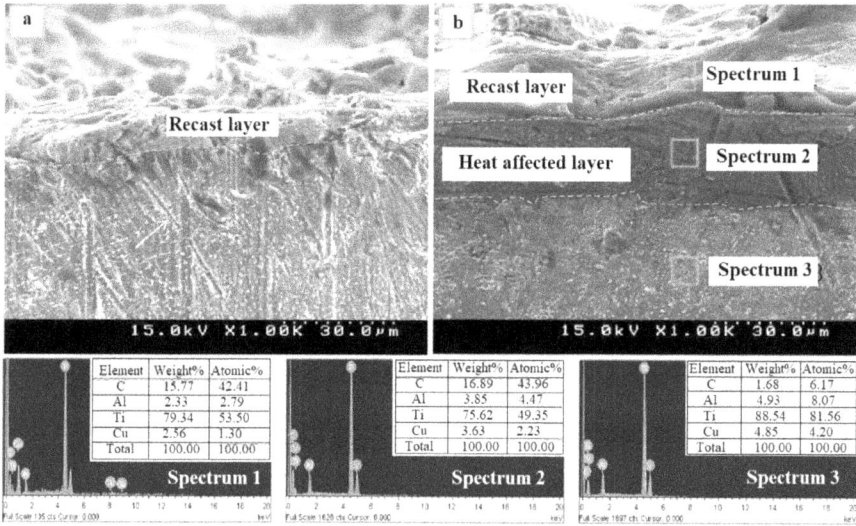

FIGURE 8.4 Cross sectional view of the EDMed surface (a) discharge time-2 s and (b) discharge time-10 s [14].

et al [14] checked the micro hardness of the cross section of the workpiece at different discharge times (2 and 10 s) with 100,000 pF capacitance. Figure 8.5 shows values of the micro hardness across the cross-section of the sample. Hardness was measured at intervals of 20 μm.

From Figure 8.5, it is evident that micro hardness of the work surface is significantly influenced by the EDM process. With both discharge times (2 and 10 s), the growth rate of micro hardness is reduced after 60 μm. This indicates that this region corresponds to the base material which was not affected by the EDM process or heat from the EDM discharge. Another point which of note from Figure 8.5 is a significant difference in the micro hardness at a depth of 40 μm with discharge times of 2 and 10 s. With 10 s discharge time, a heat affected layer will also form and this leads to the change in hardness value. However, at a depth of 20 μm the difference in hardness value is at a minimum as this region corresponds to the recast layer, which is almost the same for both discharge times.

From the above discussion, it can be seen that after the EDM process, the hardness of the heat affected layer and the recast layer, will be less than that of the matrix material. However, in the actual EDAM process, the cutting action of the milling insert will take place immediately after the EDM discharge and hence, the modified layer will not get enough time to solidify. This means that the modified layer will be in a molten state when the milling insert contacts this layer and hence, the hardness of the modified layer will be much less than the hardness shown in Figure 8.5. This lower hardness of the modified layer will certainly improve the machinability of the material, especially for hard to cut materials like Ti and Ni based alloys.

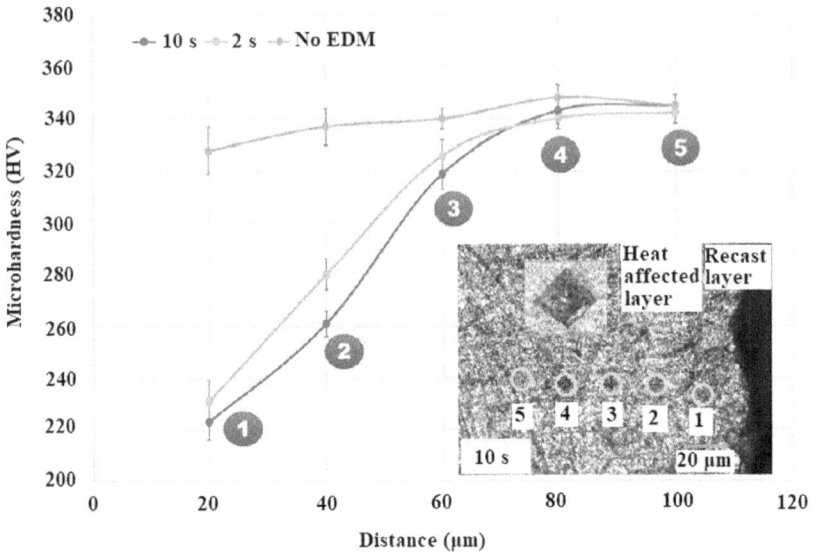

FIGURE 8.5   Variation of micro hardness with different discharge time [14]. With permission from Elsevier.

8.4   THE EFFECT OF EDAM ON SURFACE ROUGHNESS

Li et al. [14] studied the EDAM process extensively and compared the surface quality with that of the conventional milling process. Figure 8.6 shows the comparison of the surface quality of the Ti-6Al-4V under EDAM and under a conventional milling (CM) process at a feed rate of 20 μm/tooth, reported by Li et al. [14].

From Figure 8.6, it is evident that the surface roughness of the conventional milled surface is high because of the adherence of the chips to the machined surface and due to burr formation at the edges. Due to the poor thermal conductivity and to the high chemical affinity of the Ti alloys, heat accumulates around the cutting edge and this causes adherence of the chips to the tool as well as to the workpiece. However, with the EDAM process it is very clear from Figure 8.6 that chip adherence to the machined surface is minimal and also, the burr formation is less. Figure 8.7 shows the comparison of surface roughness of the machined surface under EDAM and conventional milling operations at different feed rates. It is clear that surface roughness of the EDAM machined surface is almost 5 times lower than that of the conventional milling process.

8.5   EFFECT OF EDAM ON TOOL WEAR

Li et al. [14] studied the tool wear progress of EDAM tools while machining Ti alloy and compared it with the tool wear of the conventional milling process. Figure 8.8 shows the comparison of flank wear progress of the tool insert after conventional milling and after the EDAM process when machining Ti alloy [14]. It is

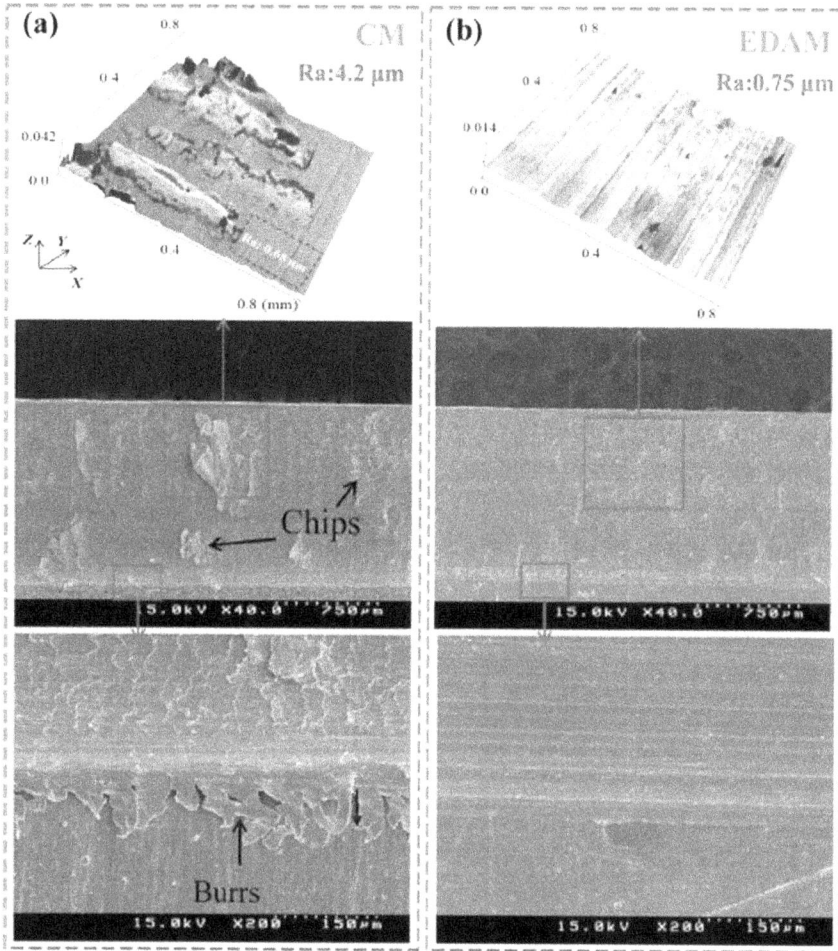

FIGURE 8.6 Surface morphology of the of Ti-6Al-4V workpiece after conventional machining and EDAM process at a feed of 20 μm/tooth [14].

clearly evident that during a conventional milling process, after 50 mm machining length, coating delamination occurred. However, with the EDAM process this was not observed during the initial stages of the machining but a small amount of built-up edge formation was noticed around the cutting edge. During conventional milling, edge chipping was observed after 150 mm length of cut. However, for the EDAM process, sticking of the chips to the flank surface and edge chipping were not found.

Li et al. [14] also studied the progress of flank wear in terms of flank wear width (VB) for tools used in the EDAM process and compared with conventional milling tools. Figure 8.9 shows the comparison between the progression of flank wear width for tools used in the EDAM process and in the conventional milling process. It is

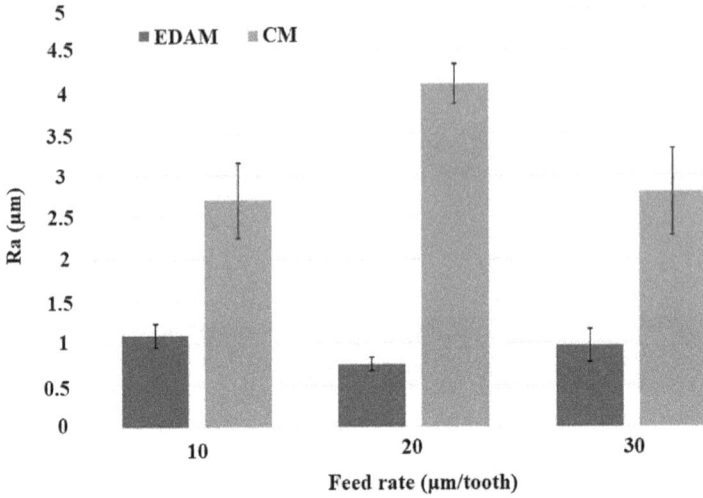

FIGURE 8.7 Comparison of surface roughness after EDAM and conventional milling of Ti alloy at different feed [14].

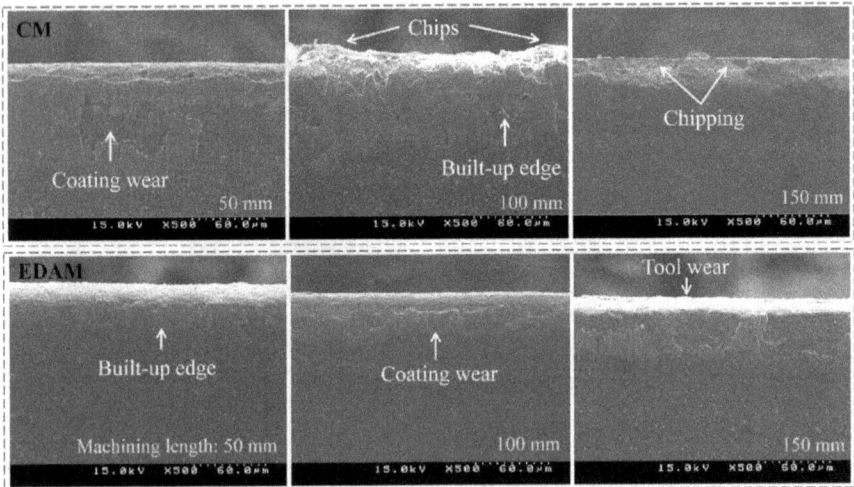

FIGURE 8.8 Tool wear of carbide insert after conventional milling and EDAM process of Ti alloy [14].

evident that the wear of tools used in the EDAM process is significantly less than those used in a conventional milling process. It can be concluded that the wear of tools used in the EDAM process is approximately three times less than that in a conventional milling process.

Li et al. [12] also compared the tool wear characteristics between conventional milling and an EDM assisted milling process based on the progress of flank wear

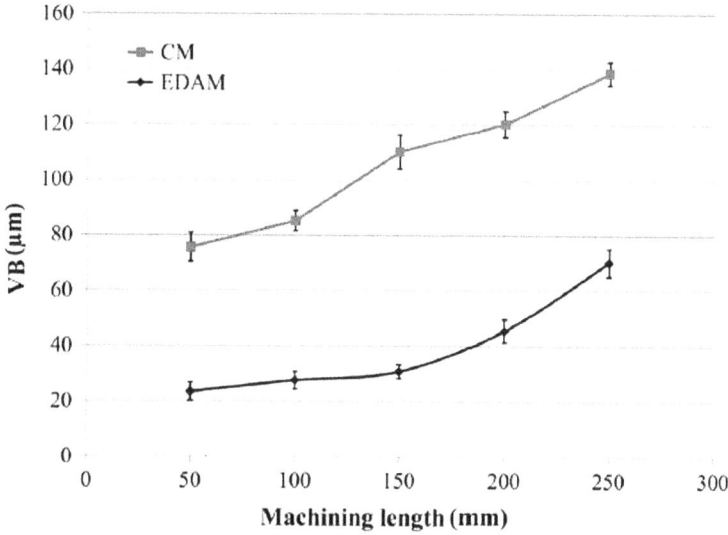

FIGURE 8.9 Comparison of flank wear progress for tools used in EDAM process and conventional milling process [14].

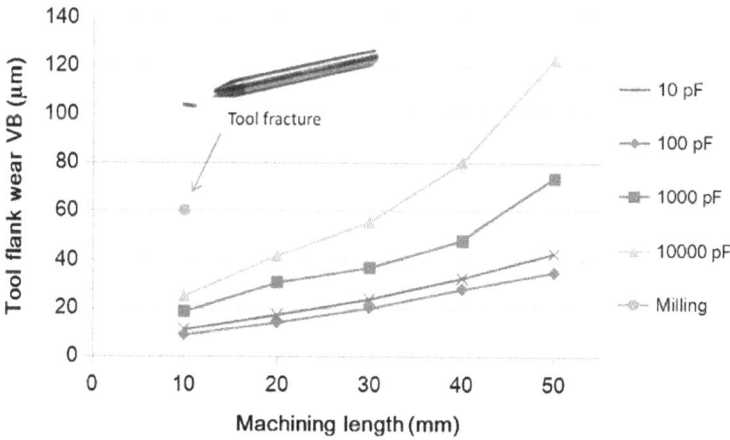

FIGURE 8.10 Comparison of progress of flank wear width (VB) with machining length with conventional milling and EDM assisted milling process [12].

width (VB) with respect to the machining length. Figure 8.10 shows the results reported by Li et al. [12]. It can be clearly seen that the EDM assisted milling process gives better tool life compared to a conventional milling process. Figure 8.11 shows the SEM images of flank face of the EDM assisted milling tool. With a small capacitance value, chipping of the cutting edge was observed as shown in Figure 8.11 (a). This could be due to insufficient melting of the workpiece material at low capacitance.

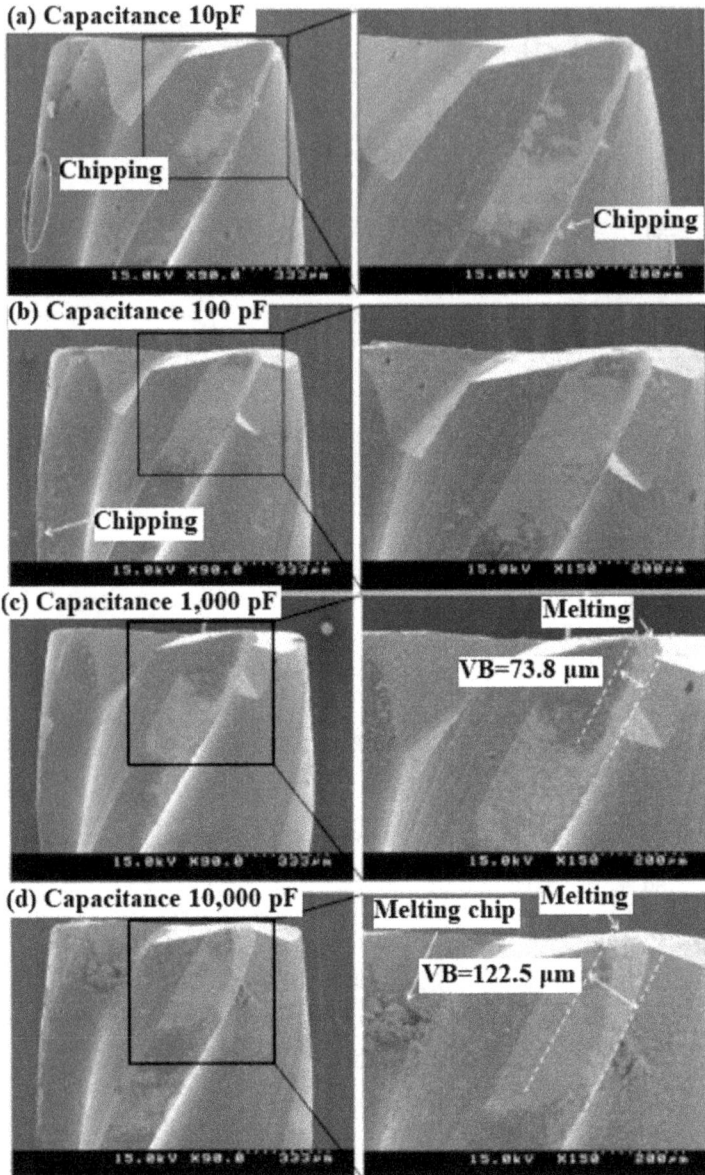

FIGURE 8.11 SEM images of flank face of the EDM assisted mill tool after machining at different capacitance value [12].

At the same time, flank wear was found to be more at higher values of capacitance as given in Figure 8.11 (c) and (d). An explanation of this is that at higher capacitance, discharge energy becomes high enough that meeting zone of the tool flank face also becomes large. Hence, it can be concluded that high discharge energy is not preferred in the EDM assisted milling process.

8.6   CHIP FORMATION IN THE EDAM PROCESS

Analysis of chip morphology and shape is always important in any machining process as it reveals much information regarding the machining mechanism, the machining temperature, and the like. Figure 8.12 shows the SEM images of the chip produced during both EDAM and a conventional milling process [14]. One of the unique characteristics of the chips produced during the EDAM process is presence of large holes as shown in Figure 8.12 (c) and (d). This could be mainly due to the craters formed during the EDM phase of the EDAM process. However, chips produced during a conventional milling process have small holes and this could be due to bending and breaking during chip formation. The edges of the chips produced during the EDAM process were found to be irregular in shape as can be seen in Figure 8.10 (c) and (d). This can also be explained on the basis of the formation of craters during the EDM phase of the EDAM process. On the other hand, the edges of the chips produced during a conventional milling process are comparatively smooth as shown in Figure 8.12 (a) and (b). Another important characteristic of chips produced during the EDAM process is the smaller size of the chips compared to the chips produced under a conventional milling process. This is due to the fact that during the EDAM process, plastic flow generated in the cutting zone is less and results in less shearing.

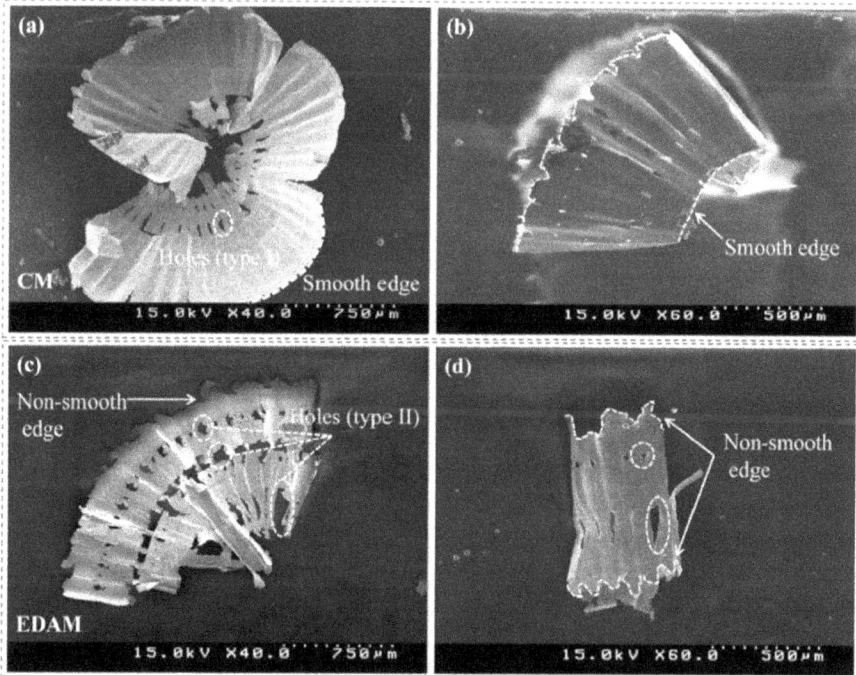

FIGURE 8.12   SEM images of the chip produced during EDAM and conventional milling processes 14].

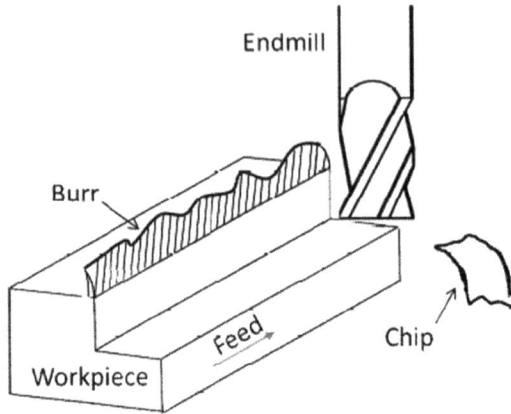

FIGURE 8.13 Top burr formation in conventional end milling process [13].

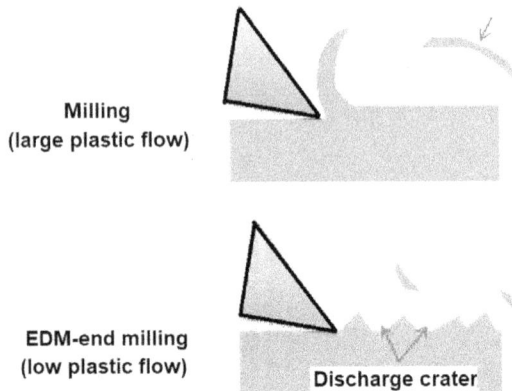

FIGURE 8.14 Schematic representation of chip formation in conventional and EDAM process [13].

8.7 BURR GENERATION IN THE EDAM PROCESS

Burr formation in a machining process is undesirable and unavoidable. The presence of burrs poses issues in the final product such as: sharp edges of the burr might cause injuries to the operator, over a period of time burrs might detach from the workpiece and later might cause damage to the product, the presence of burrs might create problems during assembly stages, on so on. Figure 8.13 shows the formation of a top burr in an end milling process. In conventional milling, a burr is generally formed due to the lack of support material as the tool edge reaches the free edge of the workpiece [15].

Figure 8.14 shows a schematic representation of the chip formation mechanism in conventional and EDAM processes. Clearly, the discharge crater produced on the surface during the EDM phase of the EDAM process results in the less plastic flow of the material at the main cutting region compared to a conventional milling process. Electric discharge generated during the EDM phase of the EDAM process means that material gets preheated before the milling cutter starts engaging with the workpiece. This can result in a reduction of fatigue strength of the material. This could be another reason for the reduction in top burr formation in the EDAM process.

8.7.1 THE EFFECT OF CAPACITANCE ON BURR FORMATION IN THE EDAM PROCESS

Kim et al. [13] studied burr formation during the EDAM process and investigated the influence of discharge energy (capacitance) on burr formation. Figure 8.15 shows the SEM images of top burr formation during conventional milling and during the EDAM process at different capacitance values [13]. Figure 8.15 (a) indicates burr formation during a conventional milling process. It can be seen that a wavy type of burr is generated under conventional milling conditions. Figure 8.15 (b)–(e) shows the top burr formation under the EDAM process at different capacitance values (10 pF, 100 pF, 1000 pF, 10000 pF respectively). Figure 8.16 shows the variation of burr height with capacitance value. It can be noticed that with an increase in the capacitance, that is, with increase in the discharge energy, burr height reduces. From Figure 8.15 (b) it can be observed that the burr has a pinnate-like morphology and as capacitance increases, burr takes a needle-like morphology. This is because at higher discharge energy the thermal effect on the workpiece will be high which would result in bigger crater dimensions and lower shear stress, resulting in reduced burr adhesion and easy fracturing of the workpiece material [13].

8.7.2 BURR MORPHOLOGY

Kim et al. [13] studied burr morphology in terms of its width. Information regarding the width of the burr is very important especially in relation to the ease of removal by any other secondary operations. Figure 8.17 shows the SEM images of the top burr morphology under conventional end milling and under the EDAM process with a capacitance value of 100 pF. It is evident from Figure 8.17 that under a conventional end milling process, burrs are thicker. However, under the EDAM process, burrs are comparatively smaller in width and have pinnate-like shapes [13].

8.8 CONCLUSION

The machining of difficult to cut materials like titanium- and nickel-based alloys through conventional tool-based operation is always a challenge. EDM assisted milling (EDAM) is an alternative solution to tackle this issue. In this process, material removal takes place in two phases, namely through EDM in the first phase and through a conventional milling process in the second phase. During the EDM phase, a modified layer is created by the electric discharge which softens the material, which

(a) Ordinary milling process

(b) EDM end-milling
capacitance: 10 pF

(c) EDM end-milling
capacitance: 100 pF

(d) EDM end-milling
capacitance: 1000 pF

(e) EDM end-milling
capacitance: 10000 pF

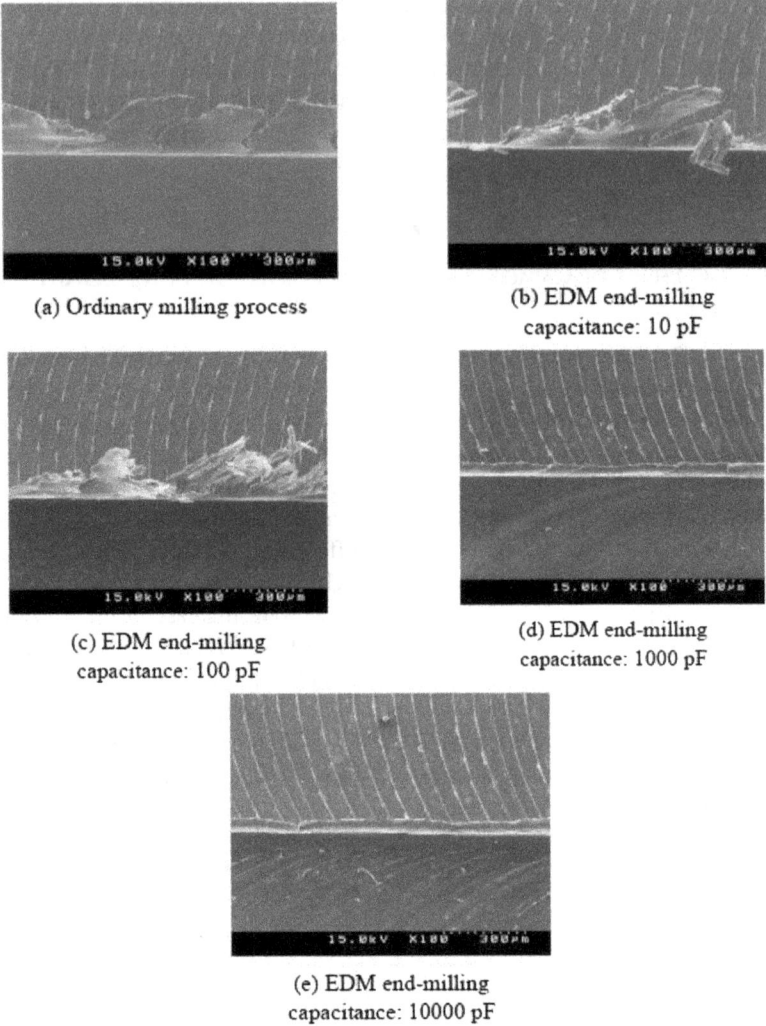

FIGURE 8.15 SEM images of the top burr formation under (a) conventional milling (b) EDAM with 10 pF (c) EDAM with 100 pF (d) EDAM with 1000 pF (e) EDAM with 10000 pF [13].

in turn eases material removal by the conventional milling process. Surface roughness after the EDAM component is found to be much better and top burr formation is also found to be reduced compared to that in a conventional milling process. Tool life also improved in the EDAM process. Chip analysis has also confirmed the presence of an EDM crater which results in small chips compared to the longer chips in a conventional milling process.

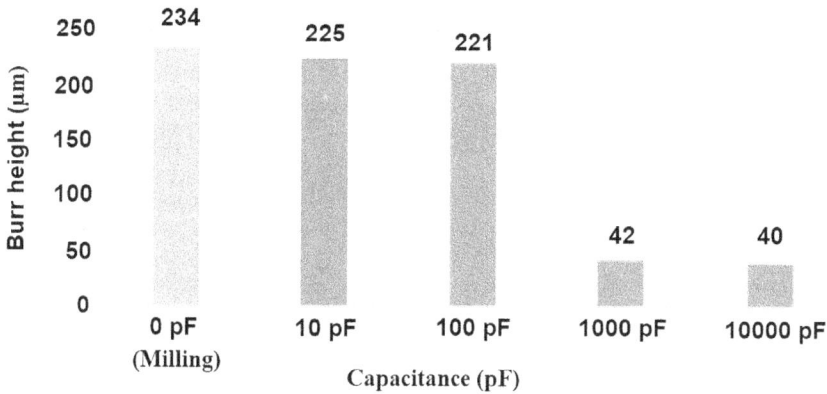

FIGURE 8.16 Variation of burr height win capacitance in EDAM process [13].

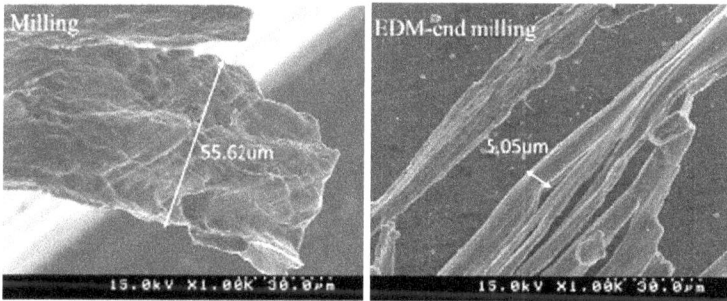

FIGURE 8.17 Comparison of burr width under conventional milling and EDAM process [13].

REFERENCES

[1] Mamedov, A., Lazoglu, I., 2016. Thermal analysis of micro milling titanium alloy Ti–6Al–4V. Journal of Material Processing Technology. 229, pp. 659–667.

[2] Huang, P.L., Li, J.F., Sun, J., Zhou, J., 2014. Study on performance in dry milling aeronautical titanium alloy thin-wall components with two types of tools. Journal of Cleaner Production. 67, pp. 258–264.

[3] An, Q., Chen, J., Tao, Z., Ming, W., Chen, M., 2020. Experimental investigation on tool wear characteristics of PVD and CVD coatings during face milling of Ti6242S and Ti-555 titanium alloys. International Journal of Refractory Metals and Hard Materials. 86, 105091.

[4] Wang, G., Liu, Z., Niu, J., Huang, W., Wang, B., 2020. Effect of electrochemical polishing on surface quality of nickel-titanium shape memory alloy after milling. Journal of Materials Research and Technology. 9, pp. 253–262.

[5] Gu, L., Li, L., Zhao, W., Rajurkar, K.P., 2012. Electrical discharge machining of Ti6Al4V with a bundled electrode. International Journal of Machine Tools and Manufacture. 53, pp. 100–106.

[6] Wang, F., Liu, Y., Zhang, Y., Tang, Z., Ji, R., Zheng, C., 2014. Compound machining of titanium alloy by super high speed EDM milling and arc machining. Journal of Materials Processing Technology. 214, pp. 531–538.

[7] Gao, C., Liu, Z., 2003. A study of ultrasonically aided micro-electrical-discharge machining by the application of workpiece vibration. Journal of Materials Processing Technology. 139, pp. 226–228.

[8] Xi, Y., Zhan, H., Rahman Rashid, R.A., Wang, G., Sun, S., Dargusch, M., 2014. Numerical modeling of laser assisted machining of a beta titanium alloy. Computational Materials Science. 92, pp. 149–156.

[9] Dandekar, C.R., Shin, Y.C., Barnes, J., 2010. Machinability improvement of titanium alloy (Ti–6Al–4V) via LAM and hybrid machining. International Journal of Machine Tools and Manufacture. 50, pp. 174–182.

[10] Shu, K.M., Tu, G.C., 2003. Study of electrical discharge grinding using metal matrix composite electrodes. International Journal of Machine Tools and Manufacture. 43, pp. 845–854.

[11] Byiringiro, J.B., Kim, M.Y., Ko, T.J., 2012. Process modeling of hybrid machining system consisted of electro discharge machining and end milling. The International Journal of Advanced Manufacturing Technology. 61, pp. 1247–1254.

[12] Li, C.P., Kim, M.Y., Islam, M.M., Ko, T.J., 2016. Mechanism analysis of hybrid machining process comprising EDM and end milling. Journal of Materials Processing Technology. 237, pp. 309–319.

[13] Kim, M.Y., Li, C.P., Kurniawan, R., Jung, S.T., Ko, T.J., 2019. Experimental investigation of burr reduction during EDM end-milling hybrid process. Journal of Mechanical Science and Technology. 33, pp. 2847–2853.

[14] Li, C., Xu, M., Yu, Z., Huang, L., Li, S., Li, P., Niu, Q., Ko T.J., 2020. Electrical discharge-assisted milling for machining titanium alloy. Journal of Materials Processing Technology. 285, pp. 116785.

[15] Aurich, J.C., Dornfeld, D., Arrazola, P.J., Franke, V., Leitz, L., Min, S., 2009. Burrs—Analysis, control and removal. CIRP Annals. 58 (2), pp. 519–542.

9 Vibration-Assisted EDM and Micro-EDM Processes

Abhimanyu Singh Mertiya[1] and
Deepak Rajendra Unune[1,2]*

[1] Department of Mechanical-Mechatronics Engineering,
The LNM Institute of Information Technology, Jaipur, India

[2] Department of Materials Science and Engineering,
INSIGNEO Institute for *in silico* Medicine, The University
of Sheffield, Sheffield, United Kingdom

*Corresponding author

CONTENTS

DOI: 10.1201/9781003202301-9

157

9.1 INTRODUCTION

Electrical discharge machining (EDM) is an electro-thermal energy based non-traditional machining process, where electrical energy is used to generate electrical sparks and where material removal occurs mainly due to thermal energy of the spark [1]. It is one of the oldest and the most widely used unconventional machining process. It can machine any material irrespective of its hardness as long as it is electrically conductive, which is also the main criterion for selecting a material for electrical discharge machining. It is mainly used to machine difficult-to-cut materials and high strength temperature resistant alloys. EDM can be used to machine difficult and intricate geometries. It is predominantly used for producing dies and moulds. Since it does not use mechanical forces for material removal, it can be used to machine even small and delicate parts without the risk of fracture of the workpiece. Hence, it is extensively used in automotive and aerospace industries for the machining of required components. EDM also finds its use in the medical industry for making surgical implants [2].

9.1.1 THE BACKGROUND OF EDM

In EDM, the material is removed from the workpiece due to erosion by rapidly recurring spark discharges taking place between the tool and workpiece (Figure 9.1), both of which must be conductive in nature. There is no contact during the entire operation. These repeated discharges generate a high temperature plasma channel. The heat generated from the plasma channel is utilized to erode the material. This plasma channel instantly vaporizes the material from both of the electrodes. The localized high temperature region melts the material from the tool and the workpiece. The

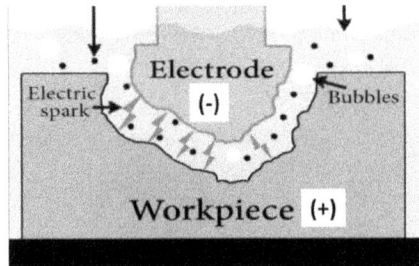

FIGURE 9.1   EDM principle [3]. Available under CC-BY- licence.

molten material is then solidified in the form of debris, and removed by flushing the dielectric fluid, from the inter-electrode gap (IEG) to maintain the sparking conditions [3]. The cavity left in the workpiece, as a result of debris removal, is called the crater. Thus, the volume of the workpiece that is removed, is symmetrically opposite to the shape of the tool. Both the tool and the workpiece are kept immersed in a dielectric fluid, which helps in concentrating the heat energy to a specified area on the work-piece [4, 5]. A thin gap is maintained between the tool and workpiece by a servo system. This gap is known as the inter-electrode gap (IEG). The IEG is maintained throughout the machining process such that its voltage difference is able to ionize the dielectric present in the region, which results in spark generation after electric breakdown. Generally, a pulsed DC of frequency 5 kHz and amplitude 80-100 volts is maintained across the IEG [6]. As every material has some irregularities present on its surface, the point where anode and cathode are closest to each other acts as the site of the discharge. The tool is the cathode and the workpiece is the anode. The reason for having the workpiece as anode is that during discharge, out of the total heat produced, about two-thirds goes to the anode. Since we want more heat at the work-piece (to enhance material removal and reduce tool wear), it is set as the anode. When the voltage across the gap becomes sufficiently high, dielectric breakdown occurs which leads to current discharge through the gap in the form of sparks in intervals of microseconds. Due to the voltage difference, an electric field is created in the IEG. Electrons from the cathode start moving due to the electromotive force of the electric field. They collide with the dielectric molecules present in the IEG, generating posi-tive ions. This ionization process creates more ions (positive and negative ions). The newly formed positive ions and electrons, along with the existing ions, then accel-erate under the influence of the electromotive force, forming a conductive discharge channel, producing a spark that causes ion-electron collisions. This collision among positive and negative ions converts their kinetic energy into thermal energy, which then results in the creation of a high temperature plasma channel. The spark that is produced generates high pressure between tool and workpiece. The combined effect of high temperature of the plasma channel and of the high-pressure melts, vaporizes and erodes the metal. The formation of the spark is quite a fast phenomenon, which results in a sudden decrement in electrical resistance in the plasma channel. This instantaneous drop significantly escalates current density causing an increment in ionization, thus producing a powerful magnetic field. Material removal is a result of vaporization and melting. However, the removal of molten metal is partial and stops when the potential difference is taken back. The termination of potential difference collapses the plasma channel, which generates shock waves that helps in expelling the molten metal thus creating a crater in that region [7].

9.1.2 Types of EDM Process

EDM processes are typically classified into various subtypes based on (i) quantity of dielectric used, (ii) machining operation, and (iii) machining dimensions used (see Figure 9.2).

EDM

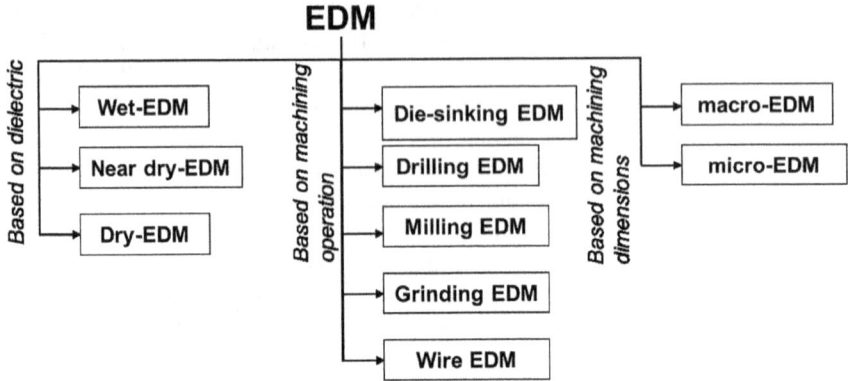

FIGURE 9.2   Types of EDM.

Based on the dielectric quantity used in EDM, it can be classified as conventional, wet, or flood-type EDM, near-dry EDM, and dry EDM. In conventional or flood-type EDM, the sparking between the electrodes takes place in the presence of a liquid dielectric fluid, or dielectric is supplied to the machining area in the form of a jet during discharge, or both. On the other hand, in the case of dry EDM, compressed air or gas is supplied in the IEG. In near-dry EDM, the air-fluid mixture is supplied in mist form in the IEG. Each of these variants has their own advantages and limitations.

Based on the type of machining operation, EDM is classified as die-sinking EDM, electrical discharge drilling (EDD), electrical discharge milling (ED milling), electrical discharge grinding (EDG), wire EDM (WEDM). Die-sinking EDM consists of a tool with a shape complimentary to that required on the newly machined surface. The electrode is generally given motion in the z-axis, while the remaining motions are applied to the workpiece. While machining, the tool sinks into the workpiece, as shown in Figure 9.3(a). In die-sink EDM, the machined hole is a replica of the shape of the tool. Electrical discharge drilling is a type of drilling where the hole is machined using the principles of spark erosion. A rotating tool traverses in the z-direction to 'drill' the hole. The newly formed hole has a circular cross-section; unlike die-sink EDM, where the cross-section is determined by the tool. ED milling is a variant of milling where spark energy is used to cut the desired shape in the workpiece. The tool is given motion in all three axes in order to form the required contour. This facilitates the formation of complex contours using simple tool geometry. Electrical discharge grinding has a conductive rotating wheel as the electrode whose base circumference participates in spark generation with the workpiece. A rotary movement of the wheel also assists in the flushing of the IEG. Wire EDM uses a pair of spools of conducting wire in place of a tool electrode [8]. The dielectric is flushed in the IEG using a dielectric jet. The wire moves continuously through the IEG, between the two spools and helps to obtain better tolerances. Figure 9.3 depicts various EDM variants based on the type of machining operations.

Apart from the above-mentioned criteria, EDM processes are also classified based on workspace dimensions or on discharge energy into macro and micro-EDM. If the

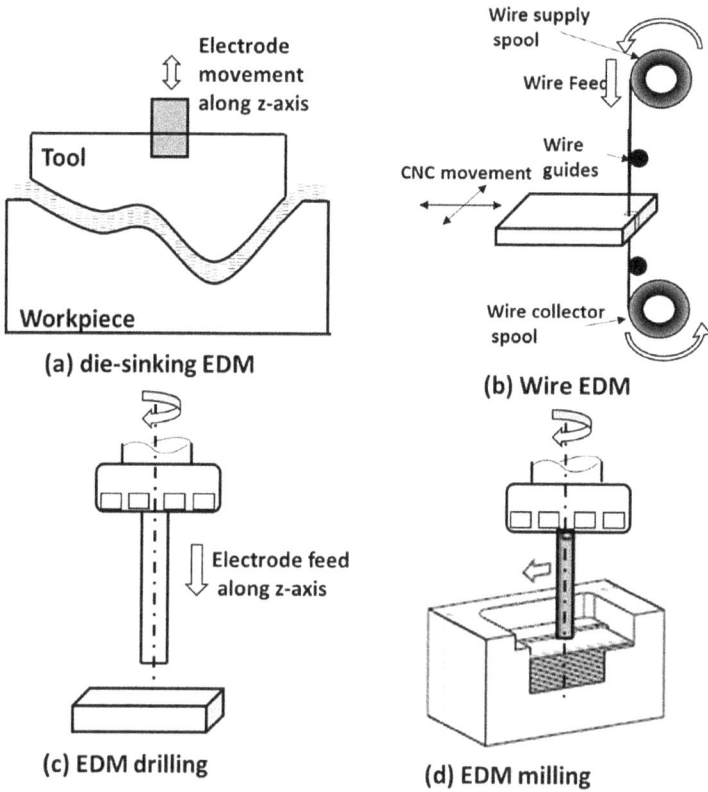

FIGURE 9.3   EDM variants.

smallest dimension of the workspace is below 1 mm or if the machining area is less than 1 mm², it is termed as micro-EDM (μEDM), otherwise it is called a macro-EDM process [9]. Figure 9.4 gives the comparison between macro-EDM and micro-EDM. Figures 9.4(a) and 9.4(b) show the crater sizes in macro and micro-EDM formed after the first spark collapse. Figures 9.4(c) and 9.4(d) give the relative comparison of the dimensions of tool and the generated spark in macro-EDM and micro-EDM.

9.1.3   MICRO VERSUS MACRO-EDM

Although macro and micro-EDM processes involve the same mechanism for material removal, in other words, material removal using the electro-thermal energy of the spark, there are significant differences between the two processes. There are differences in the power supply units, axis resolution, processing techniques, servo-control mechanisms for maintaining the gap width, tool sizes and methods of manu-facturing the tools [12–14]. As it is evident from the name of the micro-EDM process, the tool diameter must be in the range of microns. Due to the small size of the tool, the plasma channel is comparatively bigger than the tool in micro-EDM, whereas in

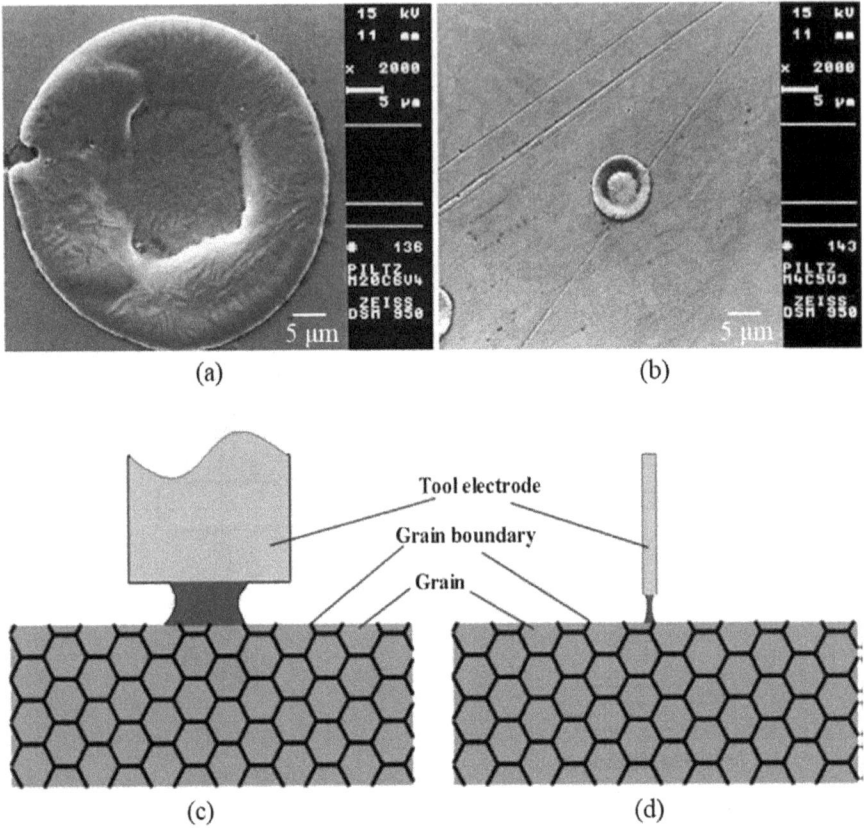

FIGURE 9.4 SEM images of the first crater formed in (a) macro-EDM and (b) micro-EDM [10] (reproduced with permission from Elsevier); dimensional comparison of the tool and the generated spark in (c) macro-EDM, and (d) micro-EDM [11] (reproduced with permission from Elsevier).

macro-EDM, the plasma channel is smaller than the tool size. Small tools also place a limitation on heat conduction and dissipation through the tool. For this reason, maximum spark energy has an upper limit in the micro-EDM process, as high energies do not dissipate quickly and tend to damage the tool (burn or break the tool), whereas no such limitation exists in the case of macro-EDM. Therefore, sparks of low energy are used in the micro-EDM process. Because of less heat conduction, heat accumulates on the tool surface, which melts and erodes the tool material. Thus, the tool in the micro-EDM process experience more wear than the tool used in the macro-EDM process. Due to the limitation on maximum spark energy in micro-EDM, craters on the surface machined using micro-EDM process are smaller in size in comparison to those machined with a macro-EDM process [10].

Low energy discharges, in the order of 10^{-7} to 10^{-6} joules, are used, with high discharge frequency (~200 Hz) in order to achieve high accuracy and precision [15].

The craters formed on the surface will be in the range of micrometers in the case of a micro-EDM process [16]. As the size of craters is quite small and the frequency of discharge is large, the surface finish of the machined surface will be good and very low values of surface roughness will be found in this process, as compared to a macro-EDM process. Thus, the surface obtained in a micro-EDM process will be a near-finished surface.

In micro-EDM, high resolution of the axes is required for the relative movement between tool and workpiece. Even a small power supply unit with relatively low capacity will be enough to create dielectric breakdown in the IEG (whose size will be in the range of microns too) [17]. The variation of flushing pressure on the electrode is more in micro-EDM in comparison with that of the macro-EDM process. In comparison to macro-EDM, the tool area exposed to the fluid pressure is smaller. However, the tool stiffness is also lower in the case of micro-EDM. Thus, even though low force is exerted by the dielectric fluid, there are chances of tool deflection due to softer tools and more pressure variations on the tools. Furthermore, due to a smaller gap between tool and machined surface during machining, high pressure drops in micro-volumes and high viscosity of the dielectric, the removal of debris is quite difficult in micro-EDM [18].

9.1.4 THE NEED FOR VIBRATION-ASSISTED EDM

The low MRR of EDM is a cause of considerable concern, and this concern is accentuated further during micro machining. Vibrations make cavitation bubbles in the dielectric present in the IEG, forming micro jets that facilitate effective debris removal from the IEG. Moreover, continuous debris removal and improved flushing in the IEG, due to vibrations, increases the effective discharge ratio. Both of these factors help in increasing the MRR substantially, when EDM is assisted by vibrations.

The gap between the electrodes (IEG) is very small, in order to induce sparks. However, this small gap results in machining instability which causes an increase in surface defects and machining time, leading to high electrode wear. Inefficient debris removal when micro-machining complex microstructures using micro-EDM creates a debris-bridge between the electrodes that causes short circuits, hampering machining efficiency. Furthermore, high debris concentration causes arcing that decreases surface quality. Vibration-assisted EDM, as discussed above, increases fresh dielectric circulation which in turn minimizes debris accumulation, thus decreasing the arcing phenomenon and increasing machining stability. This is beneficial while machining thin slots and complex contours.

In EDM, the sparking results in the melting and vaporization of the material from both the workpiece as well as from the tool electrode. The dielectric fluid present around the IEG typically takes away the debris particles that are left after every spark cycle.

In addition, production of high aspect ratio holes poses a challenge in the micro-EDM process [9]. Aspect ratio is defined as the ratio of depth to diameter of a hole [19, 20]. To achieve stable machining conditions, there should be continuous removal of debris from the IEG. But, when drilling high aspect ratio holes, the flushing conditions degrade at greater depth resulting in inefficient removal of debris and other by-products of machining, which in turn causes adhesion and decreases machining efficiency [20, 21]. Implementing vibrations in the process improves flushing in the IEG, that leads to the augmentation of molten metal and debris removal which in turn results in better machining performance [22]. Improvement in molten metal removal from the IEG also minimizes micro-cracks and the modification of the micro-structure on the machined surface [23]. It was found that a vibration-assisted EDM process has eight times more efficiency than that of EDM without any vibration assistance, with an improvement in MRR, surface characteristics and accuracy in the micro-hole dimensions [24].

9.2 METHODS OF APPLYING VIBRATIONS DURING EDM

The EDM process has three major components, namely workpiece, dielectric fluid, and the tool electrode. The workpiece undergoes erosion in order to obtain the desired shape or features. The tool helps in machining the workpiece and imparts its complementary shape to the workpiece. The dielectric fluid concentrates the spark energy on the specified region for machining and helps in cooling and debris removal after every spark cycle. Vibrations can be applied to any of the above components, to enhance their effects in the machining process. In all the cases, it is observed that vibrations applied to any of the above components resulted in an increment in the MRR, improvement in surface quality, and enhancement in machining efficiency [25–28]. Figure 9.5 shows the schematic diagram of EDM where the workpiece is undergoing vibrations [25]. Figure 9.6 gives the schematic diagram of an EDM setup where vibrations are applied to the dielectric fluid [26]. Figure 9.7 depicts the EDM setup, which includes tool vibration [27].

9.2.1 TYPES OF VIBRATION

Vibrations results in enhanced dielectric circulation in the IEG. Provision of vibrations to the tool or workpiece is the most suitable approach to enhance flushing and machining stability in EDM or micro-EDM. However, providing vibration excitations to a large component, such as the workpiece, is quite strenuous. Additionally, the design of a suitable actuation mechanism for vibrating the tool or workpiece, along with the clamping system, can prove to be another heavy task. Hence, some indirect methods are also used to impart vibrations in the IEG, such as applying vibrations to the dielectric using a sonotrode [30]. Two types of vibrations are used in EDM to improve machining output. These are classified, based on the frequency of vibrations. These are termed ultrasonic vibrations and low frequency vibrations.

9.2.1.1 Ultrasonic Vibration

The effects of ultrasonic vibrations while machining with EDM have been studied since 1985 [23]. The frequency range typically lies between 20 and 60 kHz, and amplitude varies between 2 and 10 μm. The introduction of vibrations with this range of frequency and amplitude has two remarkable effects on the EDM setup. The first effect is seen on the dielectric. Due to the introduction of vibrations, dielectric movement is increased, and high velocities and acceleration values are reached.

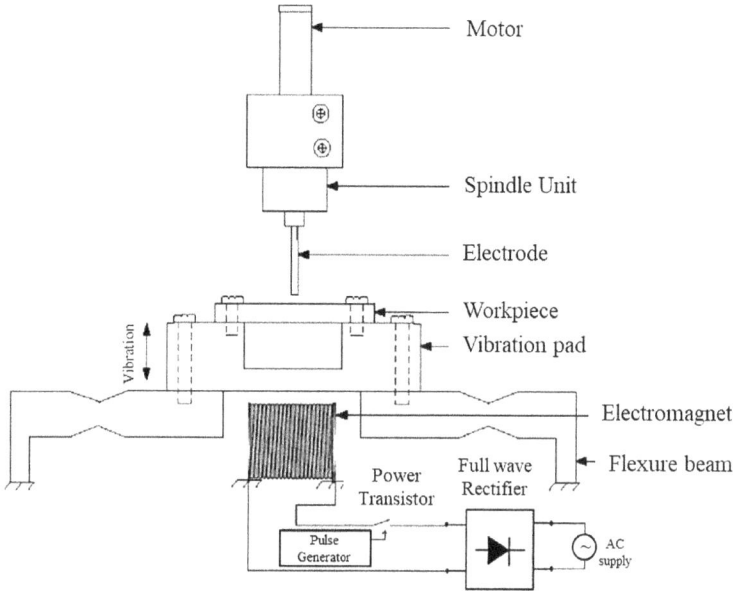

FIGURE 9.5   Vibration applying to workpiece in EDM [29]. Reproduced with permission from Elsevier.

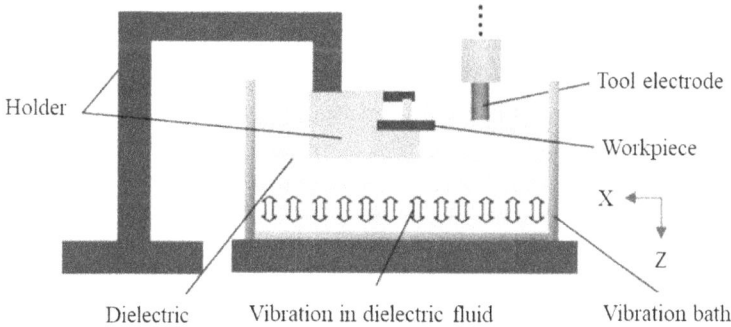

FIGURE 9.6   Vibrations applied to dielectric fluid in EDM [26]. Reproduced with permission from Elsevier.

These high values of velocity and acceleration provide an effective stirring effect in the dielectric, especially in the IEG, thus avoiding the accumulation of debris, which results in high values of the effective discharge ratio due to consistent conditions. In the absence of vibrations, an effective discharge ratio decreases after some time due to the inefficient flushing of debris. Secondly, the vibrations cause frequent changes in the width of the IEG. This alternate expansion and contraction of the IEG helps in stabilizing the process and improving dimensional accuracy, which may get distorted due to longer duration pulses. Longer duration pulses generate sparks for longer

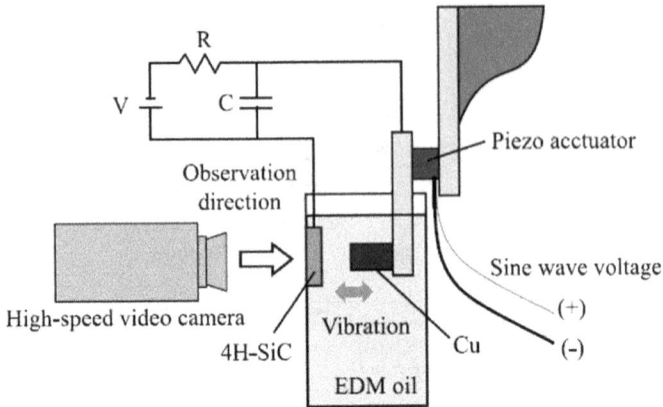

FIGURE 9.7 Vibrations applied to the tool electrode in EDM [27]. Reproduced with permission from Springer Nature.

durations, which undergo deviation after a certain time period and this leads to a degradation of dimensional accuracy. Furthermore, these longer discharges also result in increased short circuits, by developing arcs between tool and debris, instead of workpiece, and this results in decreased machine performance. The alternate expansion of the IEG collapses the long spark discharges, thus improving machining stability and eliminating geometric inaccuracies. This results in an improvement in the fabrication of intricate designs and a remarkable increment in machining speed [30]. It has been observed that the depth of a micro hole, when drilled using EDM assisted with ultrasonic vibrations, was found to be double to that drilled without any aid [31]. Moreover, less current is needed for machining when assisted with ultrasonic vibrations. Stable machining conditions are achieved when ultrasonic vibrations are used with EDM, avoiding short circuits and arcing, that lead to a better surface finish. Shorter pulse durations are possible for machining when it is assisted with ultrasonic vibrations. In fact, the range of input parameters, known as machining parameters (like current, pulse-on time, and so forth), is increased in the case of ultrasonic vibration-assisted EDM [32].

9.2.1.2 Low Frequency Vibration

The assistance of ultrasonic vibrations in EDM requires huge investment, as there is a requirement for resonant systems. Hence, the possibility of low frequency vibrations was also explored. It was found that low frequency vibrations also have similar effects to ultrasonic vibrations. The only difference lies in the frequency range. While ultrasonic vibrations have frequencies of the order of kHz, low frequency vibrations have frequencies ranging from tens to a few hundreds of Hz. The amplitude of vibrations can reach up to 20 μm. The low frequency vibrations prove quite helpful while machining difficult geometrical shapes. They help in maintaining and enhancing process stability while fabricating these geometrical shapes. They also help in decreasing dimensional and geometrical errors during machining. The vibrations also help in decreasing machining time by more than half of the machining time measured in EDM with no vibrations [30].

9.2.2 VIBRATION APPLIED TO THE TOOL OR TO THE WIRE ELECTRODE

In pursuit of achieving high machining efficiency of EDM, several experiments have been conducted with vibrations. Many researchers have applied vibrations to the tool electrode, others have used these vibrations to vibrate the dielectric fluid, while yet others have vibrated the workpiece. The use of vibrations was being reported in the late 1980s, by Kremer et al. [23]. They reported a significant increase in the MRR (up to 400%) during the finishing process using a tool vibrating at ultrasonic frequencies, with machining time decreasing to almost one-fifth of the initial time, along with stable machining conditions. They made a setup for vibrating the tool as shown in Figure 9.8. It consisted of a transducer, made of piezoelectric material, responsible for generating vibrations. These vibrations were transferred to the electrode via a tool, commonly known as sonotrode. This sonotrode was especially designed to be used in the EDM setup. It was in the form of a cross (hereby referred as 'the cross'). The design of the cross was necessary for a specific reason. Due to design restrictions, the piezoelectric transducer could not be placed between the tool electrode and the EDM head. Therefore, a medium was required, which could transfer the transducer's vibrations to the tool electrode as well as to connect the electrode with the EDM head. Moreover, since the EDM head was vertically above the tool electrode, the transducer had to be placed along the horizontal position. Hence, the vibrations generated from the transducer were in the horizontal direction, which needed to be converted to vertical vibrations before passing them to the tool. Hence, the cross was designed such that it could connect both, the transducer and the EDM head, to the tool. Additionally,

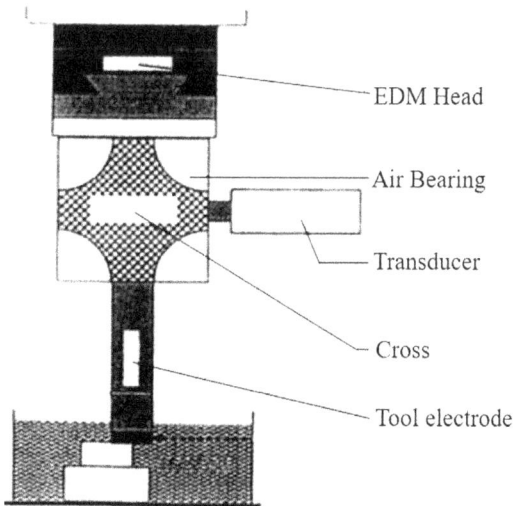

FIGURE 9.8   Setup for vibrating the tool [23]. Reproduced with permission from Elsevier.

it translated the horizontal vibrations of the transducer to vertical vibrations, and then passed them to the tool. In the conversion, although some vibrational energy was lost, frequency still remained in the kHz range, and provided amplitude up to 8 μm. Air bearings were also employed to avoid damage to the EDM machine from the ultrasonic frequency. They also constructed electrodes using two materials: a metallic part was used for effective vibration transfer and a graphite part was used for spark discharge. Apart from increasing the MRR, a significant positive change was observed in the heat affected region, the heat affected zone, that included the recast layer, the reheated layer, and the tempered layer, showed a decrease in thickness, thermal stresses, and micro cracks. The recast layer was made from the solidified metal that was melted and, using spark erosion, was more uniform. The tempered layer was the part of workpiece near to the crater, which had not melted, but reached a very high temperature (crossed the austenitic limit) and then abruptly cooled, to generate the tempered phase of the metal. The reheated layer is the transition region between tempered and recast layers. As fewer and shorter micro cracks were observed when machining a surface using EDM, it can been considered that fatigue resistance was also increased. It was also noted that surface roughness and micro hardness was not affected at low amplitudes of vibration.

Some studies have combined ultrasonic machining (USM) with electrical discharge machining (EDM) and used a DC power source, instead of a pulsed power supply [33, 34]. The vibrations caused sparks to occur, with a frequency that was equal to the ultrasonic frequency of the vibrations. During the downward movement of the tool, when the IEG had a thickness sufficient to undergo dielectric breakdown with the applied voltage, the spark started. When the tool moved up, a time came when the dielectric fluid present in the IEG was not able to undergo breakdown by the voltage imposed on it, due to the increase in the IEG size. This led to elimination of the spark. Thus, the spark started on every downward movement of the tool, and stopped on every upward motion. This caused sparks to occur at a frequency equal to that of the mechanical vibrations. One thing to be kept in mind is that while applying voltage, its magnitude should not be very high, otherwise the spark would not stop, resulting in arcing that would affect the machining adversely. The combination of USM and EDM has resulted in benefits from both of the processes. Improved surface finish of the machined surface using ultrasonic machining was combined with the high material removal rate of electrical discharge machining. A setup, as shown in Figure 9.9, was developed to do combined machining. An ultrasonic generator was used to produce high frequency electrical signals, which were converted to the same frequency mechanical vibrations using a transducer. A horn was positioned to amplify and transmit these mechanical vibrations to the EDM tool. The tool and workpiece were connected to the DC power supply, using straight polarity, to increase material removal from the workpiece. The dielectric was flushed with a jet. Sparking took place during the downward motion of the tool. The discharge mechanism caused the formation of debris, which was removed quite efficiently due to the pressure variations caused by a changing IEG width due to the vibrating tool. This setup also demonstrated a decrease in thermal residual stresses, a thin heat affected zone, less micro cracks, and highly efficient discharges (above 80%). The MRR was reported

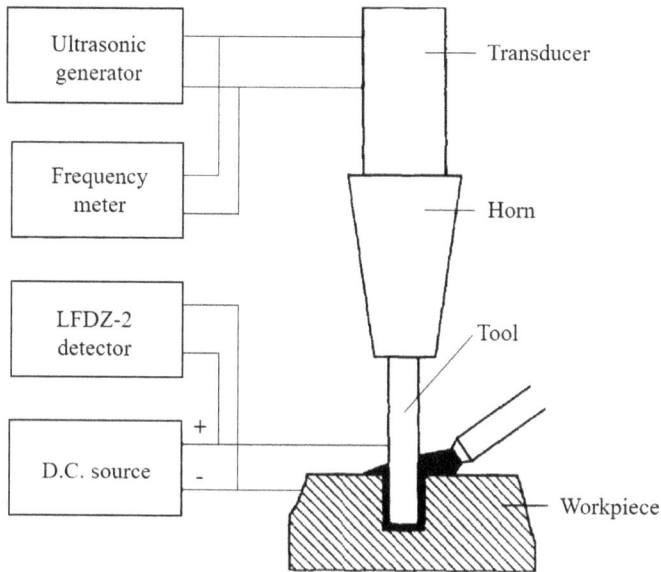

FIGURE 9.9 Setup of the combined machine [33]. Reproduced with permission from Elsevier.

to be about three times than that of USM, and twice that of EDM, used individually, without any effect on surface quality.

With the advent of miniaturization, micro-hole fabrication has become quite common and this is inevitable. However, it is quite strenuous to increase the aspect ratio along with the MRR, while maintaining the surface quality simultaneously. Many researchers have used vibrations in micro-EDM to improve its output [35–43]. Endo et al. [21] proposed a setup to be used for imparting vibrations in the micro-EDM process, as shown in Figure 9.10. A piezoelectric material, Lead Zirconate Titanate (PZT), 20 mm long in the form of laminated sheets, was used for tool vibrations. The PZT was attached to a tool holder, where the tool was installed. A unique feature was observed in this setup. Instead of placing the tool on the z-axis, the workpiece was placed there, with the help of a mandrel. The tool was given x- and y-motions. The diameter of the workpiece was in microns. While machining, a workpiece surface equivalent to the tool thickness was machined at a time, followed by the downward movement of the workpiece equal to the tool thickness for further machining. This procedure was followed until the whole length of the workpiece underwent machining. The amplitude and frequency of vibrations were measured in air. The actual vibration amplitudes in EDM oil were less than those measured. However, the difference became negligible at higher frequencies. Although the experiment was done to machine the workpiece, the same setup could be used to drill a micro hole in the workpiece, by changing the shape of tool to a cylinder with diameter in microns, and holding the workpiece on the mandrel using a clamp. The provision of dielectric flushing was made by introducing a dielectric jet. The PZT was anchored on a servo

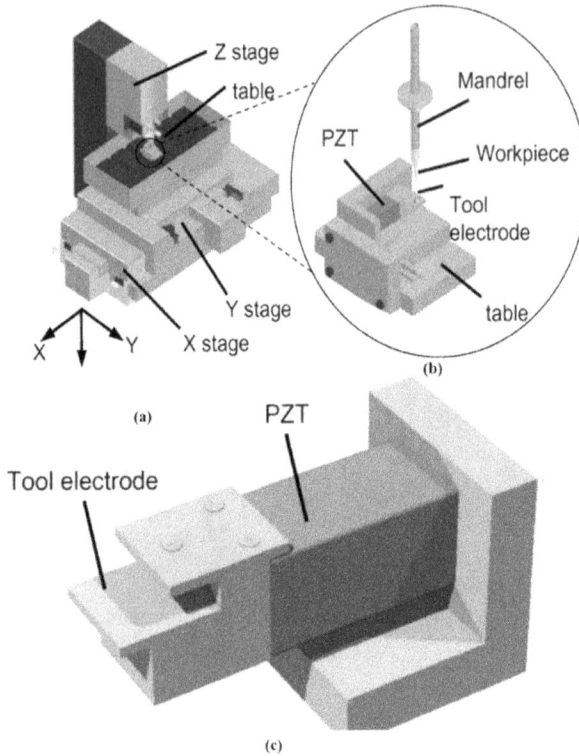

FIGURE 9.10 (a) Schematic diagram of vibration-assisted micro-EDM (machine setup); (b) enlarged image of the tool and workpiece; and (c) placement of piezoelectric transducer to vibrate the tool [21]. Reproduced with permission from Elsevier.

control mechanism to monitor the IEG. Nevertheless, use of ultrasonic frequencies bent the tool. Hence, low frequency vibrations proved quite helpful while fabricating a micro hole.

Many researchers are working to improve the efficiency of wire EDM [44–49]. In wire EDM, the tool is replaced by a wire which travels between two spools, and sparks are generated between the moving wire and the workpiece. Mohammadi et al. [48] proposed a setup for turning a workpiece using vibration assisted wire EDM. A spindle was placed in the dielectric tank, which was properly sealed to avoid seepage of dielectric in the bearing house. The cylindrical workpiece was mounted on it. A steel bar, with a wire guide engraved in it, was used to vibrate the wire. It acted as a medium to transmit vibrations to the wire from the vibration generating system. In addition, it also allowed holding the moving wire in position during the machining operation. The vibration generation system contained an ultrasonic generator to generate signals of ultrasonic frequency, a transducer to convert these signals into mechanical vibrations and a conical concentrator to concentrate and transmit the vibrations to the wire holder, which was nothing but a steel bar. As can be seen from Figure 9.11, the wire travelled through the groove in the wire holder during

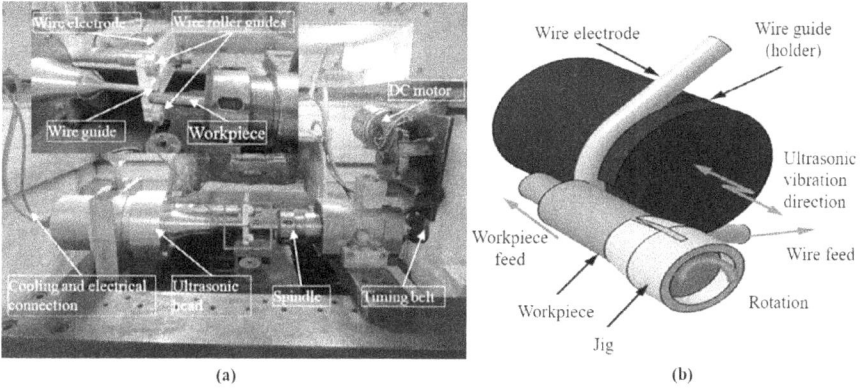

(a)   (b)

FIGURE 9.11   (a) Setup for wire vibration in wire EDM; and (b) schematic diagram of wire vibrating bar [48]. Reproduced with permission from Elsevier.

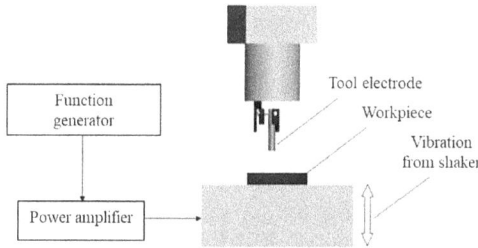

FIGURE 9.12   Schematic of the setup used in low frequency workpiece vibrations during EDM [50]. Reproduced with permission from Springer Nature.

machining. This groove helped in keeping the wire in position during machining. A concentrator transmitted the vibrations to the wire holder which vibrated in the longitudinal direction. Transverse vibrations could also be provided to the wire, by making arrangements such that the transducer vibrated in a perpendicular direction to the present direction of vibration.

9.2.3   VIBRATION APPLIED TO THE WORKPIECE

After realizing the importance of vibrations in improving the performance of the EDM process, several studies were conducted to increase machining efficiency by applying vibrations to the workpiece. Prihandana et al. [50] investigated the effect of low frequency workpiece vibrations on EDM performance. The maximum frequency of vibrations was 600 Hz, while amplitude was kept in the range of 0.75 to 1 μm. As seen in Figure 9.12, a shaker was used to impart the vibrations to the workpiece. This shaker was connected to a function generator through a power amplifier that controlled the amplitude and frequency of the vibrations. The researchers found a significant increment in MRR due to the application of low frequency vibrations, because

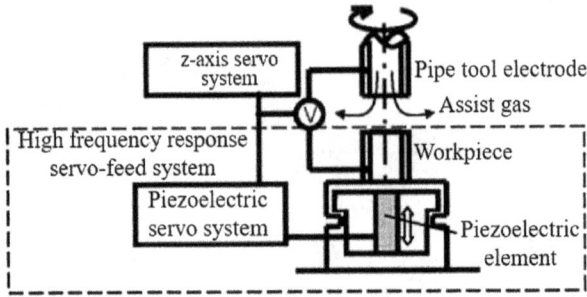

FIGURE 9.13 Schematic setup of the vibration assisted dry EDM [51]. Available under CC-BY- licence.

the vibrations had increased the frequency of attaining sufficient IEG width which increased the number of discharges. Furthermore, the improvement in dielectric flow and flushing characteristics led to better surface finish and less tool wear.

Yoshida et al. [51] developed a vibration setup for the workpiece in dry EDM. The schematic is shown in Figure 9.13. The piezoelectric transducer was placed to provide workpiece vibrations. The peculiar element of the setup was the high frequency response servo-feed system, along with the z-axis servo system. It included a piezoelectric element and a piezoelectric servo system that was responsible for workpiece vibration. This system helped in maintaining the IEG width during the machining process. In another study by Kunieda et al. [52], a performance simulator was developed that maintained the optimum IEG width during the entire machining operation by keeping a record of the difference between the present IEG width and the reference servo width, and then adjusting the present width through the z-axis servo mechanism and the piezoelectric servo mechanism, to enhance the process efficiency. They reported an increase in the MRR due to stabilization of the IEG width, and also found that there was no need for a piezoelectric servo system in die-sink EDM due to its larger IEG width, as compared to dry EDM.

Unune et al. [53] used low frequency workpiece vibrations in micro-EDM milling. The schematic of the setup is shown in Figure 9.14. The vibration device used an electromagnetic actuator to generate vibrations. An electromagnetic actuator works on the principles of electromagnetism. A time-varying electric field generates a magnetic field. Thus, if a current carrying conductor is placed in a magnetic field, it experiences a force, called an electromotive force, in a direction which is perpendicular to both, current flow and magnetic field direction. This type of actuator was used to provide vibrations. It contained an electromagnet and a ferromagnetic platform. When electricity was passed to the electromagnet, a magnetic field was generated. This magnetic field applied force to the platform and the platform became displaced from its rest position. When the current was stopped, the magnetic field collapsed, removing the force from the platform, thus bringing it back to its initial position. If a current of some frequency was passed, it would experience a force with that frequency, thus generating vibrations of that frequency. The actuator that had been developed was producing vibrations between 10 and 400 Hz, with an amplitude range of 2-15 μm.

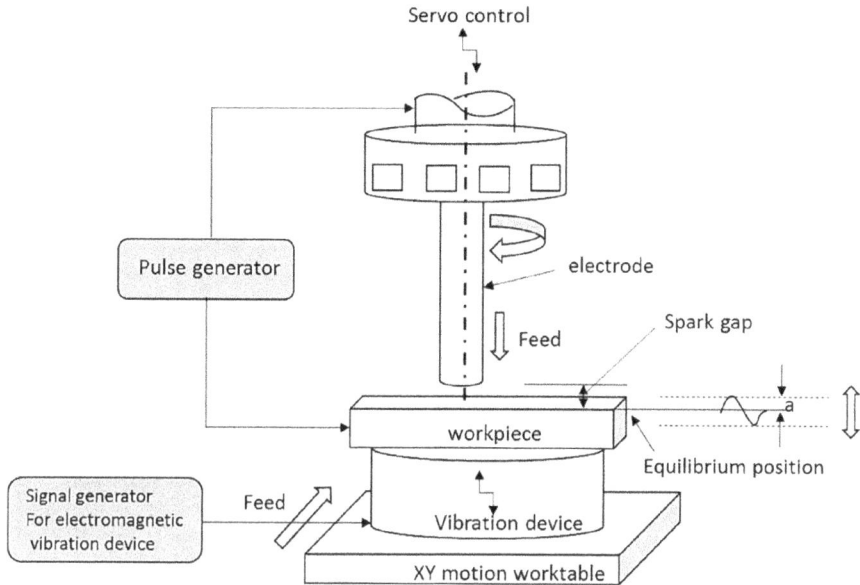

FIGURE 9.14 Low frequency vibration assisted micro-EDM setup [53]. Reproduced with permission from Taylor and Francis.

The frequency and amplitude of vibrations were changed by altering the frequency and magnitude of the applied current. The required frequency and amplitude of the applied current were controlled by a pulse generator and amplifier. Unune et al. undertook an investigation that discovered the effect of capacitance, as a prime parameter, that affected the MRR and electrode wear. They also found that high frequencies caused large sized globule formation that resulted in a rough surface of machined specimen. Hence, they suggested a low value of vibration frequency and discharge energy for good surface quality. Recently, they extended their work to investigate the influence of vibrational frequency on the dimensional accuracy and surface quality of micro-channels produced using vibration assisted micro-EDM milling [54].

Unune and Mali [47] also worked with the combination of low frequency workpiece vibration and the micro-wire EDM (micro-WEDM) process. The vibration device used here was based on the same principle that was used in the previous study of low frequency vibration assisted micro-EDM milling, in other word, it was working on electromagnetism to generate and provide vibrations to the workpiece. The tool was replaced by a wire travelling between two spools, one to provide fresh wire for the sparking phenomenon while the other was to collect the used wire coming from the IEG, as seen in Figure 9.15. This setup was able to vibrate with frequency up to 100 Hz. The researchers found that the voltage present across the IEG, capacitance and frequency of vibration primarily influence the kerf width (half of the difference between machined channel width and wire diameter). Low frequency vibrations decreased adhesion between the tool and workpiece and increased the flushing of debris out of the IEG, thus improving the efficiency of the micro-WEDM process.

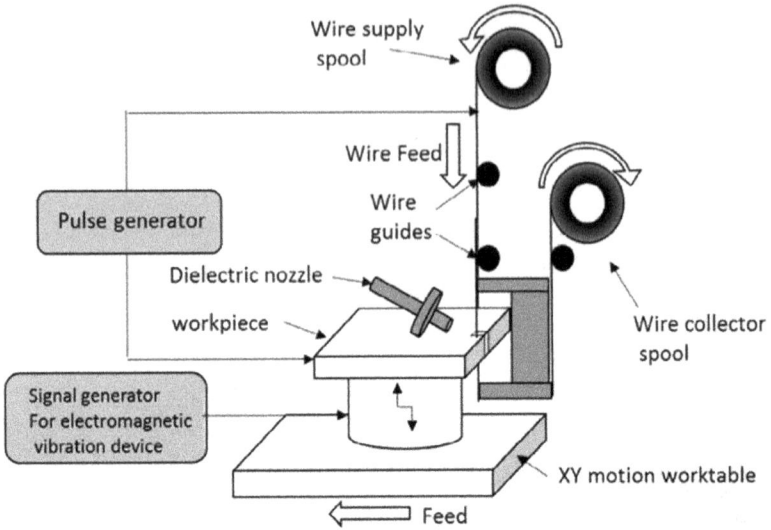

FIGURE 9.15 Schematic setup of vibration assisted micro-WEDM [47]. Reproduced with permission from Elsevier.

9.2.4 Vibration Applied to the Dielectric

As discussed earlier, vibrations applied to the tool or workpiece require complex systems and modifications in the machine, resulting in higher costs. It also helps in generating pressure variations in the IEG without affecting its width [1]. Hence, some researchers worked on applying vibrations to the dielectric fluid. Prihandana et al. [26] developed a setup for performing micro-EDM machining in the presence of dielectric vibrations. The main goal of using vibrations was to increase the velocity of debris, formed after the spark phenomenon, to enable easy debris removal from the IEG and hence, increased effective discharge ratio. As seen from Figure 9.6, a vibration bath is used to impart vibrations of frequency 43 kHz to the dielectric fluid. This frequency was chosen so as to increase the debris velocity. An increased MRR was revealed in the presence of ultrasonic vibrations of dielectric fluid, along with a decrease in adhesion of debris to the electrodes.

Schubert et al. [30] proposed a setup for vibrating dielectric fluid using a sonotrode. It involved a transducer which was capable of generating high frequency vibrations. A sonotrode was attached to the transducer and transferred the generated vibrations to the dielectric. Apart from transferring the vibrations, it also helped in amplifying the amplitude of vibrations to the desired value. The sonotrode tip was immersed in the dielectric and kept at that point where it could transfer maximum vibrations of the required range to the IEG. A schematic sketch is shown in Figure 9.16.

In another study by Ichikawa and Natsu [55], the fluid was vibrated ultrasonically using a vibrating horn, as shown in Figure 9.17, in a micro-hole drilling setup. They placed the electrode along the z-direction, and its position was changed in the z-axis,

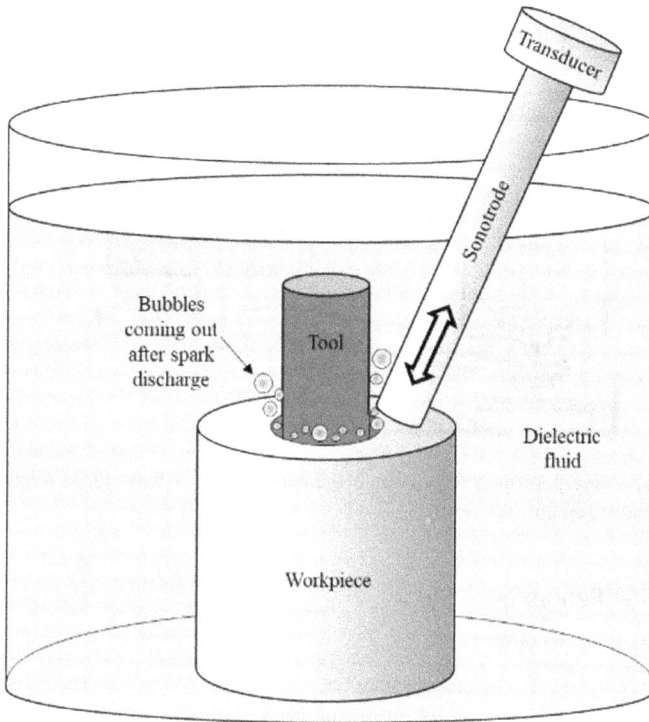

FIGURE 9.16 Schematic of dielectric vibration using sonotrode [30]. With permission.

while the x- and y-axis movement changes were given to the workpiece. A horn with a circular hole was placed such that the electrode was fed to the workpiece by passing through the circular hole present in the horn, without contacting it. The vibrations were transferred from the vibration generating setup to the dielectric using the horn. As there was no contact between the horn and the tool, vibrations were directly transferred to the dielectric without affecting the motion of the tool. With this setup in use, they observed a huge reduction in machining time, tool wear and the IEG width. In addition, the clearance gap between the tool and workpiece was found to be reduced as well, thus suggesting the possibility of machining micro-holes with tighter tolerances.

9.3   THE EFFECTS OF VIBRATION ON EDM PERFORMANCE

To achieve an improved process performance of EDM, many researchers have investigated the effects of vibration in EDM processes and analyzed the influence of the frequency and amplitude of vibrations. This section highlights the major studies incorporating investigations on vibration assisted EDM on improving process performance.

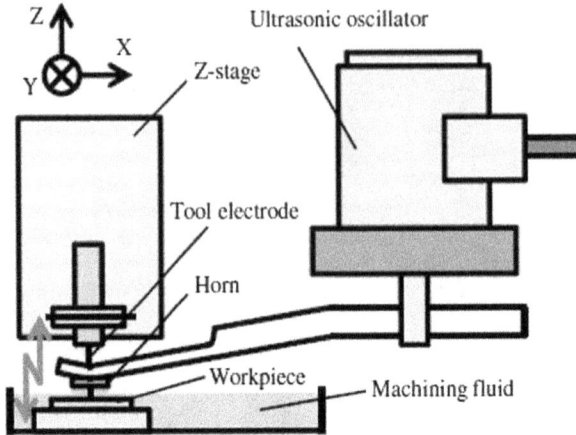

FIGURE 9.17 Setup proposed to vibrate dielectric using a horn [55]. Reproduced with permission from Elsevier.

9.3.1 EDM PROCESSES

9.3.1.1 MRR and TWR

With the application of vibration assistance, the MRR has incremented by an amount ranging from 10 to 400% depending on the type of operation, process and vibration frequency, and amplitude range used during the process. Kremer et al. [23] used a cross-shaped piezoelectric transducer for vibrating the tool to enhance the MRR in EDM and found for the roughing pass, an improvement of 10% in the MRR, whereas for the finishing pass, an increment as high as 400% was attained. The tool provided the pumping action, when aided by ultrasonic vibrations, thus enhanced debris flushing out of the IEG took place in decreased short-circuit sparks, hence improving the MRR. However, an increase in effective spark discharges also resulted in higher tool wear (10-15%).

In another study, EDM and ultrasonic machining (USM) processes were combined in a single setup by Thoe et al. [56]. The workpiece was vibrated using a piezoelectric dynamometer. The horn of the USM was connected to the tool electrode of the EDM. The tool and workpiece were connected to the closed loop force feedback system so that the system could work conveniently with an EDM servo control circuit and a USM DC servo quill assembly. The combined effect resulted in a remarkable increase in MRR. Further, the depth of hole was increased by a factor of 21 on average. Figure 9.18 shows graphically, the effect of ultrasonic vibrations on the depth of fabricated hole.

Shabgard et al. [57] used ultrasonically vibrating tool steel and found that the MRR was significantly increased for a combination of short pulses and ultrasonic vibrations at low current values (see, Figure 9.19(a)). This condition of small pulse-on time with low current values is called the finishing process. During finishing, as a high surface finish was required, the machining was performed at a low current

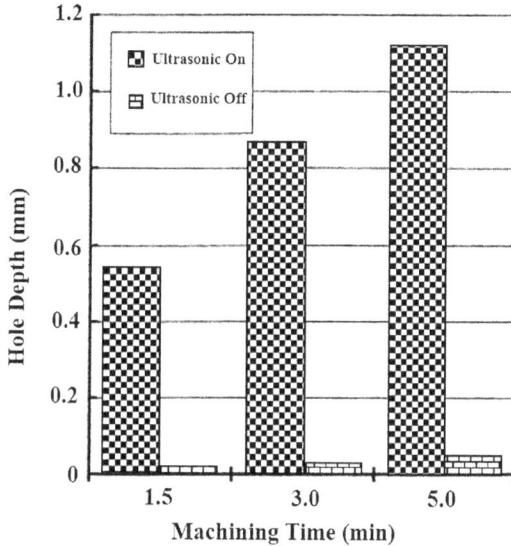

FIGURE 9.18 Comparison of hole depth at different machining time [56]. Reproduced with permission from Elsevier.

and small pulse-on time so that a lower heat flux could lead to small crater formation and thus, increasing the surface quality. This increment in the MRR was found to be caused by rapid compression and rarefaction during the tool's downward and upward movements imparted by the ultrasonic vibrations. This acted in the ejection of the newly formed debris and other particles from the IEG, reducing short-circuiting as well as arcing pulses, which stabilized the process and reduced the retraction motion. Nevertheless, a reduction in tool wear ratio (TWR) was also found during the finishing pass. Figure 9.19(b) gives the TWR trends with pulse-on time at different current values and vibration conditions. The reason for the low tool erosion ratio was attributed to the properties of the graphite tool material used in the study. Graphite underwent sublimation in the presence of spark discharge. In other words, instead of melting and then vaporizing, Graphite directly vaporizes instead of passing through a melting phase. Hence, no molten metal accumulated on the tool surface. On the other hand, due to the effects of ultrasonic vibrations, the bulk boiling of molten material present on the workpiece surface, the MRR was considerably more than the tool wear. Therefore, the TWR, which is the ratio of tool wear to the MRR, was seen to decrease in the finishing pass.

Srivastava and Pandey [58] applied vibrations to the cryogenically cooled tool electrode in the EDM process. They reported the MRR to be same as that of conventional EDM, but higher than cryogenically cooled EDM. When cryogenic cooling was applied, the temperature of the dielectric decreased below the room temperature. Hence, before the spark discharge, some of the spark energy was used in increasing the temperature of the surrounding dielectric. In addition, the introduction of cryogenic cooling also delayed the spark ignition, for the reasons stated above. Hence, less

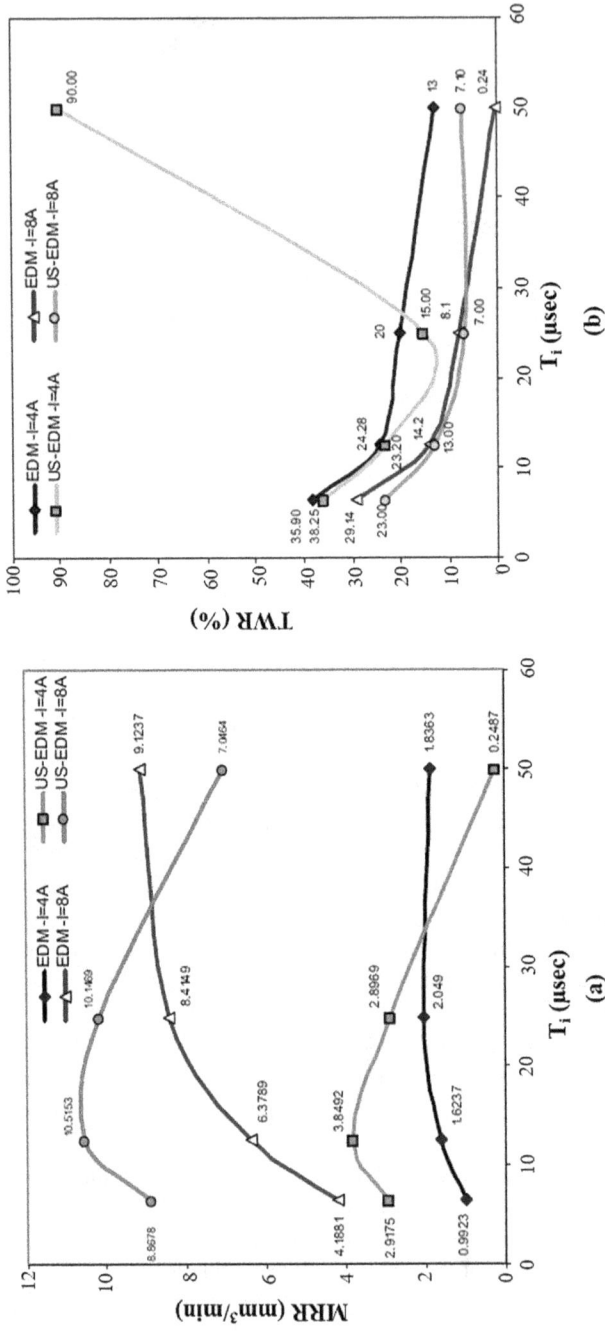

FIGURE 9.19 (a) Impact of ultrasonic vibrations (US-EDM) on MRR and (b) TWR with pulse-on time (Ti) at 4A and 8A current [57]. Available under CC-BY- licence.

energy was available for melting the material, resulting in a lower MRR than conventional EDM. Nevertheless, with the application of ultrasonic vibrations, the tool acted as a pump to thrust dielectric into the IEG. Hence, it enhanced flushing which, in turn increased the MRR. The same observation was reflected in tool wear. Cryogenically cooled EDM had least tool wear followed by vibration-assisted cryogenically cooled EDM and then conventional EDM. Apart from low erosion due to cryogenic cooling in the first process, the craters were small in size. The material removed, in the form of small debris, solidified on the surface of the tool, creating a recast layer. Thus, in this way, the total TWR was observed to be least in the cryogenically cooled EDM process. Ultrasonic vibration caused more erosion of the tool and prevented formation of a recast layer, thus increasing TWR in that case.

Teimouri and Baseri [59] used vibrations in dry EDM assisted with a rotating magnetic field. They observed that ultrasonic vibrations produce longitudinal waves in the dielectric, with regions of alternate compression and rarefaction. The half wave of rarefaction, in other words, the low-pressure phase, helps in the suction of debris from the IEG, whereas the other half, namely, the high-pressure phase, helped in expelling the debris from the IEG. This decreased the inactive pulse count. Moreover, during the removal of gases from the IEG, there was instantaneous compression and heating of gases. This instantaneous heating obstructed the heat transfer process, thus creating a localized hot spot for a short period of time. This hot spot had high temperature and pressure and produced metal atoms at higher energy levels, along with dissociating the dielectric molecules that led to an increase in ionization. Furthermore, these longitudinal waves produced the gas bubbles which were responsible for cavitation. When these bubbles collapsed, the explosion created high energy in a localized area, thus causing the cavitation. All of these factors combined on the workpiece surface to give an increased MRR. Similarly, due to increased ionization which increased the machining time, it resulted in high tool wear.

9.3.1.2 Pulse Characteristics

Rajeswari et al. [60] reported an increase in the arcing phenomenon when EDM was assisted with vibrations, with an increment in ignition delay. Arcing is mainly a result of the combined effect of contamination in the IEG and its length. Although the vibrations improved the cleaning action in the IEG, the distance between electrodes was reduced at regular intervals, causing the arcing. Further, an increase in the arcing led to an increase in recast layer thickness. However, the maximum discharge voltage in vibration-assisted EDM was reported to decrease by a quarter of that required in conventional EDM, with an approximate similar increase in maximum discharge current, with no effect on maximum discharge duration. It was found that effective sparks (that helped in material removal) and weak sparks (that did not reach the other electrode due to less energy) caused breakdown voltages resulting in ignition delay, and these sparks along with the arcs determined discharge parameters.

Murthy and Philip [61] found the existence of a pre-breakdown current. However, it was not affecting the machining, owing to its low magnitude. The voltage was found to collapse as soon as breakdown started in conventional EDM. At the same time, the

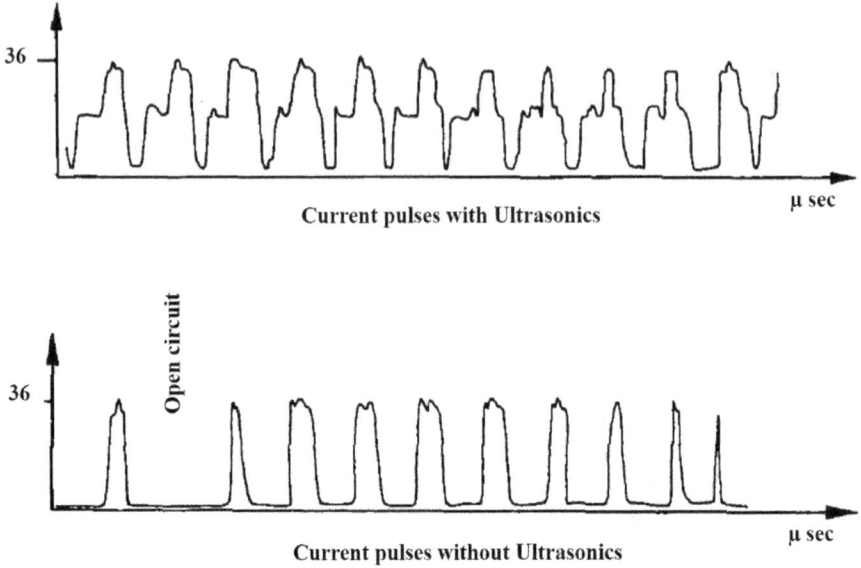

Current pulses with Ultrasonics

Current pulses without Ultrasonics

FIGURE 9.20   Change in the shape of pulses in presence of vibrations [61]. Reproduced with permission from Elsevier.

vibration-assisted EDM had a gradual voltage drop at the onset of dielectric break-down, with oscillations. Although the cavitation and acousto-electric effect helped in the rapid breakdown, the ignition delay remained unaffected due to enhanced debris removal from the IEG and frequent changes in discharge locations that balanced the increasing effect of acousto-electric current and cavitation. Acousto-electric current was produced by the superposition of ultrasonic waves, generated due to ultrasonic vibrations, on the charge carriers that were present in the spark zone. These ultrasonic waves carried the energy and momentum which, when transferred to the ions present in the plasma region increased the magnitude of current. The shape of the pulses was observed to be changing in the presence of vibrations, as seen in Figure 9.20. Here, it can be seen that there are two peaks in a pulse when vibrations are present while machining, unlike the pulse shapes where vibrations are not present. The first peak is lower in value when compared to the second peak. This is because the first peak corresponds to the pre-breakdown current while the second peak corresponds to the discharge current. Due to the absence of the pre-breakdown current in EDM sparks that are not vibration assisted, this peak is not visible here. Only the peak corresponding to the discharge current is observed.

9.3.1.3   Surface Characteristics

The surface quality of machined surfaces plays a critical role in the service life of fabricated components. The EDM process is considered to have an adverse effect on the surface quality of machined surfaces. Surfaces machined with EDM show the formation of a recast layer or white layer, cracks, globules, voids, uneven surfaces, and heat affected zone [62–66]. Thus, controlling the electrical parameters to achieve the desired surface quality has always been a challenging task for machinists. Huu et al. [67] observed that the surface peaks got shortened and more uniform with low frequency vibrations ranging from 128-512 Hz in EDM operation. The study concluded that surface roughness of the newly formed surface was decreased with a uniform surface texture. Figure 9.21 represents the microscopic images of surfaces machined with EDM, with and without frequency assistance. It is evident from Figure 9.21 that the application of vibration decreased the surface roughness. This is because vibrations assist in removing the debris and other particles from the machining region. This decreases the number of particles that get attached to the newly generated surface from machining, decreasing the surface roughness. In the absence of vibrations, the material melted from the workpiece cools rapidly and adheres to the new surface, increasing its roughness. Vibrations reduce this adhesion and improve the flushing properties of the process, which in turn increases the smoothness of the surface. Vibrations decrease the adhesion drastically (Figure 9.21(d)), which enhances the smoothness of the surface. In addition, the recast layer is also seen on the machined

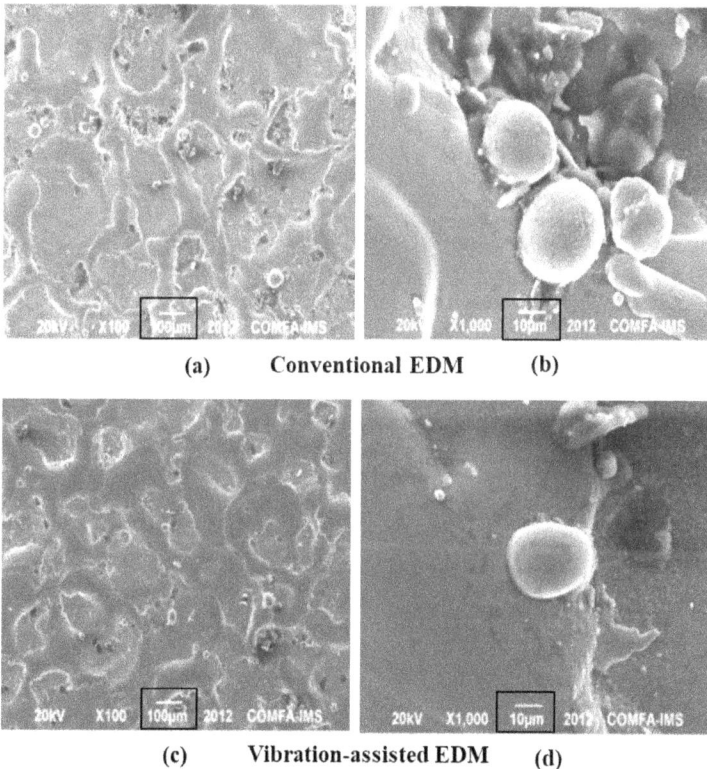

(a)   **Conventional EDM**   (b)

(c)   **Vibration-assisted EDM**   (d)

FIGURE 9.21 Topographical image of the surface generated through (a) conventional and (c) vibration-assisted EDM; debris present on the machined surface in case of (b) conventional and (d) vibration-assisted EDM [67]. Reproduced with permission from Springer Nature.

surface. Though its thickness is less in the case of vibration-assisted EDM, it is found to be thicker than the tempered layer (heat affected layer). The recast layer is formed from melted particles that are not able to flush out and stick to the machined surface due to adhesion. Note the increase in surface quality (Figure 9.21(c)) and decrease in debris concentration (Figure 9.21(d)) in vibration-assisted EDM when compared, under the same conditions, with conventional EDM (Figure 9.21(a-b) respectively).

In another study by Wang et al. [68], it was found that the surface roughness increases with vibration amplitude. As shown in Figure 9.22, the average surface roughness values for surfaces machined with conventional EDM, the vibration-assisted EDM with amplitudes 4 μm, 10 μm and 20 μm were found to be 2.4199 μm, 2.0274 μm, 1.9156 μm and 1.7013 μm, respectively. The 3D surface roughness profiles in Figure 9.22 confirmed that the surface morphology was more flattened with the increase in vibration amplitude. It was also found that EDM aided by vibrations decreased the carbon deposition on the machined surface due to the cavitation effect. Figures 9.23(a) and 9.23(b) give the extent of carbon deposition on the surface machined with conventional EDM and vibration-assisted EDM, respectively.

Lin et al. [69] compared the performance of dry EDM with vibration-assisted dry EDM, and called it hybrid EDM. They found a smooth surface formation in the case of hybrid EDM, with less micro pores, molten debris and micro cracks, which can be clearly seen in Figure 9.24(a). Moreover, the recast layer thickness in the case of

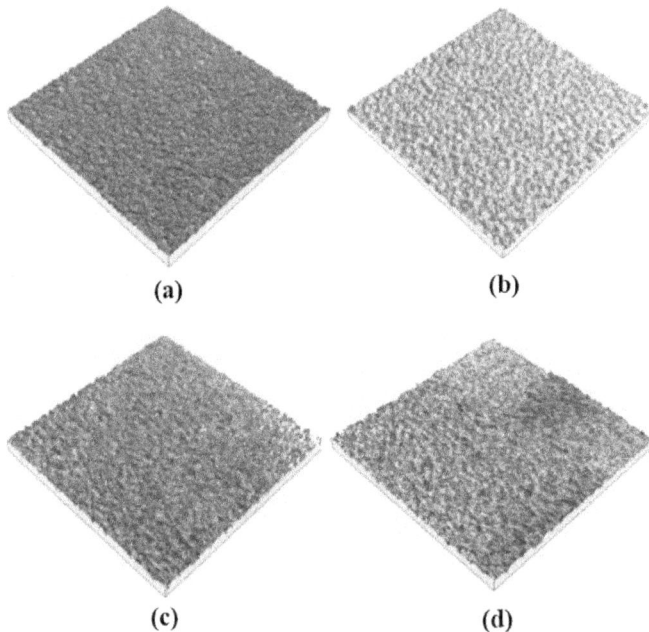

(a)   (b)

(c)   (d)

FIGURE 9.22 3D surface roughness at different EDM conditions: (a) conventional EDM; (b) at amplitude 4 μm; (c) at amplitude 10 μm; and (d) at amplitude 20 μm [68]. Reproduced with permission from Springer Nature.

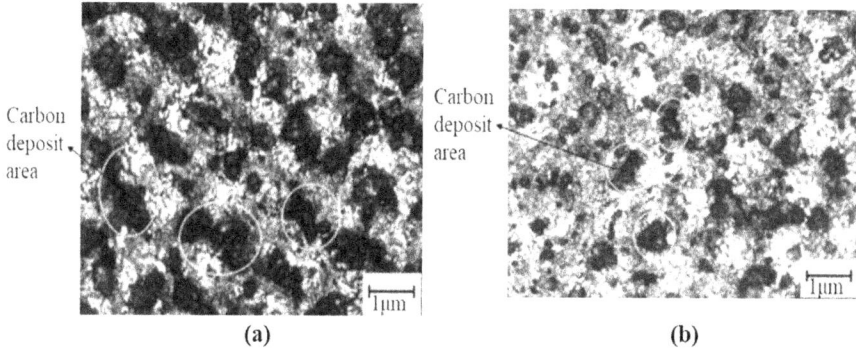

FIGURE 9.23 Comparison of carbon deposition in EDM without vibration assistance (a) and with vibration assistance and (b) in the EDM [68]. Reproduced with permission from Springer Nature.

hybrid EDM was found to be less than that of dry EDM. It was observed that when using oxygen as the gas, the recast layer thickness was found to be 19.8 μm in dry EDM, and 14.6 μm on the surface machined by hybrid EDM. Vibrations would either remove the molten material by facilitating flushing or, alternatively the cavitation due to vibrations ensured the redeposition of molten material back onto the surface, thus decreasing the recast layer, as depicted in Figure 9.24 (b).

Azhiri et al. [70] performed vibration-assisted EDM turning operation on AISI H13 steel using a copper tool and found that the presence of vibrations was beneficial in improving surface characteristics. Remarkably, the vibrations decreased the damaging discharges, thus reducing the surface irregularities. Conventional EDM turning produced many damaging discharges that destroyed surface quality quite appreciably. The SEM image (Figure 9.25(a)) shows a number of deep valleys, micro cracks and debris stuck to the machined surface, giving it an irregular texture. On the other hand, the surface topography of the surface generated with vibration-assisted EDM was relatively smoother and more even due to the restriction in the formation of damaging discharges (SEM image of Figure 9.25(b)).

9.3.1.4 Accuracy

Accuracy is another critical factor in any manufacturing process and industry always strives to achieve the highest possible machining accuracy. In EDM, not only the workpiece gets eroded due to the thermal action, but the tool electrode also gets eroded. Thus, tool wear has a significant impact on the accuracy of the produced part.

In the case of an EDM hole drilling process, it is often observed that debris and molten metal gets trapped in the holes, resulting in secondary sparking which may degrade the dimensional accuracy. It becomes prevalent in the case of holes with intricate shapes and with large aspect ratios. It is found that accuracy depends largely on tool geometry and path travelled by the tool. Tool geometry is the more significant as the tool path does not deviate and tool oscillations during machining are negligible. On the other hand, the tool geometry is severely affected due to tool wear during

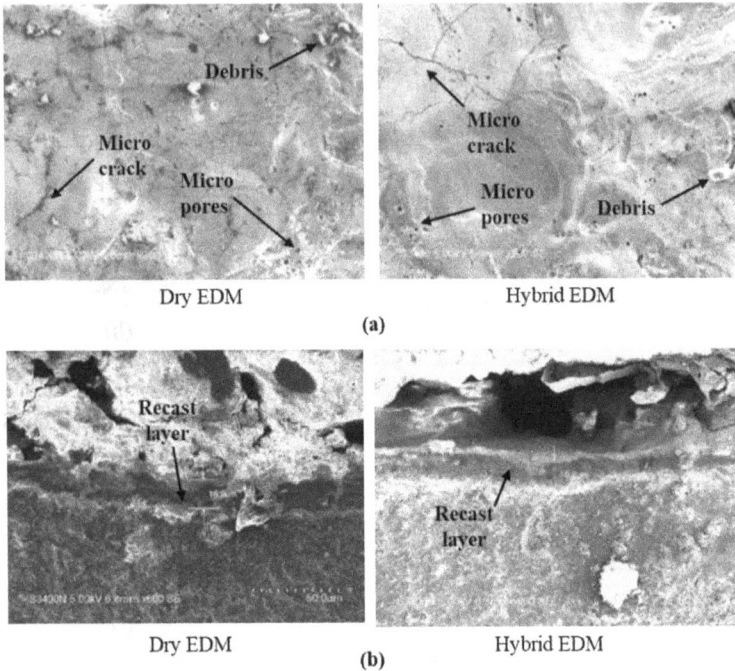

FIGURE 9.24   Comparison of surface characteristics of dry EDM with vibration-assisted dry EDM; (a) comparison of micro cracks, micro pores, and debris and (b) comparison of recast layer thickness [69]. Reproduced with permission from Springer Nature.

EDM machining. Tool wear decreases as machining proceeds due to the accumulation of carbon on its surface that gets dissociated from the dielectric fluid during spark discharge. However, tool edges are heavily eroded due to the inability of carbon to deposit at the edges. This worn edged tool causes tapered holes that affects the dimensional accuracy of newly drilled holes [71].

It is observed that the vibration-assisted EDM process significantly decreases tool wear. An electrode wear reduction of 2% was reported when axial vibrations were imparted in vibration assisted EDM [72]. This wear can be considered to be negligible, in other words, the tool is not worn out at all. Hence, no taper will occur in the machined hole, thus improving the accuracy exponentially. It has also been found that as the depth of the hole is increased during machining, the viscous resistance of the dielectric fluid and the pressure applied by the bubbles produced from the gases generated during spark discharge, cumulatively stops the movement of dielectric in the machining region. This results in an obstruction of the machining process. However, the presence of vibrations facilitates the up and down movement of the dielectric fluid. This movement continuously changes the contact angle between the dielectric fluid and the surface of the hole's wall and the tool's side wall. This is termed the wetting effect of vibrations. This effect helps in thinning the boundary

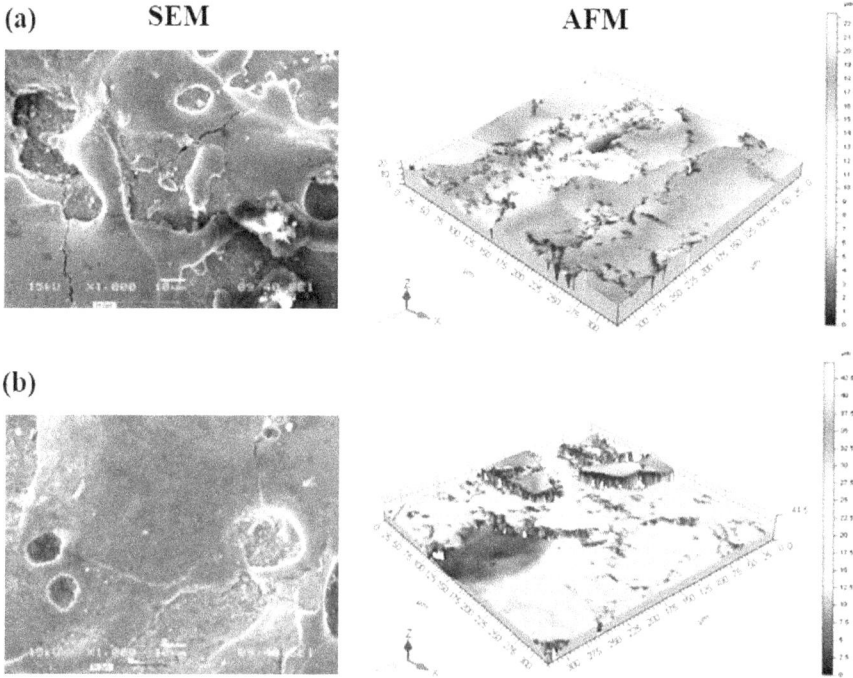

FIGURE 9.25 Pictorial comparison of EDMed surfaces using SEM and AFM imaging, (a) surface machined with conventional EDM and (b) surface machined with vibration-assisted EDM [70]. Reproduced with permission from Springer Nature.

between bubbles and dielectric fluid, which decreases viscous resistance in the IEG which ultimately facilitates debris removal, leading to a high tool feed speed in the workpiece, indicating the formation of a hole of high aspect ratio with better dimensional accuracy [73]. It was also revealed that the roundness of the tool employed in vibration-assisted EDM is lower than that of the tool used in conventional EDM. Thus, the shape projected on the hole also retains that circularity, clearly indicating improved dimensional accuracy in the presence of vibrations in EDM [58].

However, the introduction of vibrations increases the flushing of debris from the IEG and these are removed while passing between the workpiece and tool's side walls. The debris particles may cause the occurrence of sparks in this region thus creating overcut (increasing lateral thickness) more than desired, thus leading to dimensional inaccuracy [59]. Tool wear is observed to be increasing in some cases where vibrations are applied to the process. This is because of the decrease in the thickness of the recast layer that is deposited on the tool surface [74]. Thus, a thinner recast layer on the tool surface leads to the worn-out region becoming exposed to spark discharge. This exposed worn-out region then projects its shape onto the newly formed hole, thus causing a taper in the hole that results in dimensional inaccuracy.

9.3.2 Micro EDM Processes

9.3.2.1 MRR and TWR

In the case of micro-EDM, the MRR is typically observed to be increasing in the presence of low frequency vibrations, irrespective of other parameters, due to improved dielectric flushing in the IEG. In addition, electrode wear ratio (EWR) decreases when the frequency of vibrations is increased. The molten metal and debris particles, produced as a consequence of spark discharge, leads to the adhesion of tool to the workpiece. This adhesion causes short circuits which results in retraction of the tool to maintain the width of IEG. The repetitive retraction delays the machining process and reduces the MRR. Vibrations cause expansion and contraction of the IEG. In other words, vibrations facilitate relative motion between the tool and workpiece. This relative motion avoids the adhesion phenomenon. Vibrations also impart oscillatory motion to the dielectric as well as debris present in the IEG. The contraction of the IEG increases the fluid pressure which results in expulsion of the dielectric fluid and debris from the working zone, whereas the expansion of the machining zone decreases the fluid pressure in the IEG that sucks in fresh fluid [75]. This form of induced flushing from vibrations is efficient in debris removal. This effective debris removal assists in improving the MRR since the debris accumulation is reduced significantly in the IEG. Hence, less secondary sparks occur between the tool and the debris. Hence, more heat is now available on the workpiece as more sparks are occurring between the tool and the workpiece. This causes an increase in the MRR. With the increase in spark discharge between the tool and workpiece, the effective discharge ratio also increases. This increased effective discharge ratio decreases the frontal electrode wear (tool wear). Hence, with the introduction of vibrations in the micro-EDM process, tool wear is also found to be decreasing [53]. Figure 9.26 graphically depicts the increase in the MRR (Figure 9.26(a)) and the decrease in tool wear (Figure 9.26(b)).

For micro-WEDM, Unune et al. [47] reported that the MRR is directly proportional to the vibration frequency. contributing almost 11% in improved MRR. Furthermore, they also reported that the MRR is highly responsive to the combination of wire feed rate and vibration frequency, with an effect of about 22% on the MRR (see Figure 9.27). This observation was explained by considering the adhesion between the wire electrode and workpiece. During adhesion, a short circuit occurs between the wire and the workpiece. This causes the EDM machine to move backwards on its path, in the opposite direction to machining direction, in order to maintain the gap width for spark discharge to take place. This obstruction to the machining decreases the MRR. In the presence of vibrations, the adhesion between the two electrodes, namely, the wire and the workpiece, was reduced significantly, thus eliminating the chances of retraction of the machine, hence increasing the MRR. Apart from this, vibrations also changed the dielectric fluid flow in the machining zone to turbulent flow. This turbulent flow helped in improving debris evacuation from the IEG, thus reducing the number of ineffective discharges which in turn, again increased the MRR. An improvement of 66.2% was seen in the MRR when EDM was assisted with low frequency vibrations in the range 0–80 Hz.

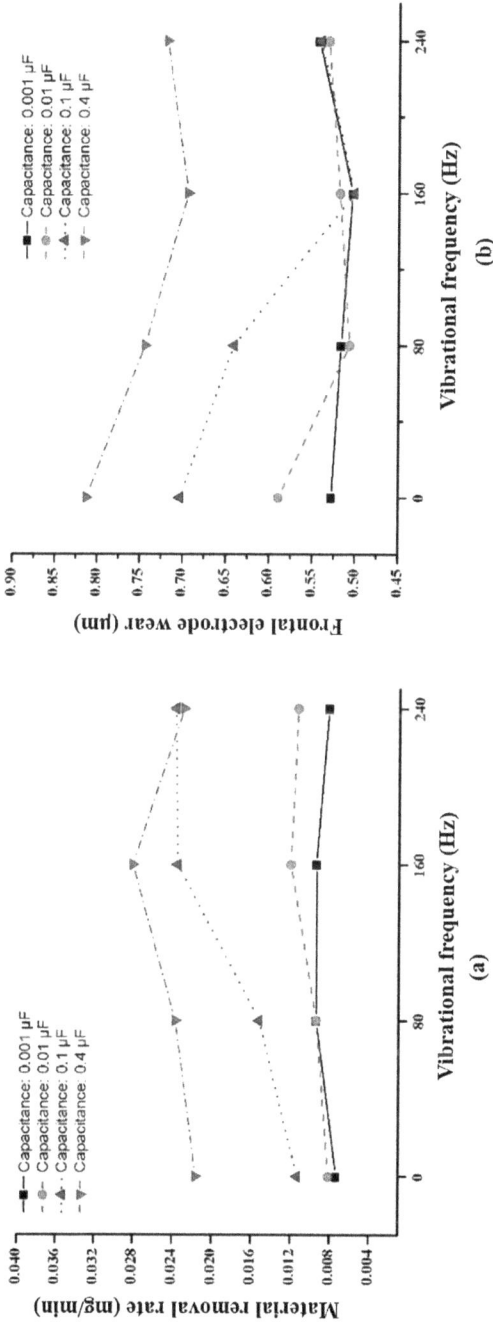

FIGURE 9.26 Effect of vibrational frequency on (a) Material Removal Rate (MRR) and (b) frontal electrode wear (tool wear) [53]. Reproduced with permission from Taylor & Francis.

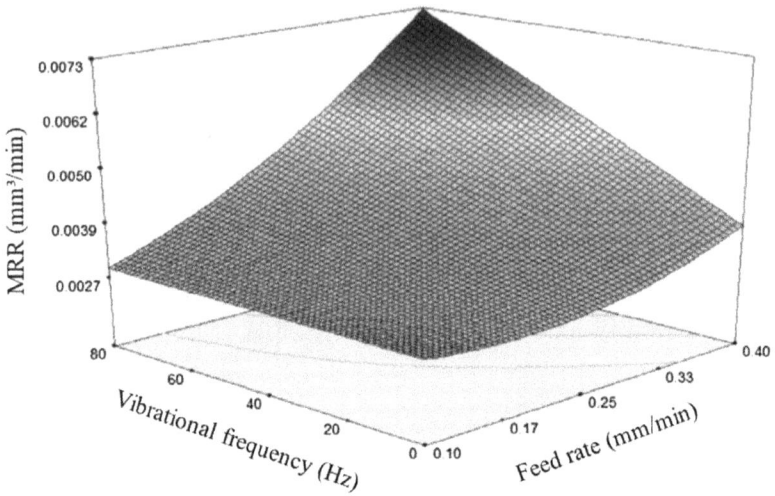

FIGURE 9.27 Variation of MRR with vibrational frequency and feed rate [47]. Reproduced with permission from Elsevier.

9.3.2.2 Pulse Characteristics

Jahan et al. [29] proposed a new variable to evaluate the effectiveness of low frequency vibrations in the micro-EDM process. They denoted this parameter by K_v and defined it as the ratio of the maximum value of acceleration in the direction of gravitational field to the acceleration due to gravity. They proposed that the value of K_v must be greater than one in order to drill micro holes with high aspect ratios. The percentage of normal pulses was also found to be more in the case of $K_v > 1$. In addition, machining efficiency was also found to be improving remarkably when $K_v > 1$. Hence, they examined the pulse characteristics for this case. In the presence of low frequency vibrations, they found four types of pulses, as shown in Figure 9.28.

Stable machining conditions are achieved in the case of a greater number of normal sparks and effective arcs, called effective pulses. On the other hand, if the number of transient short and complex sparks, which are collectively termed, short pulses is greater, this condition leads to unstable machining. The introduction of vibrations increased the percentage of normal sparks and effective arcs by improving flushing that eliminates the chances of unstable machining. In fact, the machining of high aspect ratio holes was found to occur only for $K_v > 1$, suggesting a difficulty in machining such holes in the absence of vibrations. Figure 9.29 gives the percentage of various pulse types at different aspect ratios for discrete values of K_v. It is observed that effective pulses have increased to more than 75% while decreasing short pulses to less than 5% at the same time, when K_v is found to be greater than one, which leads to machining stability and this, in turn, increases process efficiency. Thus, it can be inferred that vibrations increase the effective pulses in the process, improving machine performance.

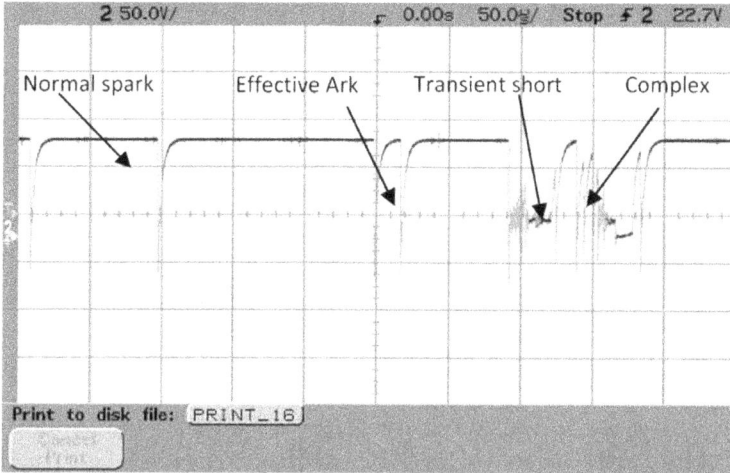

FIGURE 9.28 Different types of pulses during machining from low frequency vibration-assisted EDM setup [29]. Reproduced with permission from Elsevier.

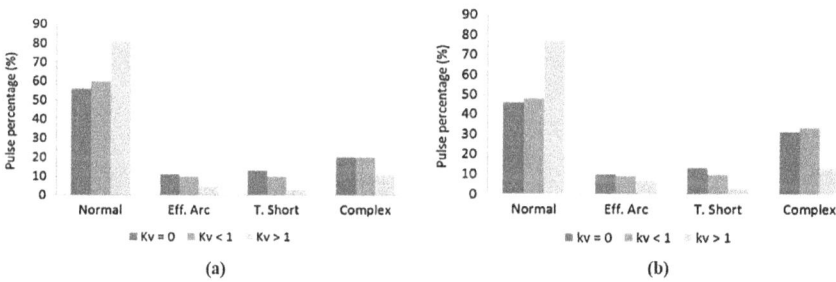

FIGURE 9.29 Percentage of individual pulses at different aspect ratios for distinct values of K_v, hole-aspect ratio: (a) 5 and (b) 7.5 [29]. Reproduced with permission from Elsevier.

9.3.2.3 Surface Characteristics

Figure 9.30 shows the surface characteristics of micro-holes machined with micro-EDM that is aided with low frequency workpiece vibrations. A poor surface finish was obtained in the absence of vibrations during machining. Burrs were not seen outside the machined hole when low frequency vibrations were employed during the process. This is because vibrations decrease the adhesion effect, along with facilitating the removal of debris by improving flushing characteristics. This also helps in enhancing the surface quality. Figure 9.31 presents a cross-sectional view of the machined hole. It can be seen that the surface of the hole machined using a micro-EDM setup, assisted with low frequency vibrations, was smoother than that machined without the assistance of vibrations. The presence of carbon deposited regions (black spots) was visible on the hole surface when machined without vibrations. In addition, the debris is also seen sticking on the surface. These were not present on the surface of holes

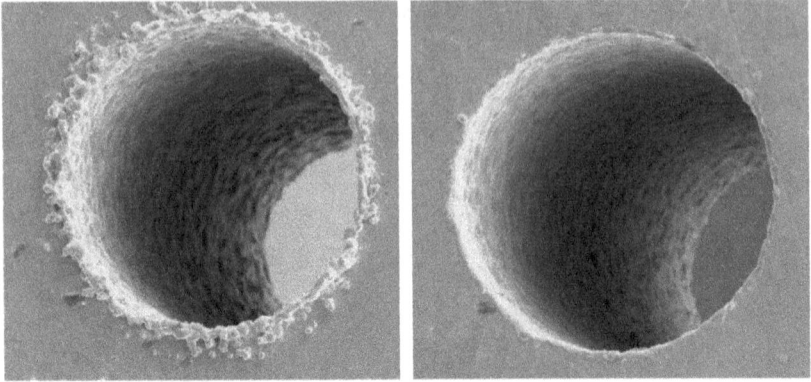

FIGURE 9.30   Microholes produced in Inconel 718 using micro-EDM, hole surface in absence of vibration (left); and in presence of vibration (right) [76]. Reproduced with permission from Elsevier.

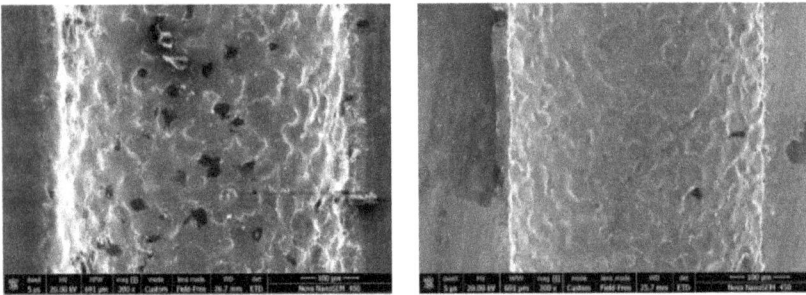

FIGURE 9.31   Cross-sectional view of micro-hole with micro-EDM; Surface characteristics in absence of vibration (left); and in presence of vibrations (right) [76]. Reproduced with permission from Elsevier.

machined in the presence of low frequency vibrations. The reason is that vibrations do not allow debris particles to adhere to the surface and improved dielectric flushing expels them from the IEG before re-solidification. Furthermore, due to vibrations which result in more flushing, carbon deposition is reduced significantly [76].

9.3.2.4   Accuracy

Dimensional accuracy is a major challenge in machining high aspect ratio micro holes using the micro-EDM process. The inaccuracy in the dimensions of the machined hole is basically initiated by tool wear, followed by interruption of the feed control system in the machining process due to this high tool erosion. Due to an accumulation of debris in the IEG, the dielectric strength of the medium reduces, which leads to arcing which increases tool wear and decreases material removal, hence resulting in the loss of dimensional accuracy. In addition, secondary sparks

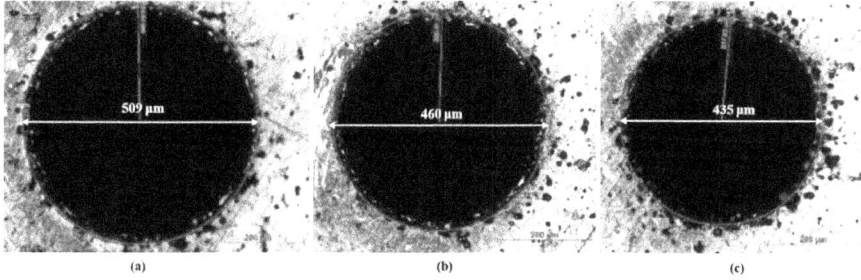

FIGURE 9.32   Hole dimensions when machined in absence of vibrations (a); vibrations with frequency of 80 Hz (b); and frequency 160 Hz (c) [76]. Reproduced with permission from Elsevier.

also start occurring due to debris between the newly formed surface of the hole and the side walls of the tool. Secondary sparks are those discharges that occur between the tool and the debris suspended in the machining zone. Some of the debris particles, when coming out of the IEG, become stuck between the side walls of the tool and the hole surface. The debris stuck in this region are the cause of secondary sparks occurring in this region, between the tool edges and the newly generated surface of the hole. These sparks are responsible for an increase in overcut (lateral width) and a tapered hole [77].

It is apparent that vibrations decrease the unwanted overcut (lateral width). Figure 9.32 shows the effect of vibrations on the dimensional accuracy of the machined hole. The desired diameter for the machined holes was 300 μm. As the frequency of vibrations is increased, it can be seen that deviation is reduced. This is aided by the process of effective flushing. In the absence of vibrations, debris becomes stuck between the side walls of the tool and the newly formed surface of the micro hole. This leads to more spark discharges and an expansion in the plasma channel which causes the increment in the overcut. Conversely, vibrations result in reduced tool and workpiece adhesion along with enhanced expulsion of debris due to improvement in flushing conditions, which creates turbulence along with the cycle of alternate suction and ejection of dielectric fluid that improves debris removal. However, it is also noted that with the rise in vibration frequency, hole taper is also increased. This is because of the debris that is now flushed out in a larger quantity due to the vibration effect resulting in enhanced flushing. This debris generates sparks while moving out of the hole along the side walls of the tool. The generated sparks release heat that leads to the melting of material, which leads to an increase in the diameter of the hole at the top, thus causing taper [76].

9.4   RESEARCH TRENDS

This section discusses the present research trends in the field of vibration-assisted EDM. It also discusses some of the main effects of vibrations when coupled with the EDM setup. It also gives a glimpse of future scope in this field.

9.4.1 VIBRATION-ASSISTED ELECTRIC DISCHARGE MILLING

Although, electrical discharge milling is good for milling simple contours, the conventional EDM process is not recommended for machining complex three-dimensional geometry because the ineffective flushing at point corners may lead to the accumulation of debris. This debris later results in tapering of edges and other problems which cause unnecessary deviation of the generated surface from the desired contour. An EDM setup aided by vibrations can overcome such problems since vibrations help in improving flushing characteristics which leads to enhanced debris removal from the machining region and thus, increased ability to form intricate three-dimensional surfaces effectively. When milling micro-channels in the workpiece using the micro-EDM milling setup and providing low frequency vibrations to the workpiece, debris removal is observed to be improved due to oscillations in the dielectric fluid caused by the vibrations that increase the supply of fresh dielectric in IEG. This also helps in debris evacuation from the machining zone [53]. It is observed that the MRR is improved by approximately 14% in the presence of vibrations. The contribution of vibrations to electrode wear is found to be about 4%. Apart from this, it has also been revealed that low frequency vibrations help in the formation of smooth contours, which match the desired contour profile, due to the formation of craters of a smaller size [62]. It is also seen that instead of applying vibrations to the tool, workpiece vibrations improve machining performance significantly in the electric discharge milling process. The ultrasonic vibrations can accelerate the flow of dielectric fluid which helps in effective debris removal from the IEG. Additionally, periodic contraction and expansion of the IEG eliminates the possibility of arcing, thus leading to stable machining of the workpiece with high dimensional accuracy [78].

9.4.2 POWDER MIXED VIBRATION ASSISTED EDM

Recently, a positive effect has been seen when vibrations are employed in a powder mixed EDM setup and improved process performance can be achieved with the aid of both powder and vibration in EDM. When dielectric fluid involves some suspended particles of additive to improve the machining performance, it is termed as powder mixed EDM. However, it is a big challenge to prevent agglomeration or settling of the suspended particles on the base as this adversely affects machining stability. This problem can be solved by the use of vibrations. The disturbance caused by vibrations in the dielectric fluid induces a stirring effect that helps in activating the additive particles and keeps them suspended in the dielectric fluid, thus ensuring uniform machining conditions throughout the process without any effect on the stability during the process. This process has a wide potential in the surface treatment industry, facilitating the process of metal deposition uniformly and with high speeds [1]. It is found that if the suspended particles have lubricating properties, it can help in improving the surface characteristics of the machined surface. The suspended particles increase the number of spark discharges. The high density of suspended particles also helps in increasing material removal. Due to vibrations, the particles get charged and acquire some acceleration. When they collide from the workpiece surface after this acceleration, it produces an impact which facilitates workpiece erosion. The particles that remain suspended due to vibrations, help in providing the current path over the whole surface, instead of concentrating the current only at the center of the tool. This helps in achieving a flat surface, with smooth surface texture and the absence of black spots that are caused due to deposition of carbon that gets decomposed from the dielectric fluid at very high temperatures. However, due to uniform distribution and not letting the current concentrate at a single point (namely, the center of the tool), such high temperatures are not reached, thus avoiding the formation of black spots. Along with helping the additive particles to remain suspended in the dielectric fluid, vibrations also help in reducing adhesion and improve flushing which enhances material removal significantly. The combination of vibrations and additive suspension results in significant reduction in machining time, thus speeding up the process of machining [26]. It can also be seen that the combined effect of additives and vibrations improves dimensional accuracy and enhances the ability of the machine to produce holes of high aspect ratio. It simplifies the upward movement of cavitation bubbles which speeds up the process of debris removal, a phenomenon termed as cloud cavitation [79].

9.4.3 ULTRASONIC-ASSISTED DRY AND NEAR-DRY EDM

Dry EDM and near-dry EDM were introduced to overcome the problems of environmental pollution caused by the disposal of dielectric fluid. However, use of gas or mist has its own limitations, including low process stability and low material removal. Hence, vibrations are employed to negate some of the disadvantages of using gas as the dielectric medium. It was observed that the use of vibrations in near-dry EDM gives a MRR more than the MRR experienced when vibrations are provided in a conventional EDM process, and double than that of the dry EDM process in the absence of vibrations [22]. In addition, it was found that tool wear is almost zero in vibration assisted near-dry EDM. This implies that dimensional accuracy is quite high in the process. Moreover, the change in the values of surface roughness is negligible in vibration assisted near-dry EDM when compared with that of conventional EDM [80]. It has been proven experimentally that the use of vibrations in dry EDM has helped in stabilizing the machining process. This stability brings an added advantage with it. Due to the generation of stable sparks, more material removal occurs that ultimately increases the material removal rate [52]. When ultrasonic vibrations are applied to the workpiece, it starts acting as a pump which pushes the debris out of the machining area. As the density of the debris is less in the region, a greater number of normal discharges are seen that increase the material removal rate as well as the process stability [59].

9.5 CONCLUSION

The new-age metals such as Nickel and Titanium superalloys are being extensively used in many industries, particularly aerospace applications, due to their remarkable mechanical and chemical properties, such as high temperature resistance, extreme hardness, and high corrosion resistance. However, such properties also create a barrier to the machinability of the metals using conventional processes. Hence, unconventional processes are employed for machining. The electrical discharge machining process is widely used for machining any conductive materials, irrespective of their hardness. The ability to machine any complex shape with high accuracy makes the process quite popular amongst the various unconventional machining processes. However, EDM still has some challenges such as low cutting rate, degeneracy of the accuracy of parts and poor surface integrity. One of the possible solutions for such challenges is the use of vibrations in EDM and micro-EDM. The literature suggests that the use of vibration results in speeding up material removal, decreasing tool wear and that it improves the surface integrity of the newly formed surface after machining. The vibration-assisted EDM process is in the development phase even at this point of time and has a broad scope for research. There is a need to devise measurement techniques that can be used during the ongoing machining process to monitor the dimensional accuracy and other parameters which would be of advantage to the setup. There is a need to devise some cost-effective methods of producing the vibrations.

REFERENCES

[1] B. C. Khatri, P. Rathod, and J. B. Valaki, "Ultrasonic vibration–assisted electric discharge machining: A research review," *Proceedings of the Institution of Mechanical Engineers, Part B: Journal of Engineering Manufacture,* vol. 230, pp. 319–330, 2015.

[2] K. H. Ho and S. T. Newman, "State of the art electrical discharge machining (EDM)," *International Journal of Machine Tools and Manufacture,* vol. 43, pp. 1287–1300, 2003.

[3] J. E. Abu Qudeiri, A. Saleh, A. Ziout, A. I. Mourad, M. H. Abidi, and A. Elkaseer, "Advanced Electric Discharge Machining of Stainless Steels: Assessment of the State of the Art, Gaps and Future Prospect," *Materials (Basel),* vol. 12, 2019.

[4] M. R. Shabgard and F. Kabirinia, "Effect of Dielectric Liquid on Characteristics of WC-Co Powder Synthesized Using EDM Process," *Materials and Manufacturing Processes,* vol. 29, pp. 1269–1276, 2014.

[5] I. A. Bucklow and M. Cole, "Spark-machining," *Metallurgical Reviews,* vol. 14, pp. 103–118, 1969.

[6] M. P. Jahan, "Micro-electrical discharge machining," in *Nontraditional machining processes,* ed: Springer, 2013, pp. 111–151.

[7] V. K. Jain, *Advanced machining processes*: Allied publishers, 2009.

[8] V. Lalwani, P. Sharma, C. I. Pruncu, and D. R. Unune, "Response Surface Methodology and Artificial Neural Network-Based Models for Predicting Performance of Wire Electrical Discharge Machining of Inconel 718 Alloy," *Journal of Manufacturing and Materials Processing,* vol. 4, p. 44, 2020.

[9] U. Maradia, M. Boccadoro, J. Stirnimann, I. Beltrami, F. Kuster, and K. Wegener, "Die-sink EDM in meso-micro machining," *Procedia Cirp,* vol. 1, pp. 166–171, 2012.

[10] E. Uhlmann, S. Piltz, and U. Doll, "Machining of micro/miniature dies and moulds by electrical discharge machining—recent development," *Journal of Materials Processing Technology,* vol. 167, pp. 488–493, 2005.

[11] Q. Liu, Q. Zhang, M. Zhang, and J. Zhang, "Review of size effects in micro electrical discharge machining," *Precision Engineering,* vol. 44, pp. 29–40, 2016.

[12] Y. S. Wong, M. Rahman, H. S. Lim, H. Han, and N. Ravi, "Investigation of micro-EDM material removal characteristics using single RC-pulse discharges," *Journal of Materials Processing Technology,* vol. 140, pp. 303–307, 2003.

[13] M. Rahman, A. Asad, T. Masaki, T. Saleh, Y. Wong, and A. S. Kumar, "A multiprocess machine tool for compound micromachining," *International Journal of Machine Tools and Manufacture,* vol. 50, pp. 344–356, 2010.

[14] T. Masuzawa, "Micro-EDM," *Proceedings of ISEM XIII, 2001,* 2001.

[15] E. Gentili, L. Tabaglio, and F. Aggogeri, "Review on micromachining techniques," in *AMST'05 advanced manufacturing systems and technology,* ed: Springer, 2005, pp. 387–396.

[16] E. C. Jameson, "Description and development of electrical discharge machining (EDM)," *Electrical discharge machining, society of manufacturing engineers, Dearbern, Michigan,* vol. 16, 2001.

[17] T. Masuzawa, "State of the art of micromachining," *Cirp Annals,* vol. 49, pp. 473–488, 2000.

[18] Z. Katz and C. Tibbles, "Analysis of micro-scale EDM process," *The International Journal of Advanced Manufacturing Technology,* vol. 25, pp. 923–928, 2005.

[19] Z. Zhang, H. Zhang, X. Pan, J. Das, and J. J. P. M. L. Eckert, "Effect of aspect ratio on the compressive deformation and fracture behaviour of Zr-based bulk metallic glass," vol. 85, pp. 513–521, 2005.

[20] R. Singh, A. Dvivedi, and P. Kumar, "EDM of high aspect ratio micro-holes on Ti-6Al-4V alloy by synchronizing energy interactions," *Materials and Manufacturing Processes,* vol. 35, pp. 1188–1203, 2020.

[21] T. Endo, T. Tsujimoto, and K. Mitsui, "Study of vibration-assisted micro-EDM—the effect of vibration on machining time and stability of discharge," *Precision Engineering,* vol. 32, pp. 269–277, 2008.

[22] Q. H. Zhang, R. Du, J. H. Zhang, and Q. B. Zhang, "An investigation of ultrasonic-assisted electrical discharge machining in gas," *International Journal of Machine Tools and Manufacture,* vol. 46, pp. 1582–1588, 2006.

[23] D. Kremer, J. L. Lebrun, B. Hosari, and A. Moisan, "Effects of Ultrasonic Vibrations on the Performances in EDM," *CIRP Annals,* vol. 38, pp. 199–202, 1989.

[24] D. R. Unune and H. S. Mali, "Current status and applications of hybrid micro-machining processes: A review," *Proceedings of the Institution of Mechanical Engineers, Part B: Journal of Engineering Manufacture,* vol. 229, pp. 1681–1693, 2014.

[25] M. Jahan, M. Rahman, Y. Wong, and L. Fuhua, "On-machine fabrication of high-aspect-ratio micro-electrodes and application in vibration-assisted micro-electrodischarge drilling of tungsten carbide," *Proceedings of the Institution of Mechanical Engineers, Part B: Journal of Engineering Manufacture,* vol. 224, pp. 795–814, 2010.

[26] G. S. Prihandana, M. Mahardika, M. Hamdi, Y. S. Wong, and K. Mitsui, "Effect of micro-powder suspension and ultrasonic vibration of dielectric fluid in micro-EDM processes—Taguchi approach," *International Journal of Machine Tools and Manufacture,* vol. 49, pp. 1035–1041, 2009.

[27] T. Shitara, K. Fujita, and J. Yan, "Direct observation of discharging phenomena in vibration-assisted micro-electrical discharge machining," *The International Journal of Advanced Manufacturing Technology,* vol. 108, pp. 1125–1138, 2020.

[28] M. T. Shervani-Tabar, K. Maghsoudi, and M. R. Shabgard, "Effects of Simultaneous Ultrasonic Vibration of the Tool and the Workpiece in Ultrasonic Assisted EDM," *International Journal for Computational Methods in Engineering Science and Mechanics,* vol. 14, pp. 1–9, 2013.

[29] M. P. Jahan, Y. S. Wong, and M. Rahman, "Evaluation of the effectiveness of low frequency workpiece vibration in deep-hole micro-EDM drilling of tungsten carbide," *Journal of Manufacturing Processes,* vol. 14, pp. 343–359, 2012.

[30] A. Schubert, H. Zeidler, M. H. Oschätzchen, J. Schneider, and M. Hahn, "Enhancing Micro-EDM using Ultrasonic Vibration and Approaches for Machining of Nonconducting Ceramics," *Strojniški vestnik – Journal of Mechanical Engineering,* vol. 59, pp. 156–164, 2013.

[31] O. Hitoshi, N. Teruo, and M. Iwao, "Study of micro machining of metals by EDM with high frequency vibration," *Takushima Prefectual Industrial Technology Center,* 2001.

[32] N. Mohd Abbas, D. G. Solomon, and M. Fuad Bahari, "A review on current research trends in electrical discharge machining (EDM)," *International Journal of Machine Tools and Manufacture,* vol. 47, pp. 1214–1228, 2007.

[33] J. Zhixin, Z. Jianhua, and A. Xing, "Study on a new kind of combined machining technology of ultrasonic machining and electrical discharge machining," *International Journal of Machine Tools and Manufacture,* vol. 37, pp. 193–199, 1997.

[34] J. Zhang, T. Lee, W. Lau, and X. Ai, "Spark erosion with ultrasonic frequency," *Journal of Materials Processing Technology,* vol. 68, pp. 83–88, 1997.

[35] Y. S. Liao and H. W. Liang, "Study of Vibration Assisted Inclined feed Micro-EDM Drilling," *Procedia CIRP,* vol. 42, pp. 552–556, 2016.

[36] K. Maity and R. K. Singh, "An optimisation of micro-EDM operation for fabrication of micro-hole," *The International Journal of Advanced Manufacturing Technology,* vol. 61, pp. 1221–1229, 2012.

[37] Z. Weiliang, W. Zhenlong, and D. Desheng, "A new micro-EDM reverse copying technology for microelectrode array fabrication," 2006.

[38] Z. Wansheng, W. Zhenlong, D. Shichun, C. Guanxin, and W. Hongyu, "Ultrasonic and electric discharge machining to deep and small hole on titanium alloy," *Journal of Materials Processing Technology,* vol. 120, pp. 101–106, 2002.

[39] M. Mahardika and K. Mitsui, "A new method for monitoring micro-electric discharge machining processes," *International Journal of Machine Tools and Manufacture,* vol. 48, pp. 446–458, 2008.

[40] D. J. Kim, S. M. Yi, Y. S. Lee, and C. N. Chu, "Straight hole micro EDM with a cylindrical tool using a variable capacitance method accompanied by ultrasonic vibration," *Journal of Micromechanics and Microengineering,* vol. 16, p. 1092, 2006.

[41] S. A. Mastud, N. S. Kothari, R. K. Singh, J. Samuel, and S. S. Joshi, "Analysis of debris motion in vibration assisted reverse micro electrical discharge machining," in *ASME 2014 International Manufacturing Science and Engineering Conference collocated with the JSME 2014 International Conference on Materials and Processing and the 42nd North American Manufacturing Research Conference,* 2014.

[42] E. Bamberg and S. Heamawatanachai, "Orbital electrode actuation to improve efficiency of drilling micro-holes by micro-EDM," *Journal of materials processing technology,* vol. 209, pp. 1826–1834, 2009.

[43] H. Huang, H. Zhang, L. Zhou, and H. Zheng, "Ultrasonic vibration assisted electro-discharge machining of microholes in Nitinol," *Journal of micromechanics and microengineering,* vol. 13, p. 693, 2003.

[44] Z. N. Guo, T. C. Lee, T. M. Yue, and W. S. Lau, "Study on the machining mechanism of WEDM with ultrasonic vibration of the wire," *Journal of materials processing technology,* vol. 69, pp. 212–221, 1997.

[45] Z. N. Guo, T. C. Lee, T. M. Yue, and W. S. Lau, "A Study of Ultrasonic-aided Wire Electrical Discharge Machining," *Journal of materials processing technology,* vol. 63, pp. 823–828, 1997.

[46] K. T. Hoang and S. H. Yang, "A study on the effect of different vibration-assisted methods in micro-WEDM," *Journal of Materials Processing Technology,* vol. 213, pp. 1616–1622, 2013.

[47] D. R. Unune and H. S. Mali, "Experimental investigation on low-frequency vibration assisted micro-WEDM of Inconel 718," *Engineering Science and Technology, an International Journal,* vol. 20, pp. 222–231, 2017.

[48] A. Mohammadi, A. F. Tehrani, and A. Abdullah, "Introducing a New Technique in Wire Electrical Discharge Turning and Evaluating Ultrasonic Vibration on Material Removal Rate," *Procedia CIRP,* vol. 6, pp. 583–588, 2013.

[49] A. Mohammadi, A. F. Tehrani, and A. Abdullah, "Investigation on the effects of ultrasonic vibration on material removal rate and surface roughness in wire electrical discharge turning," *The International Journal of Advanced Manufacturing Technology,* vol. 70, pp. 1235–1246, 2013.

[50] G. S. Prihandana, M. Mahardika, M. Hamdi, and K. Mitsui, "Effect of low-frequency vibration on workpiece in EDM processes," *Journal of Mechanical Science and Technology,* vol. 25, pp. 1231–1234, 2011.

[51] M. Yoshida, "Improvement of material removal rate of dry EDM using piezoelectric actuator coupled with servo-feed mechanism," *CAPE'98, Tokyo, Japan,* 1998.

[52] M. Kunieda, T. Takaya, and S. Nakano, "Improvement of dry EDM characteristics using piezoelectric actuator," *CIRP Annals,* vol. 53, pp. 183–186, 2004.

[53] D. R. Unune and H. S. Mali, "Experimental investigation on low-frequency vibration-assisted μ-ED milling of Inconel 718," *Materials and Manufacturing Processes,* vol. 33, pp. 964–976, 2017.

[54] D. R. Unune and H. S. Mali, "Dimensional accuracy and surface quality of micro-channels with low-frequency vibration assistance in micro-electro-discharge milling," *Advances in Materials and Processing Technologies,* pp. 1–12, 2020.

[55] T. Ichikawa and W. Natsu, "Realization of Micro-EDM under Ultra-Small Discharge Energy by Applying Ultrasonic Vibration to Machining Fluid," *Procedia CIRP,* vol. 6, pp. 326–331, 2013.

[56] T. B. Thoe, D. K. Aspinwall, and N. Killey, "Combined ultrasonic and electrical discharge machining of ceramic coated nickel alloy," *Journal of Materials Processing Technology,* vol. 92, pp. 323–328, 1999.

[57] M. R. Shabgard, B. Sadizadeh, and H. Kakoulvand, "The Effect of Ultrasonic Vibration of Workpiece in Electrical Discharge Machining of AISIH13 Tool Steel," *World Academy of Science, Engineering and Technology,* vol. 3, pp. 332–336, 2009.

[58] V. Srivastava and P. M. Pandey, "Effect of process parameters on the performance of EDM process with ultrasonic assisted cryogenically cooled electrode," *Journal of Manufacturing Processes,* vol. 14, pp. 393–402, 2012.

[59] R. Teimouri and H. Baseri, "Experimental study of rotary magnetic field-assisted dry EDM with ultrasonic vibration of workpiece," *The International Journal of Advanced Manufacturing Technology,* vol. 67, pp. 1371–1384, 2012.

[60] R. Rajeswari and M. Shunmugam, "Comparative evaluation of powder-mixed and ultrasonic-assisted rough die-sinking electrical discharge machining based on pulse characteristics," *Proceedings of the Institution of Mechanical Engineers, Part B: Journal of Engineering Manufacture,* vol. 233, pp. 2515–2530, 2019.

[61] V. S. R. Murthy and P. K. Philip, "Pulse train analysis in ultrasonic assisted EDM," *International Journal of Machine Tools and Manufacture,* vol. 27, pp. 469–477, 1987.

[62] D. R. Unune and H. S. Mali, "Parametric modeling and optimization for abrasive mixed surface electro discharge diamond grinding of Inconel 718 using response surface methodology," *International Journal of Advanced Manufacturing Technology*, vol. 93, pp. 3859–3872, 2017.

[63] D. R. Unune, V. P. Singh, and H. S. Mali, "Experimental Investigations of Abrasive Mixed Electro Discharge Diamond Grinding of Nimonic 80A," *Materials and Manufacturing Processes*, vol. 31, pp. 1718–1723, 2015.

[64] D. R. Unune, C. K. Nirala, and H. S. Mali, "ANN-NSGA-II dual approach for modeling and optimization in abrasive mixed electro discharge diamond grinding of Monel K-500," *Engineering Science and Technology, an International Journal*, vol. 21, pp. 322–329, 2018.

[65] U. Aich, "Investigation for the presence of chaos in surface topography generated by EDM," *Tribology International*, vol. 120, pp. 411–433, 2018.

[66] K. Wang, Q. Zhang, G. Zhu, and J. Zhang, "Effects of tool electrode size on surface characteristics in micro-EDM," *The International Journal of Advanced Manufacturing Technology*, vol. 96, pp. 3909–3916, 2018.

[67] P. N. Huu, L. B. Tien, Q. T. Duc, D. P. Van, C. N. Xuan, T. N. Van, et al., "Multi-objective optimization of process parameter in EDM using low-frequency vibration of workpiece assigned for SKD61," *Sādhanā*, vol. 44, 2019.

[68] Y. Wang, Z. Liu, J. Shi, Y. Dong, S. Yang, X. Zhang, et al., "Analysis of material removal and surface generation mechanism of ultrasonic vibration–assisted EDM," *The International Journal of Advanced Manufacturing Technology*, vol. 110, pp. 177–189, 2020.

[69] M. Mohammadijoo, L. Collins, H. Henein, and D. G. Ivey, "Evaluation of cold wire addition effect on heat input and productivity of tandem submerged arc welding for low-carbon microalloyed steels," *The International Journal of Advanced Manufacturing Technology*, vol. 92, pp. 817–829, 2017.

[70] R. B. Azhiri, A. S. Bideskan, F. Javidpour, and R. M. Tekiyeh, "Study on material removal rate, surface quality, and residual stress of AISI D2 tool steel in electrical discharge machining in presence of ultrasonic vibration effect," *The International Journal of Advanced Manufacturing Technology*, vol. 101, pp. 2849–2860, 2019.

[71] S. Kumar, S. Grover, and R. S. Walia, "Analyzing and modeling the performance index of ultrasonic vibration assisted EDM using graph theory and matrix approach," *International Journal on Interactive Design and Manufacturing (IJIDeM)*, vol. 12, pp. 225–242, 2016.

[72] M. Iwai, S. Ninomiya, and K. Suzuki, "Improvement of EDM properties of PCD with electrode vibrated by ultrasonic transducer," *Procedia CIRP*, vol. 6, pp. 146–150, 2013.

[73] Z. Y. Yu, Y. Zhang, J. Li, J. Luan, F. Zhao, and D. Guo, "High aspect ratio micro-hole drilling aided with ultrasonic vibration and planetary movement of electrode by micro-EDM," *CIRP Annals*, vol. 58, pp. 213–216, 2009.

[74] A. Abdullah, M. R. Shabgard, A. Ivanov, and M. T. Shervanyi-Tabar, "Effect of ultrasonic-assisted EDM on the surface integrity of cemented tungsten carbide (WC-Co)," *The International Journal of Advanced Manufacturing Technology*, vol. 41, pp. 268–280, 2008.

[75] D. R. Unune and H. S. Mali, "Experimental Investigations on Low Frequency Workpiece Vibration in Micro Electro Discharge Drilling of Inconel 718," Presented at the 6th International & 27th All India Manufacturing Technology, Design and Research Conference (AIMTDR-2016), College of Engineering, Pune, 2016.

[76] D. R. Unune, C. K. Nirala, and H. S. Mali, "Accuracy and quality of micro-holes in vibration assisted micro-electro-discharge drilling of Inconel 718," *Measurement,* vol. 135, pp. 424–437, 2019.

[77] S. Kumar, S. Grover, and R. S. Walia, "Optimisation strategies in ultrasonic vibration assisted electrical discharge machining: a review," *International Journal of Precision Technology,* vol. 7, pp. 51–84, 2017.

[78] J. M. Jafferson, P. Hariharan, and J. Ram Kumar, "Effects of Ultrasonic Vibration and Magnetic Field in Micro-EDM Milling of Nonmagnetic Material," *Materials and Manufacturing Processes*, vol. 29, pp. 357–363, 2014.

[79] P. J. Liew, J. Yan, T. J. I. J. o. M. T. Kuriyagawa, and Manufacture, "Fabrication of deep micro-holes in reaction-bonded SiC by ultrasonic cavitation assisted micro-EDM," *International Journal of Machine Tools and Manufacture*, vol. 76, pp. 13–20, 2014.

[80] C. Zhang, R. Zou, Z. Yu, and W. Natsu, "Micro EDM aided by ultrasonic vibration in nitrogen plasma jet and mist," *The International Journal of Advanced Manufacturing Technology*, vol. 106, pp. 5269–5276, 2020.

10 Magnetic Field Assistance in the EDM Process

Mahavir Singh, Vyom Sharma and Janakarajan Ramkumar

Department of Mechanical Engineering, Indian Institute of Technology Kanpur, Uttar Pradesh, India

CONTENTS

DOI: 10.1201/9781003202301-10

10.1   INTRODUCTION

Electric discharge machining (EDM) is an advanced machining process, which has shown its utility in the machining of hard and difficult-to-cut materials [1]. It has undergone several advancements in machining performance over the years. Still, the process has a low machining rate compared to traditional machining techniques and some advanced machining processes [2,3]. The machining yield, namely, unit removal rate (mass material rate, volumetric material removal rate) of the EDM process is very much determined by the energy of a unit pulse (discharge energy), as the size of the crater formed is directly related to the discharge energy. However, discharge energy cannot be increased any further as the broader or deeper crater would degrade the relative tolerance and surface roughness of the machined feature. During any of the EDM operations, the conversion of input electrical energy into thermal energy takes place, which in turn melts and evaporates the material from both tool and workpiece electrodes. A fraction of molten metal is flushed away from the inter-electrode gap (IEG) by the pressurized flow of dielectric fluid and the explosive force generated by the collapse of the plasma during the pulse-off period, whereas the remaining molten liquid either re-solidifies on the nascent surface to form the recast layer or converts into small metallic particles, called debris, due to the cooling effect of the dielectric fluid [4]. Figure 10.1 shows schematically: the EDM operation, the formation of debris and the bubbles in the discharge gaps. The amount of EDM debris is enormously high as machining happens at high frequency (kHz or MHz). After each discharge, the debris has to be removed to facilitate favorable conditions for the next discharge. If the debris is less, it might expedite an early discharge (at low discharge voltage or at a higher IEG) or reduce the discharge delay time in a somewhat similar way to the mechanism of powder mixed dielectric. However, if an appropriate flushing technique is not used, the debris starts accumulating in the IEG and becomes severe after a certain machining duration [5]. This debris, if not removed successfully, results in various abnormal discharges, which eventually reduces machining efficiency. The abnormal discharges include short-circuit (direct metal to metal contact), secondary discharges (discharge between debris and electrode), or higher-order discharges (discharges between debris), which ultimately reduces machining efficiency [6]. The success of the EDM process, namely, a higher machining rate or unit removal rate, and closer geometric tolerances, is obtainable with the efficacious elimination of the machining by-product, that is, debris. Thus, debris flushing and appropriate dielectric circulation in the IEG have been major issues that lead to the development

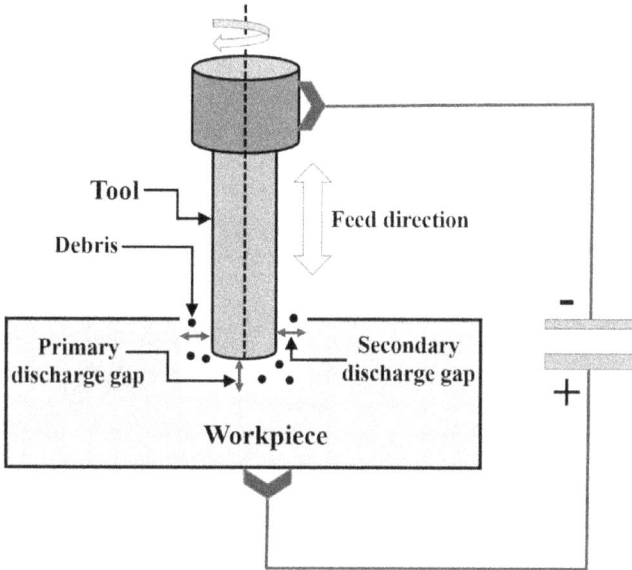

FIGURE 10.1   Representation of the EDM and debris dispersal in the IEG.

of a number of variants as well as hybrid machining techniques of EDM. There have been a number of techniques that augment the evacuation of debris from the IEG in the EDM process. These include: use of ultrasonic vibration to the tool or workpiece (sometimes dielectric fluid too) [7], pressurized jet flushing, the orbital motion of the tool electrode [8], amended tool electrodes [9], the use of gravity assistance for debris removal (workpiece positioned over the tool) [10], powder mixed dielectric [11], and magnetic field assistance [12]. Among the techniques mentioned above, magnetic field assistance in the EDM process has been success- fully implemented for the cleaning of IEG by providing the necessary forces on the debris of magnetic materials. In addition to debris cleaning, it confines the plasma channel as well as applying a normal force on the melt pool to alter the material removal phenomenon. Electric discharge micromachining (EDMM) is a variation of the traditionally used EDM process whereby the minimization of discharge energy per pulse creates a crater of relatively smaller size to facilitate higher dimensional accuracy of the miniaturized features [13]. However, the IEG is reduced significantly in EDMM, which makes the expulsion of debris more crit- ical than with conventional EDM. The MFAEDM process is a hybridization of the two processes which integrates the magnetic field and the EDM processes simul- taneously. The magnets are so arranged that the magnetic field force acts in the melt pool in the upward direction. Ferromagnetic debris is expelled from the IEG. Moreover, the plasma characterization, which includes electron density, plasma temperature, mean free path of electrons and current density are equally influenced by the Lorentz force in the magnetic field.

10.2 DEBRIS FLUSHING TECHNIQUES IN EDM

A number of debris flushing techniques are available to the EDM process, some of which are related to gap enlargement, utilizing the mechanical motion of the electrodes, and others are related to discharging at higher IEG by modifying the dielectric fluid. A few of them are briefly introduced in the subsequent sections.

10.2.1 JET FLUSHING

Jet flushing is the conventional method used for the removal of debris during the machining operation. It can be accomplished by dispensing the jet through a tube tool (internal hole in the tool) or can be supplied externally to the machining zone [14]. The dielectric fluid velocity and its pressure determine how effectively the machining products are removed from the IEG. Pressurized jet flushing appears to be valuable in the conventional EDM process where higher IEGs and tool electrodes of larger diameters are employed. Increasing dielectric pressure slightly increases the slit depth and reduces the width. However, fluid pressure cannot be increased beyond a specific range as it degrades the machining accuracy [15]. Moreover, it deteriorates the tool electrodes, especially in EDMM, where delicate microtools of dimensions ranging from 50–500 microns are used. Instead, it is a common practice to use multiple jets from different directions to enhance the flushing process. Jet flushing with a fixed nozzle position is less effective than dynamic jet flushing in which the position of the jet dispensing nozzle changes continuously according to the machining position [16].

10.2.2 ULTRASONIC VIBRATION ASSISTANCE

Ultrasonic vibrations of high frequency, but low amplitude have been implemented in the EDM process to create a hybrid process, called ultrasonic vibration-assisted EDM [17]. In this technique, the vibrations is given to workpiece electrodes (generally) during the pulse-off period, which enhances the IEG for the successful elimination of the machining debris [17]. It is a common practice to apply ultrasonic vibrations to the workpiece electrode owing to its positioning on the fixture where an ultrasonic vibration attachment can be installed easily. Moreover, if it is provided to the tool, there is a possibility of tool deflection and, interaction of the tool with the workpiece.

10.2.3 ORBITAL MOTION OF THE TOOL ELECTRODE

It is possible to decouple the tool from the machined feature by actuating the tool in a controlled trajectory. In orbital EDM, there are three possibilities for the machining of a circular hole of a specific depth [8]. In the first case, the depth of the hole can be divided into a number of cylindrical units of a certain thickness. The tool is initially inserted into the workpiece up to the depth of the first cylindrical unit, followed by an outward linear motion and, finally, a circular motion to remove the first cylindrical unit. This procedure continues until the predetermined depth of the hole is achieved. In the second case, initially, the tool is inserted up to the entire depth of the hole; subsequently, there is a linear outward motion and then a circular motion of the tool. Here the distance of the linear outward motion is a small radius (a segment of the intended radius of the hole). In subsequent steps, the linear outward motion (the radius of the hole) is increased to achieve the final desired radius. In the third case, along with the circular motion of the electrode, simultaneous feed in the direction of hole depth (vertically downward) can be provided to actuate the tool on a spiral path. In all three scenarios, the tool is decoupled from the feature size, which provides an additional path for dielectric flushing, and debris ejection, thus improving machining efficiency.

10.2.4 MODIFIED TOOL ELECTRODES

Conventionally used cylindrical and foil tool electrodes are modified by machining certain micro-profiles, namely μ-holes, μ-slots along the axis and inclined to the axis of a cylindrical tool [9]. On plate and thin foil tool electrodes, the micro-features are machined on the foil surface (perpendicular to the direction of tool feed) [18]. These micro-features assist in the evacuation of debris particles and allow the entry of fresh dielectric, as they provide additional pathways. However, the modification in the tool electrodes requires additional machining processes such as ED-drilling or wire-EDM processes. An alternative to the above mentioned modified cylindrical tool is to employ a helical tool electrode similar to a micro-drill or end-mill used in conventional machining. The helixes on the tool provide ways for the evacuation of debris particles and assist in the circulation of fresh dielectric fluid to the machining zone [19]. Tilting of the workpiece at an angle to the horizontal direction has also shown better utility in reducing debris accumulation in the IEG [20].

10.2.5 GRAVITY-ASSISTED EDM

Contrary to the conventional orientation of the electrodes in EDM, namely, with the tool positioned above the workpiece, in the gravity-assisted EDM, the orientation of the electrodes is reversed (tool at the bottom of the workpiece) so that debris can be separated due to the force of gravity. Considerable enhancement in machining rate and the aspect ratio of the feature is obtained by means of gravity-assisted debris removal techniques [10]. This is possible due to the prevention of abnormal discharges as maximum material removal occurs from normal (effective) discharges. Gravity-assisted EDM has not been used extensively, with only a few initial attempts having been made hitherto. The technique is limited by the weight of the workpiece and the additional rotational arrangements required for tool rotation. The heavy workpiece may pose some difficulty as it is directly mounted on the spindle.

10.2.6 POWDER MIXED DIELECTRIC

Powder mixed dielectric in the EDM is a method whereby some electrically conductive or non-conductive particles (having relative permittivity higher than the liquid dielectric) are mixed in the host (liquid) dielectric. The resultant dielectric under the action of the electric field modifies the discharge conditions and facilitates the formation of sparks at a higher IEG. The powder particles form a chain along with the location of the highest electric field in the IEG, thus bridging the gap, which allows spark formation at the larger IEG. The higher the discharge gap, the greater is the possibility of debris removal and the smaller are the chances of short circuits and abnormal discharges. Powder mixed dielectric has shown its usefulness in enhancing machining yield, reducing the surface roughness of the feature and thus bringing stability to the machining process [11].

10.3 THE NEED FOR MAGNETIC FIELD ASSISTANCE IN EDM

From the brief introduction of the techniques mentioned above, primarily used for the evacuation of debris, it is evident that all of the methods are equally effective in enlarging the IEG over which discharges can occur. Still, debris needs to be expelled from the IEG by inducing some kind of driving force, which is a pressurized jet of dielectric in conventional EDM. None of the techniques discussed above directly influence the debris particle using a driving force imposed on the particle. Magnetic field assistance for debris removal is a technique whereby the magnetic force directly acts on the debris particle and helps in evacuating it from the machining zone. Moreover, it does not require any configurational change to the machining setup except for the installation of magnets. Debris expulsion during the EDM operation becomes of the utmost significant owing to the fact that accumulation of debris in the IEG consequences in various abnormal discharges such as secondary discharge (between debris and electrode), tertiary discharge (amongst the debris) and, short circuit (debris ultimately bridging the IEG, resulting in metal–to–metal contact). These discharges are relatively less effective in material removal than the normal discharges due to the release of less discharge energy during the abnormal discharges. Pulse-discrimination techniques (pulse classification based on the discharge waveform) in the EDM can determine the nature of discharges in conventional EDM and MFAEDM. With the implementation of the magnetic field force in the process, it has been established that the number of normal discharges is substantially higher than those in the absence of a magnetic field. This can be simply understood in terms of the effective debris removal during the machining operation. The accumulation of debris becomes more intense as machining progresses as compared to the initial phase of machining. Thus, it is essential to eject the debris effectively, especially at higher machining depth; otherwise, it either reduces the machining rate or completely stops further machining due to the manifestation of frequent short-circuits.

10.4 FUNDAMENTALS OF MAGNETIC FIELD-ASSISTED EDM

The magnetic field influences the motion of debris particles, plasma channel dynamics and the melt pool. Thus, it is indispensable to understand them individually for a better insight of the MFAEDM. In the following subsections, the fundamentals of the role of MFA in EDM is highlighted.

10.4.1   THE INFLUENCE ON A DEBRIS PARTICLE

When magnets (permanent or electromagnet) are placed adjacent to the workpiece surface, the magnetic field lines leave from the North pole and enter from the South pole within a single magnet. At the same time, the field lines extend from the North pole of one magnet to the South pole of another magnet generating inter-magnetic field lines. Figure 10.2 shows the schematic illustration of MFAEDM. Electric field lines are along the direction of the flow of electrons in the plasma channel, whereas the magnetic field lines are extending from North pole of one magnet to the South pole of another magnet, namely, perpendicular to the flow of electrons. A debris particle in the IEG can be considered as a dipole on which two forces are acting simultaneously, that is, the magnetic field force (F_B) and the centrifugal force (F_C, due to rotation of the tool electrode) as shown in Figure 10.3. The position of the magnet is so arranged that the resultant of these two forces (F_R) acts in the direction of the field lines (a slightly upward direction) when the machining of a hole is under way [12]. The resultant force expels debris from the hole during the cutting operation.

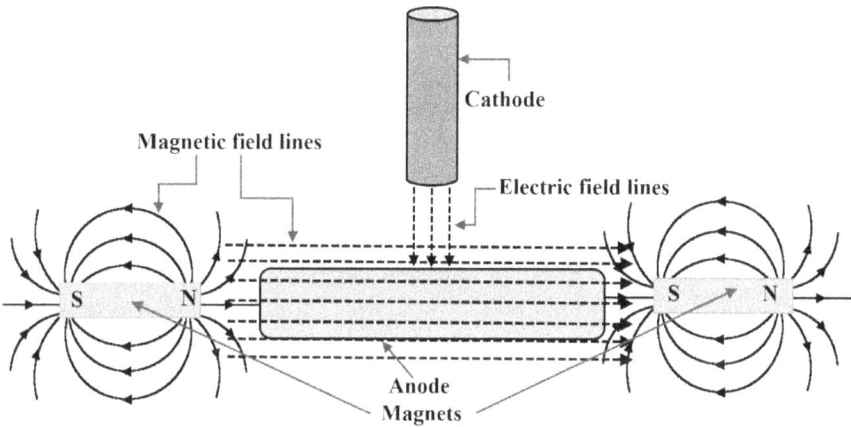

FIGURE 10.2   Schematic of the MFAEDM process.

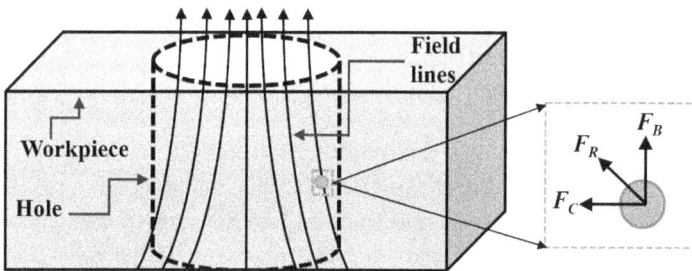

FIGURE 10.3   Forces acting on a debris particle inside a machined hole. Adapted from [12].

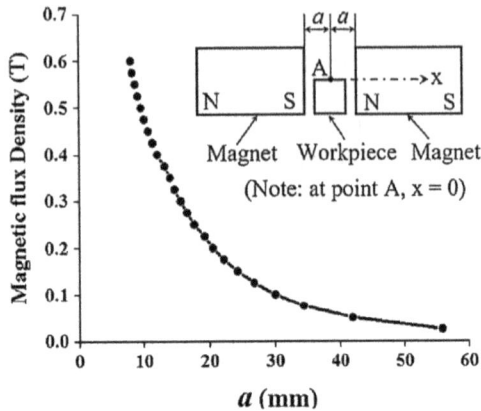

FIGURE 10.4 Variation of MFD with an increase in distance between magnet and workpiece [22].

The distance of each magnet from the center of the workpiece directly influences the magnetic flux density (MFD). As the distance between each magnet and workpiece increases, there is an exponential decrease in the MFD, as shown in Figure 10.4. In the figure, "*a*" represents the distance between a magnet and the center of the workpiece (A) on either side. With an increase in distance (*a*) from 10 mm to 40 mm, a substantial decline in MFD (0.6 T to 0.05 T) is seen. Therefore, for higher machining performance, magnets should be placed in close proximity to the workpiece to take advantage of the maximum MFD. Thus, magnet positioning with respect to the workpiece is of utmost importance for obtaining the maximum advantages from MFAEDM. In addition to the simple configuration whereby two magnets are placed on either side of the workpiece, magnets can also be placed in other configurations, such as four magnets placed at an angle of 90° [21], or a greater number of magnets, as shown in Figure 10.5.

10.4.2 THE INFLUENCE ON THE PLASMA CHANNEL

Besides a magnetic field force acting on the debris particles in the influence of the external magnetic field, plasma parameters which predominantly affect the machining rate in EDM are also modified. Three major aspects of the plasma (plasma parameters) that highly influence the effectiveness of EDM are influenced by the external magnetic field. These aspects are explained below.

10.4.2.1 The Confinement of the Plasma Channel

The confinement of the plasma channel refers to the reduction in radius of the heat source incident on the workpiece electrode. A higher energy density is achievable with the plasma channel compressed to a smaller area, as plasma is the primary heat source in the EDM process. Due to the application of an external electric field, an

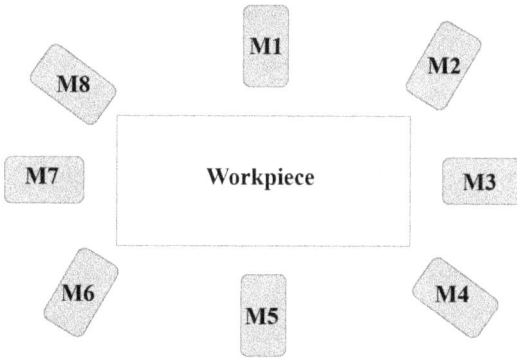

FIGURE 10.5 Installation of magnets around the workpiece.

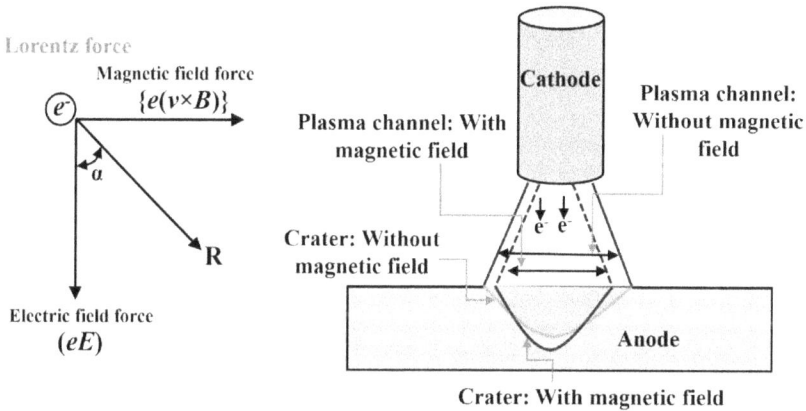

FIGURE 10.6 Schematic of confinement of plasma channel in the existence of the magnetic field (inset: representation of Lorentz force). Adapted from [23].

electric force acts on the charged particles in the plasma channel. In MFAEDM, as the magnetic field is applied (static or pulsating), the magnetic force acts in the transverse direction to the electric field. This force acts radially inward (towards the center of plasma) in the plasma channel to confine it to a smaller radius. The combined effect of electric force and magnetic force, in other words, Lorentz force, minimizes the plasma channel, and thus a smaller diameter crater is formed on the surface [23]. A self-induced Lorentz force also acts on the plasma channel, but it gets nullified as it is present in both the conditions whether or not the magnetic field is functional. Figure 10.6 depicts a representation of the plasma column, the crater formed on the workpiece in traditional EDM (no magnetic field), and MFAEDM. It has also been established that the pulsed magnetic field reduces the losses in magnetic flux associated with a permanent magnet, and is thus found to be more effective in augmenting machining rate, producing better surface integrity and causing insignifi-cant tool wear in the EDM [24].

10.4.2.2 Reduction in the Mean Free Path of Electrons

The mean free path (MFP) of electrons in the plasma channel determines the number of collisions per unit time. It has been established that the MFP of electrons is inversely proportional to the plasma channel pressure. Higher plasma pressure reduces the mean free path of the electrons. As the magnetic field is applied in the EDM process, due to the combined effect of electrostatic and Lorentz forces, the plasma is confined, which in turn increases the plasma pressure. Therefore, the MFP of the electrons in the plasma channel reduces as compared to that without any magnetic field [23, 25]. The MFP of electrons is given as follows [23]:

$$= \frac{RT}{2^{0.5} d^2 NP} \tag{10.1}$$

where, R and T are the gas constant and plasma temperature, respectively. N is Avogadro's number, and P is the plasma pressure. From the equation of the MFP, it is also evident that it is proportional to plasma temperature.

10.4.2.3 Increased Current Density in the IEG

Current density rises as the plasma column is confined, and the MFP of electrons decreases. Current density varies in inverse proportion to that of the plasma channel area, and it increases due to the higher number of collisions (in the lesser MFP). Thus, the combined effect of plasma confinement and the decrease in MFP results in an increased current density. A higher current density creates a crater with a smaller diameter, but of higher depth because energy is concentrated at a smaller point on the workpiece electrode.

It has been observed that there is a significant reduction in crater diameter in the existence of the magnetic field when machining is performed in dry conditions [26]. However, the decline in crater diameter with the magnetic field in the liquid dielectric is relatively small. This can be explained by the fact that in liquid dielec-tric due to higher inertia and viscosity, the plasma channel compresses to a smaller area. When the magnetic field is imposed in it, further confinement of plasma occurs whereas, in the dry EDM process, due to the absence of any high viscosity fluid, the maximum reduction in the plasma channel radius happens solely due to the applied magnetic field. Therefore, the maximum decrease in crater diameter is obtained in dry EDM in the magnetic field. Similarly, the crater depth increases significantly owing to higher energy density caused by the plasma channel con-finement in MFAEDM [26]. Figure 10.7 shows the SEM micrograph of a single crater formed in dry-EDM conditions with and without the magnetic field of MFD 0.1 T. It can be clearly seen from the SEM image, that the crater diameter is suf-ficiently large when there is no magnetic field in both dry and liquid dielectric. However, a slight increase in the crater depth is observed when the magnetic field is added.

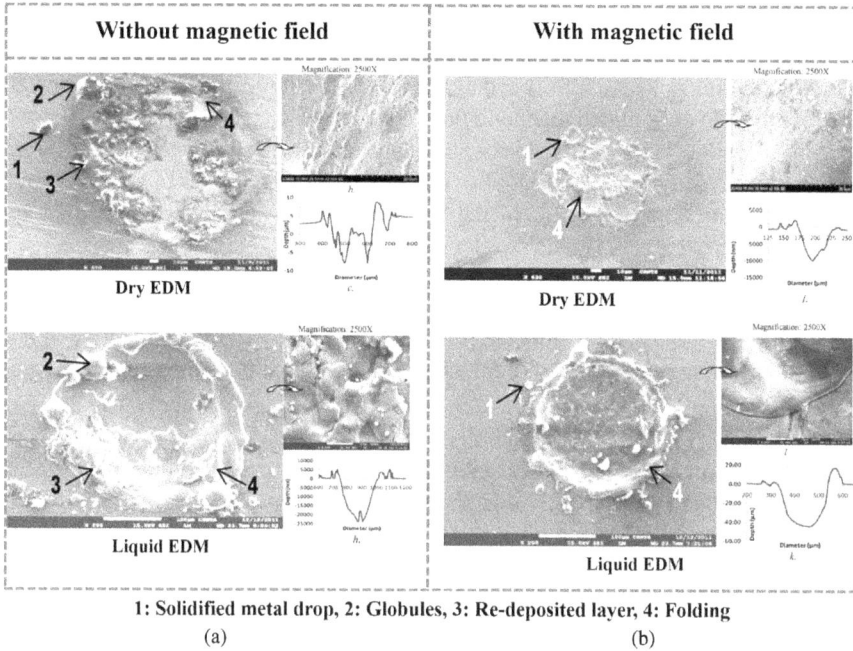

1: Solidified metal drop, 2: Globules, 3: Re-deposited layer, 4: Folding

(a)   (b)

FIGURE 10.7   SEM image of (a) dry-EDM and liquid EDM without a magnetic field (80 V, 20 A, 4 A, 60 μs) and (b) dry-EDM and liquid EDM with magnetic field (80 V, 20 A, 4 A, 60 μs, 0.1 T) [26].

10.4.3   THE INFLUENCE ON NON-MAGNETIC MATERIALS

Inherently non-magnetic materials do not experience any force in the presence of a magnetic field. However, if the non-magnetic workpiece and magnets are oriented to promote a directional current in the workpiece material such that the current and magnetic field are perpendicular to each other, they generate a Lorentz force on the plasma channel and the tool electrode [27]. The Lorentz force has an influence on EDM plasma, as has been discussed in the preceding sections. Due to the confinement of the plasma channel, its pressure increases enormously, which contributes to increasing current density. Moreover, plasma pressure also has a dominant role in determining the ejection of molten metal from the generated crater during the pulse-off period. The ejection of molten metal occurs due to the collapse of the plasma channel or of the vapor bubble encapsulating it, as soon as the energy source is withdrawn. The higher the pressure of plasma channel, the greater is the force of the explosion, and the higher is the ejection of molten metal, thus augmenting removal efficiency by preventing the formation of the recast layer. The plasma channel exerts a force on the melt pool in ejecting the molten liquid metal. Thus, the Lorentz force can confine the plasma channel, to generate higher exploding pressure on the melt pool. Lorentz force is defined as follows [28]:

$$F = J \times B \qquad (10.2)$$

where, J is the current density (A/m²), B is the magnetic field (T), and $\times$ represents the cross-product. The Lorentz force is generated only if both current and magnetic field vectors are not parallel. The magnitude of this force is maximum when the directions of the current and magnetic field are normal to each other [12].

Another variation in MFAEDM for non-magnetic material is to generate a Lorentz force acting directly in the melt pool to eject the molten material. In a typical EDM process, due to flow of current in the plasma channel (J_2) and applied magnetic field (B), a Lorentz force (F_2) is developed along the workpiece surface as shown in Figure 10.8 (a). As soon as current reaches the workpiece, it scatters isotropically, and thus there is no Lorentz force on the melt pool (Figure 10.8 (b)). However, if an additional current (J_1) can be permitted to flow through the workpiece along its surface by arranging a low path electrode, it will induce an additional Lorentz force (F_1), which acts in the direction normal to the workpiece surface (in the melt pool). The resultant (F_R) of both the Lorentz forces, namely, F_1 and F_2 acts on the melt pool as depicted in Figure 10.8 (c), which ejects the extra molten material [28, 29]. An experimentally determined single crater volume was shown to have increased by 28% whereas the

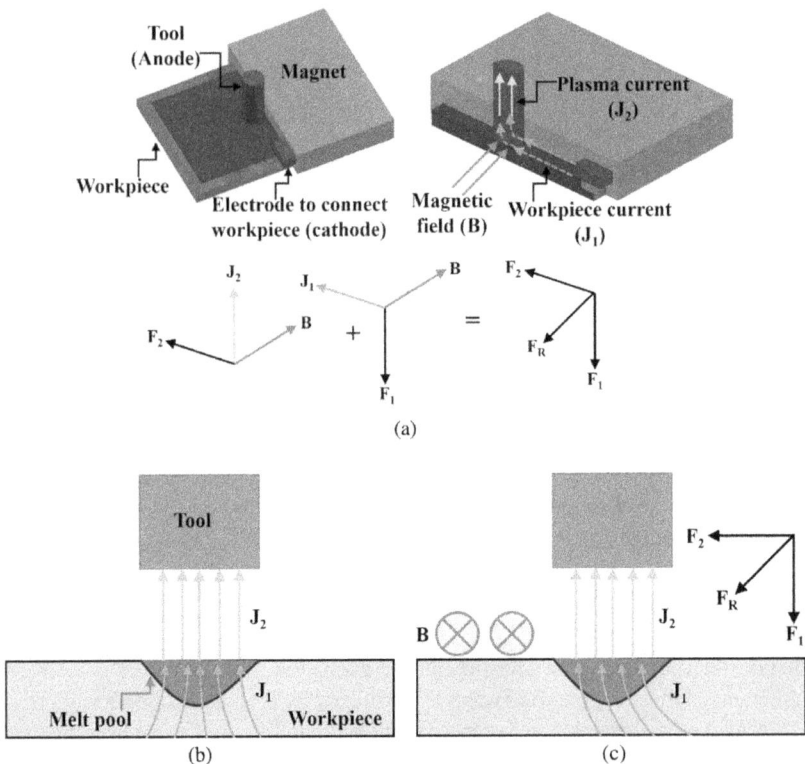

FIGURE 10.8   (a) Schematic diagram showing directional flow of current and different forces acting, (b) isotropic distribution of discharge current in the molten pool, and (c) directional discharge current in the melt pool. Adapted from [28].

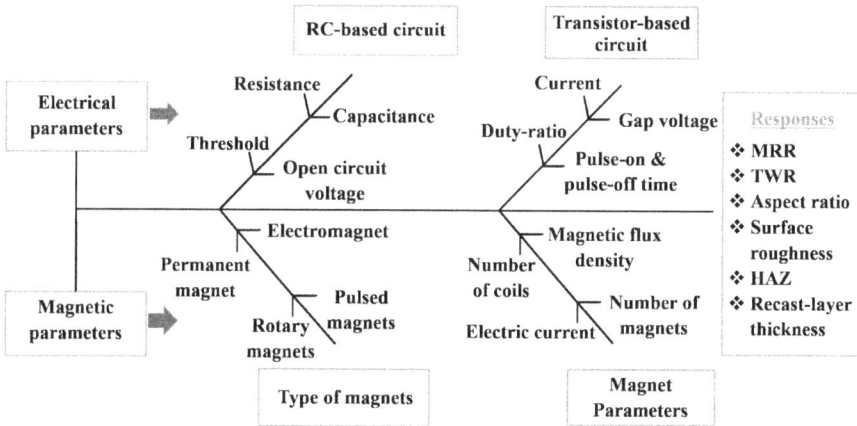

FIGURE 10.9   Fishbone diagram of the MFAEDM process. Adapted from [30].

height of bulge around the crater decreased by 50% when the Lorentz force acted in the melt pool and the magnetic field intensity was increased by 100% (0.33 T to 0.66 T) [28].

10.5   INPUT VARIABLES AND TYPICAL RESPONSES IN MFAEDM

The majority of the parameters related to a typical EDM process and its variants are common to MFAEDM, except for the parameters due to a magnetic field. The common input parameters include electrical parameters based on the resistance-capacitance circuit and transistor-based circuit. Moreover, non-electrical parameters such as parameters related to tool electrode, dielectric, and the like, are not introduced in this chapter. Typical input parameters in MFAEDM due to the addition of the magnetic field are the kinds of magnets being used (permanent, electromagnet, pulsed and rotary magnets). Magnetic flux density, the number of magnets, the magnitude of the electric current and the number of coils in the electromagnets are the major variables which affect the performance of the machining process. The output responses are similar to any of the EDM variants, namely MRR, TWR, the aspect ratio of the features, their surface integrity, and so forth. Figure 10.9 illustrates the cause–and–effect diagram (fishbone diagram) depicting some of the major input and output parameters pertaining to the MFAEDM process.

10.6   THE INFLUENCE OF MAGNETIC FIELD ASSISTANCE ON TYPICAL OUTPUT RESPONSES IN EDM

In the subsequent subsections, the influence of the addition of the magnetic field in the EDM process is discussed. The typical output responses, which collectively determine the productivity, geometric accuracy and surface integrity of the EDM process, are discussed below.

10.6.1 THE MATERIAL REMOVAL RATE

One of the critical output responses, which determines the productivity of the EDM process is the material removal rate (MRR). It is merely the rate at which material is removed from the workpiece (mass, volume, or linear) per unit time. The typical input process variables that influence the MRR include discharge energy per pulse, tool reed-rate, tool rotational speed, workpiece material, and so forth. Besides these controllable variables, debris flushing from the IEG is a factor that largely influences the overall performance of the EDM process. The addition of magnetic field force in the EDM process mainly improves the MRR in comparison with the case where machining is performed without a magnetic field, as given in Figure 10.10 (a). This

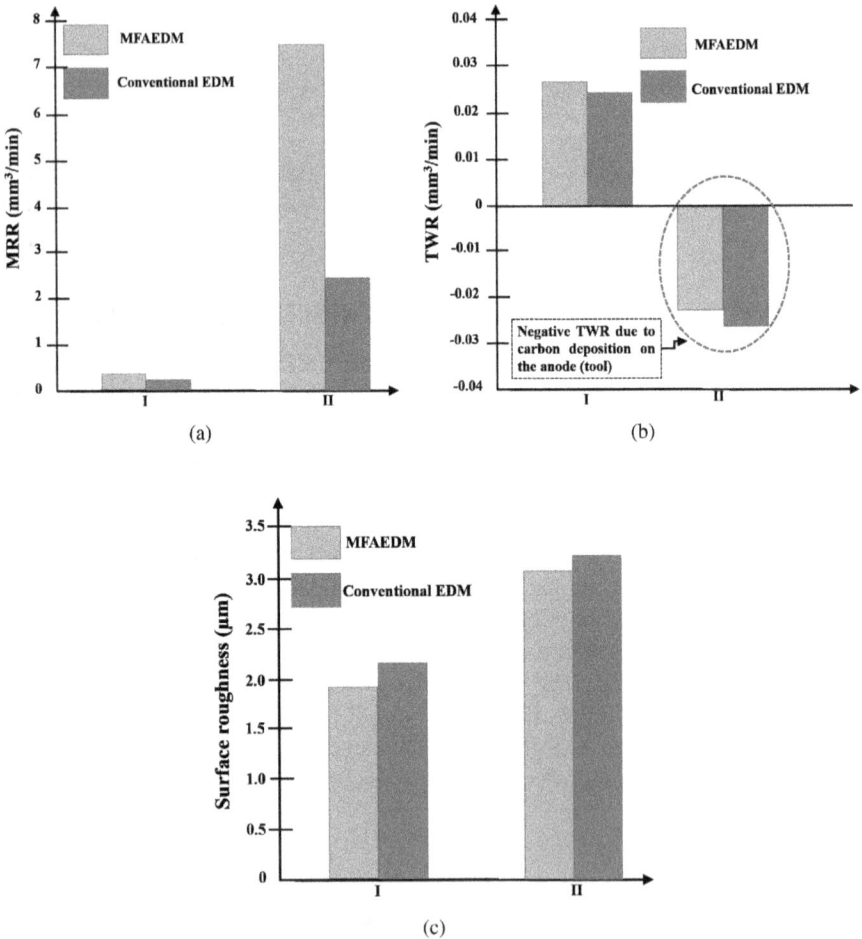

FIGURE 10.10 Comparison of (a) MRR, (b) TWR, and (c) surface roughness (R_a), attained with conventional EDM and MFAEDM (constant parameters: no-load voltage =120V, duty factor = 0.5, polarity: reverse), I (current=5A, pulse-on time= 20µs), II (current=20 A, pulse-on time= 350 µs). Adapted from [31].

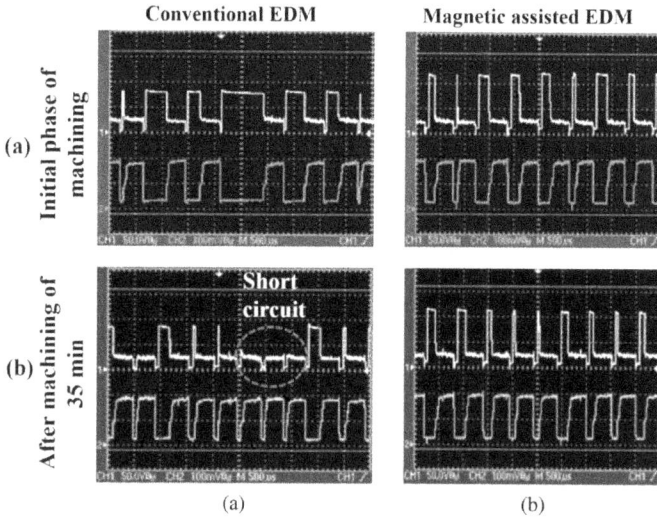

FIGURE 10.11   EDM discharge waveform (a) during the initial phase of machining and (b) after machining of 35 mins [31].

has been attributed to the successful removal of debris particles, which eliminates secondary discharges, arcing and short circuits (responsible for abnormal discharges, thus lowers removal efficiency).

Figure 10.11 represents the discharge waveform captured during EDM machining. At the beginning of machining (Figure 10.11a) the number of short circuits is least, and as such, debris accumulation is not a big problem. However, as machining progresses, it can be clearly seen from the waveform (Figure 10.11b) that the number of short circuits is relatively higher as debris is accumulating in the machining zone due to the higher machining depth. However, the short circuits are reduced in MFAEDM. Thus, it is essential to remove that debris for smooth and efficient machining operations. A similar effect from the magnetic field coupled with electrode rotation, on the mass removal rate (mg/min) has been shown in [32]. Conversely, the MRR has shown a declining tendency with an increase in magnetic flux density. This has been explained due to a change in the path of electrons from linear to cycloid as the Lorentz force is integrated with the electric field. The curvature of the path of the cycloid motion is greater than the path of the linear motion. Thus, the ions are slightly defocussed as the magnetic field increases. For this reason, the MRR decreases with an increase in magnetic flux, but it is still higher than that without a magnetic field. The general conclusion is that the machining rate in MFEDM is relatively higher than with conventional EDM. Theoretically, it has been ascertained that the MRR in MFAEDM is higher than that of conventional EDM, by a factor, which is the ratio of the total energy supplied to the anode due to the electric and magnetic fields to the energy supplied to the anode due to the electric field alone [33]. Plasma flushing efficiency (PFE), which is generally defined as the ratio of the machined volume of the crater to the theoretical volume of the crater, has shown an increasing trend when the magnetic field is applied in EDM [34]. This trend of increasing PFE is observed in almost all energy conditions. Besides higher machining yield, effective removal of debris from the side machining gap also helps in minimizing the radial overcut. Hence, better dimensional accuracy is obtained [35, 36].

10.6.2 The Tool Wear Rate

As the number of normal discharges is higher with a MFAEDM process, similar to the higher removal amount from the workpiece surface, the tool material also erodes at a higher rate. Therefore, a considerable improvement in tool wear rate is difficult to obtain. In some machining conditions (low machining current and short pulse-on time), the tool wear rate is found to be slightly higher using magnetic field-assisted EDM. However, with an increase in discharge current and pulse-on time, the decomposition of hydrocarbon dielectric results in the formation of a carbon layer on the tool (anode) surface (in the case of the reversed polarity of electrodes, the tool is the anode and the workpiece is the cathode), which, owing to high electrical resistivity and boiling temperature, reduces erosion of the tool surface [37]. Thus, negative tool wear (deposition on the tool) is possible in the EDM process. The carbon deposition on the tool surface is affected by the application of the magnetic field as the debris removal agitates the dielectric fluid, which eventually reduces the formation of the carbon layer. Thus, at higher discharge rate, negative tool wear is reduced in MFAEDM, as depicted in Figure 10.10(b) [31]. There are some other experimental works that have reported a slight decrease in relative electrode wear (ratio of the MRR to the TWR) due to a decrease in abnormal discharges [38]. They have used straight polarity of electrodes (with the workpiece as an anode and the tool as a cathode), due to which carbon layer formation did not take place on the tool surface. However, a more conclusive analysis is required in order to assertain the consequences of the magnetic field on tool wear behaviour under a magnetic field.

10.6.3 The Surface Roughness and Surface Integrity

The surface roughness of the electric discharge machined (EDMed) surface, machined under a magnetic field is found to be smaller than that obtained without a magnetic field. Adhesion of debris particles on the machined surface deteriorates its surface finish, which, due to the application of the magnetic field, is reduced to a greater extent. Thus, it is evident that a smoother surface is possible in magnetic field-assisted EDM, which is shown in Figure 10.10(c) [31]. Interestingly, the hardened debris particles also remove the peaks on the workpiece, similar to the magnetic abrasive finishing process, which assists in reducing the surface roughness of the component [32]. Figure 10.12 shows the SEM image of the machined surface in conventional EDM and EDM machining under a magnetic field (MFAEDM). It can be seen that debris is attached to the surface. Micro-cracks are also present on the surface in conventional EDM. However, the number of micro-cracks is minimized, probably due to the absence of debris particles. Thus, improved surface integrity with the application of a magnetic field in the EDM process has been reported [39].

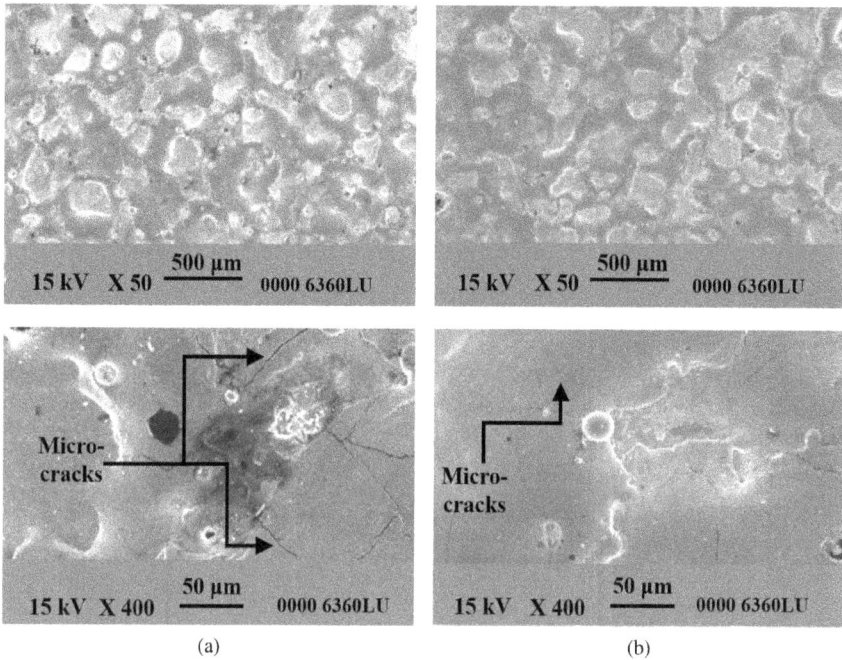

FIGURE 10.12 SEM image of the machined surface: (a) conventional EDM and (b) magnetic force-assisted EDM [31].

The recast layer is formed on the machined surface due to redeposition of molten liquid. It mostly depends on the extent of the molten metal ejection. If the molten metal is not effectively removed, it resolidifies on the machined surface due to the cooling effect of the liquid dielectric. It has been discussed earlier that as plasma pressure increases due to compressed plasma in MFAEDM (which, on explosion generates a larger force on the melt pool), this results in the efficient evacuation of molten metal, thus preventing/minimizing the creation of the recast layer, as shown in the SEM image in Figure 10.13 [31, 40]. In addition to the higher pressure of the plasma channel, debris is not allowed to adhere to the machined surface in the existence of a magnetic field, and it reduces the thickness of the recast layer [41].

10.6.4 THE ASPECT RATIO OF THE FEATURE

A significant increase in the aspect ratio of the feature (defined as the ratio of the depth of the feature to its width or hole depth to its diameter) has been reported by implementing a magnetic field force in the EDM process. It is explained by the fact that the machining progresses as long as the machining products, namely, the debris is expelled effectively from the narrow IEG, and fresh dielectric is supplied to the IEG for subsequent sparks. As soon as the gap bridges due to poor debris removal, the tool encounters short-circuits, requiring frequent retraction and in-feed of the tool, which

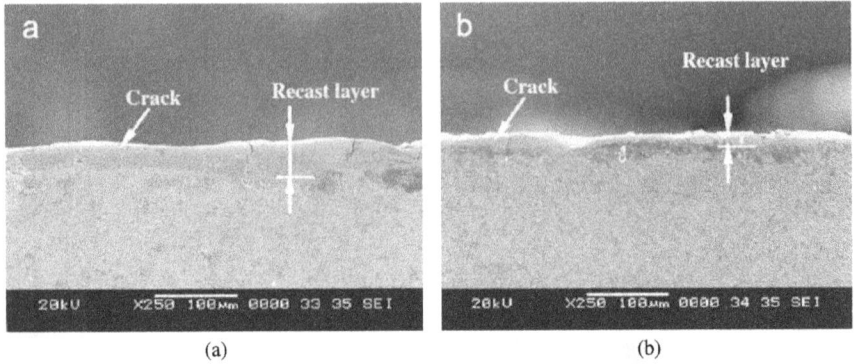

FIGURE 10.13 Micro-cracks and recast layer under (a) conventional EDM and (b) magnetic force-assisted EDM [31].

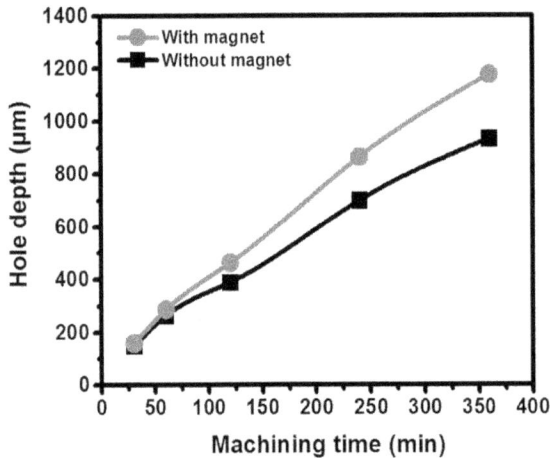

FIGURE 10.14 Comparative assessment of hole depth in the magnetic field and without magnetic field. Adapted from [12].

suppresses machining rate and if short circuits persist, further machining operation stops. Thus, there is a limit to the maximum depth up to which the machining process occurs successfully. Figure 10.14 depicts a comparative assessment of hole depth at different machining time with and without a magnetic field [12]. It is evident that a relatively higher hole depth is obtained at almost all machining times.

10.7 MAGNETIC FIELD ASSISTANCE IN EDM VARIANTS

A brief introduction is presented in the subsequent sub-sections pertaining to the influence of magnetic field assistance on the variants of the EDM process. Three major variants, namely, ED-drilling, wire-EDM, and ED-Milling, are considered.

10.7.1   ELECTRIC DISCHARGE DRILLING AND DIE-SINKING EDM

Electric discharge drilling (ED-drilling) unarguably is one of the most commonly used variants of the EDM process for the fabrication of micro-sized (macro also) blind as well as through cut holes. These holes have specific utility in various applications, namely, cooling holes in turbine blades, holes for inkjet printer heads, micro-sized holes for the fuel injectors in diesel engines, and an array of micro-holes in the filter and textile industries [42]. A cylindrical electrode rotating about its axis is generally used as a tool electrode whereby the rotational motion is the primary means of the expulsion of debris due to centrifugal action and agitation in the dielectric fluid. However, for a high aspect ratio (the ratio of the depth of a hole to its diameter), the rotational motion of the tool alone becomes incapable of ejecting the debris particles. As machining depth keeps increasing, the amount of debris in the frontal IEG (between tool end face and workpiece) and the side IEG (between the cylindrical surface of the tool and the workpiece) also increases. The increased concentration of debris at higher depth creates frequent short circuits, which continuously cause the tool to retract from the machining direction. This accumulated debris either creates short circuits in the IEG or generates secondary or higher level discharges, which are less effective. Therefore, the machining efficiency decreases drastically for high aspect ratio machining. Magnetic-field assistance, in addition to high-speed tool rotation, enhances the debris flushing rate and thus reduces the likelihoods of short-circuits and abnormal discharges. The depth of hole that can be machined, increases substantially in magnetic-assisted EDM. Die-sinking EDM utilizes a formed tool to generate a reverse replica of the tool on the workpiece whereby rotational motion is not provided. Thus, the flushing becomes more critical in die-sinking EDM than that in ED-drilling (tool rotation in ED-drilling assists in dielectric circulation and debris ejection). Therefore, MFAEDM can also be integrated with the die-sinking EDM process.

10.7.2   WIRE-ELECTRIC DISCHARGE MACHINING

Wire-electric discharge machining (wire-EDM) is the most versatile variation of the EDM process, which utilizes a wire of size ranging from 30–300 µm as a tool electrode to cut materials, fabricate micro-tools of different cross-sections and to manufacture various complex micro/macro components. The wire-EDM process is characterized by the width of machining (kerf-loss) as it determines the accuracy of the process. Kerf-loss is the loss of material due to the cutting action. It includes the wire diameter and the radial overcut that is generated between the wire and the workpiece on either side of the wire [43]. This kerf is unwanted as it reduces the machining accuracy but, at the same time, it is necessary for the flushing of fresh dielectric and debris from the machining zone [44]. Discharges in the secondary spark gap (between the wire electrode and the workpiece in the direction normal to the wire motion) which are observed due to the presence of machining debris, increases the side gap and kerf-loss. Moreover, the presence of debris in the frontal gap retracts the wire electrode, which contributes to non-productive time, and thus lowering the cutting rate. Magnetic field integrated in the wire-EDM setup can significantly reduce the chances of debris retention in the frontal and secondary discharge gap. This eventually results in higher machining accuracy and enhanced productivity associated with the wire-EDM process. Conversely, other parameters such as energy consumption for cutting and thermal deformation in the machined features are observed to be reduced significantly as magnetic field force confines the plasma channel to increase the plasma stability and increases electron density [45].

10.7.3 ELECTRIC DISCHARGE MILLING

Electric discharge milling (ED-Milling) utilizes the simultaneous motion of a simple cylindrical tool in three directions in addition to its rotation about its own axis to fabricate 3-D micro/macro features [30]. Thus, ED-Milling eliminates the need for a preformed tool for 3-D machining. The process has some distinctive limitations such as low MRR, wear of the tool electrode on the end and on cylindrical surfaces, and on the taper of machined features. Although dimensional accuracy (taper) is a function of tool wear in ED-Milling, a low machining rate is attributed to poor flushing of the dielectric fluid and debris. The problem is somewhat similar to the wire-EDM process. However, unlike the wire-EDM process whereby the debris can be expelled from the machining zone lead by the gravitational force as it is a through cut machining operation, the problem of debris accumulation is severe in the ED-Milling operation when it is performed for a blind cut (blind channel, blind pocket). There is a possibility of bringing productivity augmentation to the ED-milling via the application of magnetic field assistance. A bulk machining approach whereby the intended depth of the feature is machined in a single pass has a higher rate of machining, but the depth of machining is restricted by the flushing of debris and side wear of the tool electrode. Simultaneous motion of the tool electrode in three directions makes it difficult to induce a magnetic force in the current machining location. This can be resolved by providing motion to the magnet fixture along with tool motion.

10.8 LIMITATIONS

Although magnetic field assistance has been instrumental in bringing productivity enhancement to the EDM, it is primarily restricted to magnetic materials for debris removal. However, aspects of EDM plasma are influenced by both magnetic and non-magnetic materials. The modified plasma pressure results in increased machining rate due to higher current density, but it does not exert any force on the debris particles. Thus, debris accumulation in the IEG and its detrimental effect on machining stability (in terms of the nature of the discharge) remains problematic for non-magnetic materials. When the magnetic field is coupled in powder mixed dielectric in the EDM, additional filtration provision is required to segregate the ferromagnetic powder particles; otherwise, the powder particles concentration would change as machining progresses.

10.9 RESEARCH OPPORTUNITIES

Despite the fact that MFAEDM is a promising hybridization of the EDM process for overall performance augmentation in EDM, the number of research papers published in the domain of MFAEDM is relatively less. This leads to possibilities for extensive research activity into further development of the process. Some of the research opportunities that can be accomplished in the realm of MFAEDM are as follows:

1. Magnetic field assistance has not been explored much in relation to the variants of EDM, such as ED-drilling, wire-EDM, ED-Milling, and so forth.
2. A varying magnetic field such as electromagnets can be implemented to develop adaptive machining whereby the intensity of the magnetic field can be varied according to machining conditions. At higher machining depth, debris removal becomes indispensable as compared to the initial machining, which demands higher magnetic field force at higher machining depth. This is possible with a varying magnetic field.
3. Numerical simulation and theoretical understanding can strengthen the understanding of plasma confinement, of melt pool hydrodynamics, and of the crater formed in MFAEDM.
4. A conclusive analysis is required in order to understand the effect of the magnetic field on tool wear in the EDM process. Due to a reduction in abnormal discharges as machining rate increases, a similar effect can be predicted for tool material wear (higher tool wear). However, owing to a decreased in the number of irregular discharges and short circuit, reduction in tool wear has also been reported. Thus, it is essential to comprehend the role of the magnetic field on tool electrode wear in EDM.

10.10   CONCLUSION

MFAEDM, a well-established hybridization of the EDM process, has been instrumental in achieving productivity enhancement and bringing stability in EDM. This has been possible via the improved expulsion of machined debris particles and by modifying the plasma parameters, which eventually leads to enhanced machining efficiency. There have been a limited number of experimental, theoretical, and optimization studies pertaining to MFAEDM, which have concluded that a significant improvement in machining efficiency under the stimulus of the magnetic field is possible. However, the influence of a magnetic field in the variants of the EDM process, namely die-sinking EDM, wire-EDM, ED-Milling, and microtool fabrication techniques, has to be analyzed prior to the widespread utilization of the MFAEDM. Adaptive control of magnetic field intensity with respect to machining depth can be a useful variable in the currently used constant magnetic field, which can be influential in high aspect ratio machining.

ACKNOWLEDGMENTS

Many thanks to Mr. Pawan Singh (B. Tech, Institute of Technology Gopeshwar, Chamoli, Uttarakhand India) for extending help in drawing the figures.

ACRONYMS

EDM: Electric discharge machining
MFA: Magnetic field assistance
MFAEDM: Magnetic field-assisted EDM
EDMM: Electric discharge micromachining
IEG: Inter-electrode gap
MFD: Magnetic flux density
MFP: Mean free path
MRR: Material removal rate
TWR: Tool wear rate
REW: Relative electrode wear
μ: Micro
ED-drilling: Electric discharge drilling
ED-milling: Electric discharge milling
Wire-EDM: Wire-electric discharge machining

REFERENCES

[1] Ho KH, Newman ST (2003) State of the art electrical discharge machining (EDM). Int J Mach Tools Manuf 43:1287–1300. https://doi.org/10.1016/S0890-6955(03)00162-7

[2] Singh M, Saxena P, Ramkumar J, Rao R V. (2019) Multi-spark numerical simulation of the micro-EDM process: an extension of a single-spark numerical study. Int J Adv Manuf Technol 108:2701–2715. https://doi.org/https://doi.org/10.1007/s00170-020-05566-6

[3] Huang R, Yi Y, Guo G, Xiong X (2020) Investigation of multielectrode multiloop with series capacitance pulse generator for EDM. Int J Adv Manuf Technol 109:143–154. https://doi.org/10.1007/s00170-020-05652-9

[4] Kunieda M, Lauwers B, Rajurkar KP, Schumacher BM (2005) Advancing EDM through Fundamental Insight into the Process. CIRP Ann. – Manuf. Technol. 54:64–87.

[5] Rajurkar KP, Pandit SM (1986) Formation and ejection of ed debris. J Manuf Sci Eng Trans ASME 108:22–26. https://doi.org/10.1115/1.3187036

[6] Roy T, Datta D, Balasubramaniam R (2020) Debris based discharge segregation in reverse micro EDM. Measurement 153. https://doi.org/10.1016/j.measurement.2019.107433

[7] Huang H, Zhang H, Zhou L, Zheng HY (2003) Ultrasonic vibration assisted electro-discharge machining of microholes in Nitinol. J Micromechanics Microengineering 13:693–700. https://doi.org/10.1088/0960-1317/13/5/322

[8] Bamberg E, Heamawatanachai S (2009) Orbital electrode actuation to improve efficiency of drilling micro-holes by micro-EDM. J Mater Process Technol 209:1826–1834. https://doi.org/10.1016/j.jmatprotec.2008.04.044

[9] Kumar R, Singh I (2018) Productivity improvement of micro EDM process by improvised tool. Precis Eng 51:529–535

[10] Murali M, Yeo SH (2004) A novel spark erosion technique for the fabrication of high aspect ratio micro-grooves. Microsyst Technol 10:628–632. https://doi.org/10.1007/s00542-003-0341-8

[11] Marashi H, Jafarlou DM, Sarhan AAD, Hamdi M (2016) State of the art in powder mixed dielectric for EDM applications. Precis Eng 46:11–33. https://doi.org/10.1016/j.precisioneng.2016.05.010

[12] Yeo SH, Murali M, Cheah HT (2004) Magnetic field assisted micro electro-discharge machining. J Micromechanics Microengineering 14:1526–1529. https://doi.org/10.1088/0960-1317/14/11/013

[13] Mujumdar SS, Curreli D, Kapoor SG, Ruzic D (2015) Modeling of Melt pool Formation and Material Removal in Micro-Electrodischarge Machining. J Manuf Sci Eng 137:0310071–9. https://doi.org/10.1142/S0217979212440079

[14] Erden A (1983) ROLE OF DIELECTRIC FLUSIDNG ON ELECTRIC DISCHARGE MACHINING. Proc Twenty-third Int Mach Tool Des Res Conf 283–284.

[15] Wang AC, Yan BH, Tang YX, Huang FY (2005) The feasibility study on a fabricated micro slit die using micro EDM. Int J Adv Manuf Technol 25:10–16. https://doi.org/10.1007/s00170-003-1831-7

[16] Masuzawa T, Cui X, Taniguchi N (1992) Improved Jet Flushing for EDM. CIRP Ann - Manuf Technol 41 :239–242. https://doi.org/10.1016/S0007-8506(07)61194-9

[17] Singh P, Yadava V, Narayan A (2018) Parametric study of ultrasonic-assisted hole sinking micro-EDM of titanium alloy. Int J Adv Manuf Technol 94:2551–2562. https://doi.org/10.1007/s00170-017-1051-1

[18] Flaño O, Ayesta I, Izquierdo B, et al (2018) Improvement of EDM performance in high-aspect ratio slot machining using multi-holed electrodes. Precis Eng 51:223–231.

[19] Plaza S, Sanchez JA, Perez E, et al (2014) Experimental study on micro EDM-drilling of Ti6Al4V using helical electrode. Precis Eng 38(4):821–827.

[20] Zhou T, Zhou C, Liang Z, Wang X (2018) Machining mechanism in tilt electrical discharge milling for lens mold. Int J Adv Manuf Technol 95:2747–2755. https://doi.org/10.1007/s00170-017-1408-5

[21] Kesava Reddy C, Manzoor Hussain M, Satyanarayana S, Murali Krishna MVS (2018) Experimental Investigation – Magnetic Assisted Electro Discharge Machining. IOP Conf Ser Mater Sci Eng 346 . https://doi.org/10.1088/1757-899X/346/1/012062

[22] Beravala H, Pandey PM (2018) Experimental investigations to evaluate the effect of magnetic field on the performance of air and argon gas assisted EDM processes. J Manuf Process 34:356–373. https://doi.org/10.1016/j.jmapro.2018.06.026

[23] Gupta A, Joshi SS (2017) Modelling effect of magnetic field on material removal in dry electrical discharge machining. Plasma Sci Technol 19:1–10. https://doi.org/10.1088/2058-6272/19/2/025505

[24] Joshi S, Govindan P, Malshe A, Rajurkar K (2011) Experimental characterization of dry EDM performed in a pulsating magnetic field. CIRP Ann - Manuf Technol 60:239–242. https://doi.org/10.1016/j.cirp.2011.03.114

[25] Rouniyar AK, Shandilya P (2019) Fabrication and experimental investigation of magnetic field assisted powder mixed electrical discharge machining on machining of aluminum 6061 alloy. Proc Inst Mech Eng Part B J Eng Manuf 233:2283–2291. https://doi.org/10.1177/0954405419838954

[26] Govindan P, Gupta A, Joshi SS, et al (2013) Single-spark analysis of removal phenomenon in magnetic field assisted dry EDM. J Mater Process Technol 213:1048–1058. https://doi.org/10.1016/j.jmatprotec.2013.01.016

[27] Zhang Z, Huang H, Ming W, et al (2016) Study on machining characteristics of WEDM with ultrasonic vibration and magnetic field assisted techniques. J Mater Process Technol 234:342–352. https://doi.org/10.1016/j.jmatprotec.2016.04.007

[28] Heinz K, Kapoor SG, DeVor RE, Surla V (2011) An investigation of magnetic-field-assisted material removal in micro-EDM for non-magnetic materials. J Manuf Sci Eng Trans ASME 133:1–9. https://doi.org/10.1115/1.4003488

[29] Asad A, Yu Z, Zou R, Zhang C (2020) Magnetic Field Assisted Micro EDM in Nitrogen Plasma Jet and HVAJ. Int J Mater Mech Manuf 8:27–33. https://doi.org/10.18178/ijmmm.2020.8.1.479

[30] Singh M, V. K. Jain, Ramkumar J (2019) Micro-electrical Discharge Milling Operation. In: Golam Kibria, Muhammad P. Jahan, Bhattacharyya B (eds) Micro-electrical Discharge Machining Processes Technologies and Applications. Springer Nature Singapore Pte Ltd., pp. 23–51.

[31] Lin YC, Lee HS (2008) Machining characteristics of magnetic force-assisted EDM. Int J Mach Tools Manuf 48:1179–1186. https://doi.org/10.1016/j.ijmachtools.2008.04.004

[32] Teimouri R, Baseri H (2012) Effects of magnetic field and rotary tool on EDM performance. J Manuf Process 14:316–322. https://doi.org/10.1016/j.jmapro.2012.04.002

[33] Beravala H, Pandey PM (2020) Modelling of material removal rate in the magnetic field and air-assisted electrical discharge machining. Proc Inst Mech Eng Part C J Mech Eng Sci 234:1286–1297. https://doi.org/10.1177/0954406219892297

[34] Shabgard MR, Gholipoor A, Mohammadpourfard M (2019) Investigating the effects of external magnetic field on machining characteristics of electrical discharge machining process, numerically and experimentally. Int J Adv Manuf Technol 102:55–65. https://doi.org/10.1007/s00170-018-3167-3

[35] Renjith R, Paul L (2019) Machining characteristics of micro-magnetic field assisted EDM (μ-MFAEDM). Mater Today Proc. 27(3):2000–2004. https://doi.org/10.1016/j.matpr.2019.09.047

[36] Mukhopadhyay P, Adhikary S, Samanta AK, et al (2019) External force assisted electro discharge machining of SS 316. Mater Today Proc 19 :626–629. https://doi.org/10.1016/j.matpr.2019.07.743

[37] Xia H, Kunieda M, Nishiwaki N (1993) Removal Amount Difference between Anode Cathode in EDM Process. Int J Electr Mach 1:45–52.

[38] Lin YC, Chen YF, Wang DA, Lee HS (2009) Optimization of machining parameters in magnetic force assisted EDM based on Taguchi method. J Mater Process Technol 209:3374–3383. https://doi.org/10.1016/j.jmatprotec.2008.07.052

[39] Azhiri RB, Jadidi A, Teimouri R (2020) Electrical discharge turning by assistance of external magnetic field, part II: Study of surface integrity. Int J Light Mater Manuf 3:305–315.

[40] Lin YC, Lee HS (2009) Optimization of machining parameters using magnetic-force-assisted EDM based on gray relational analysis. Int J Adv Manuf Technol 42:1052–1064. https://doi.org/10.1007/s00170-008-1662-7

[41] Bains PS, Sidhu SS, Payal HS (2018) Magnetic Field Assisted EDM: New Horizons for Improved Surface Properties. Silicon 10:1275–1282. https://doi.org/10.1007/s12633-017-9600-7

[42] Zhang L, Tong H, Li Y (2015) Precision machining of micro tool electrodes in micro EDM for drilling array micro holes. Precis Eng 39:100–106. https://doi.org/10.1016/j.precisioneng.2014.07.010

[43] Singh M, Ramkumar J, Rao RV, Balic J (2019) Experimental investigation and multi-objective optimization of micro-wire electrical discharge machining of a titanium alloy using Jaya algorithm. Adv Prod Eng Manag 14:251–263.

[44] Singh M, Singh A, Ramkumar J (2019) Thin-wall micromachining of Ti – 6Al – 4V using micro-wire electrical discharge machining process. J Brazilian Soc Mech Sci Eng 41:338:1–12

[45] Zhang Y, Zhang Z, Zhang G, Li W (2020) Reduction of Energy Consumption and Thermal Deformation in WEDM by Magnetic Field Assisted Technology. Int J Precis Eng Manuf Green Technol 7:391–404. https://doi.org/10.1007/s40684-019-00086-5

11 Sequential Laser and Electrical Discharge Machining

P. S. Suvin and *Ranjeet Kumar Sahu*

Department of Mechanical Engineering, National Institute of Technology Karnataka, Mangalore, India

*Corresponding author

CONTENTS

11.1   INTRODUCTION

The conventional machining processes have met the requirements of industry over the decades. However, to become successful in today's global market, the industry must meet the various customer demands arising in the market. In particular, there is a demand for the processing of exotic materials (e.g., nitroalloy, hastalloy, waspalloy, nimonics, carbides, ceramics, and the like.) to obtain features of desired shape and size, for the machining of intricate 3D shapes, for ultrafine surface finish products, for miniaturized products to increase the overall efficiency of various systems and for high quality products. All of this is needed with less cost, increased productivity and further reductions in the wastage of material. However, conventional machining techniques can hardly meet these demands. Industrial researchers have made continuous efforts to advance machining techniques. Eventually, to meet the above demands, advanced machining techniques emerged as efficient and economic alternatives to conventional techniques.

DOI: 10.1201/9781003202301-11

Advanced machining techniques are those techniques that are used to create features or products by removing material, usually from the workpieces irrespective of their sizes using precisely controlled energy sources. The energy sources can be either in the form of mechanical/thermal/chemical or electrochemical sources. The principle of material removal in mechanical energy based advanced machining techniques is the erosion of the workpiece mechanically. In thermal techniques, the mechanism of material removal is melting and evaporation using repetitive thermal energy discharges, usually on the workpiece. The principle of material removal in chemical/electrochemical energy based advanced machining techniques is ionic displacement.

The capabilities of advanced machining techniques are found to be boundless in comparison with conventional techniques. Advanced machining techniques are non-contact in nature and involve the machining of complex shapes on different conducting and non-conducting materials with no barrier to their hardness and fragility. The precise shaping of super alloys, carbides, ceramics, and the like, with better quality and less cost could be achieved using these techniques. Nowadays, since these techniques are automatically controlled, they became simple, and exhibit high reliability and repeatability. In order to accomplish better output results, various process parameters can be adjusted. For these reasons, the demand for advanced machining techniques has become increasingly strong and is widely recognized across the global manufacturing industry. Each of these techniques has its own advantages and limitations and one process cannot substitute for the other. The advanced machining techniques differ from each other in their characteristic features, operation and fields of application.

Moreover, in recent times, attention has been paid by researchers on hybrid advanced machining techniques. With growing trend towards the arrival of exotic materials with exceptional characteristics and with the need for ever improving accuracy in the machining of intricate parts of definite shape, hybrid advanced machining techniques (HAMTs) have become increasingly important. HAMTs deal with the integration of various energy source based advanced machining techniques or with supporting a particular process using process energy. In HAMTs, various techniques are integrated within a single machining platform [1-5]. Table 11.1 shows the classification of single energy source based advanced machining techniques as well as HAMTs and their mechanisms for material removal [1-7].

Among HAMTs, the integration of electrical discharge machining (EDM) and laser beam machining (LBM) based on thermoelectric and radiation sources of energy has become considerable important for complexity of workpiece shape, size and surface integrity. The application of these combined energy sources in developing advanced machining techniques has greatly helped in achieving the economically viable machining of exotic materials and has satisfied other additional needs. Prior to the discussion of the assistance of laser process energy to EDM which is responsible for material removal, it is imperative to realize the conceptual overview of EDM and LBM.

TABLE 11.1

Classification of advanced machining techniques as well as HAMTs

Energy type	Mechanisms of material removal	Energy source	Process	Applications
Mechanical	Mechanical erosion of workpiece	Mechanical motion	USM	Round and irregular holes, impressions, etc.
		Pneumatic	AJM	Drilling, cutting, deburring, etc.
		Hydraulic	WJM	Paint removal, cleaning, cutting frozen meat, etc.
		Hydraulic	AWJM	Peening, cutting, textile, leather industry, etc.
Thermal	Melting and evaporation of workpiece	Electric spark	EDM	Holes in nozzles and catheters, channel cutting, curved surfaces, etc.
		High speed electrons	EBM	Drilling fine holes, cutting contours in sheets, etc.
		Powerful radiation	LBM	Drilling fine holes, cladding, etc.
		Ionized substance	IBM	3D patterning in IC circuits, micro tools and micro dies fabrication, etc.
		Ionized substance	PAM	Cutting plates, etc.
Chemical	Corrosive reaction	Corrosive agent	CHM	Pockets, contours, MEMS, etc.
Electro-chemical	Ion displacement	Electric current	ECM	Blind holes, cavities, etc.
Hybrid Machining Processes				
Electro-chemical and thermal	Melting and evaporation, and chemical etching	Electrical discharges	ECSM	Holes, grooves, channels, complex shape contours, etc.
Chemical and mechanical	Chemical reaction and mechanical abrasion	Abrasive slurry with chemicals	CMP	Polishing of Au and Ti, optoelectronic components, etc.
Electro-chemical and mechanical	Electrolytic reaction and grinding action	Electric current and mechanical motion	ELID	Small hole grinding, fine finish of hard and brittle materials, etc.
Thermal	Ablation	Powerful radiation and electric spark	LAEDM	High quality drilled holes and channels, etc.
USM	Ultrasonic Machining	IBM		Ion Beam Machining

TABLE 11.1   (Continued)
Classification of advanced machining techniques as well as HAMTs

Energy type	Mechanisms of material removal	Energy source	Process Applications
AJM	Abrasive Jet Machining	PAM	Plasma Arc Machining
WJM	Water Jet Machining	CHM	Chemical Machining
AWJM	Abrasive Water Jet Machining	ECM	Electro Chemical Machining
EDM	Electrical Discharge Machining ECSM		Electro Chemical Spark Machining
EBM	Electron Beam Machining	CMP	Chemo-Mechanical Polishing
LBM	Laser Beam Machining	ELID	Electrolytic In-process Dressing
LAEDM	Laser-assisted Electrical Discharge Machining		

11.2   THE CONCEPTUAL REALIZATION OF ELECTRICAL DISCHARGE MACHINING

Electrical discharge machining (EDM) is a thermo-electric machining process which erodes material from the workpiece (usually the anode) and tool (cathode) using repetitive spark discharges between the two electrodes immersed in a dielectric medium. The high frequency sparks remove the material from both the electrodes efficiently in the form of debris by melting and evaporation. It can be considered as machining of the components with no constraint on their size, creating features at macro/micro level. The mechanism of the EDM technique is discussed in various literature.

EDM is the most promising and cost-effective advanced machining technique owing to the contactless nature of the machining. The technique involves low input energy resulting in a lower unit removal of material. It has the ability to produce complex features on conducting and semiconducting materials at a relatively low cost and taking relatively less time with a high degree of accuracy and minimizing damage to the properties of the material. Moreover, this technique is viable for the machining of ceramic materials due to its ability to produce flexible machined ceramic products with high accuracy. In EDM, the spark gap must be maintained constant at the required position in order to ensure stable machining and better accuracy. The greater the uniformity of the spark gap, the better the machining accuracy, and thus the accuracy in the dimensional features. This was achieved using gap controlled tool feed systems. The gap control tool feed systems used in the EDM process include servo DC motors, stepper motors and other electro-mechanical servo drives. Further, EDM has found widespread applications in various industrial sectors, such as in the production of fuel injection automotive nozzles, mold and die manufacturing, small and burr-free hole drilling in turbine blades, holes in filters for food processing and textile industries, holes for catheters, needles and other medical device components, spinneret holes for synthetic fibers, making spool valves and keyways for cylinder, channels in nuclear reactors and electronic components, and so forth.

11.3 THE CONCEPTUAL REALIZATION OF LASER-BEAM MACHINING

Laser is an acronym for light amplification by stimulated emission of radiation. Using an optical lens, the laser beam can be focused. Further, this focused laser, owing to its high-power density in excess of $1MW/mm^2$ can be used for the removal of material through ablation. Laser power density will be of the order of $\sim 10^7$ W/cm^2. Laser-beam machining uses a high heat energy coherent beam of light to remove a tiny bit of material in the form of debris from any metallic/non-metallic components through melting and vaporization. LBM is mainly used for the machining of holes in a precision manner.

There are a variety of lasers used in machining including CO_2 Laser, Nd:YAG Lasers, Diode Lasers, Excimer Lasers and Fiber Lasers. The CO_2 laser is a gas laser which has a pulsed or continuous wave type mode of operation. It emits light in the infrared region. The Nd:YAG laser is short for the Neodymium ion Nd^{3+} doped Yttrium Aluminum Garnet laser. This laser has a pulsed or continuous wave type mode of operation and exhibits a high divergence beam quality. The pulsed mode can be accomplished by exciting either Xenon flash lamp pumped or diode pumped optical lasers. The efficiency for lamp pumped lasers is in the range 2-5% and for diode pumped lasers, 15-20%.

The schematic for laser beam machining is given in Figure 11.1. One end has a partially reflective mirror and another end has a fully reflective mirror. Photons coming out will create an intense laser beam, which is coherent and highly directional. The

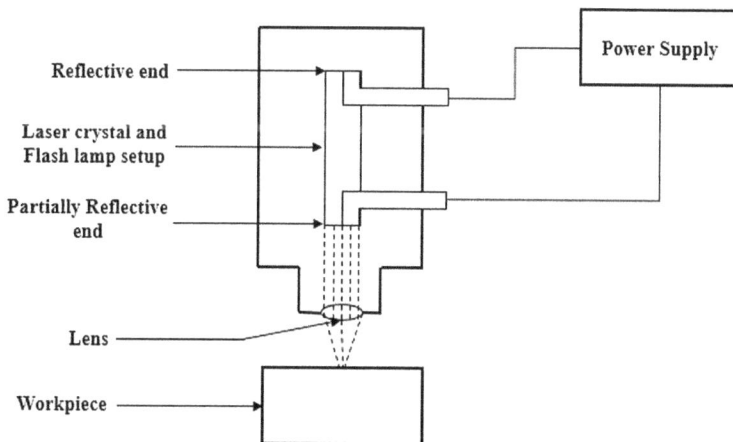

FIGURE 11.1   Schematic of a solid state laser.

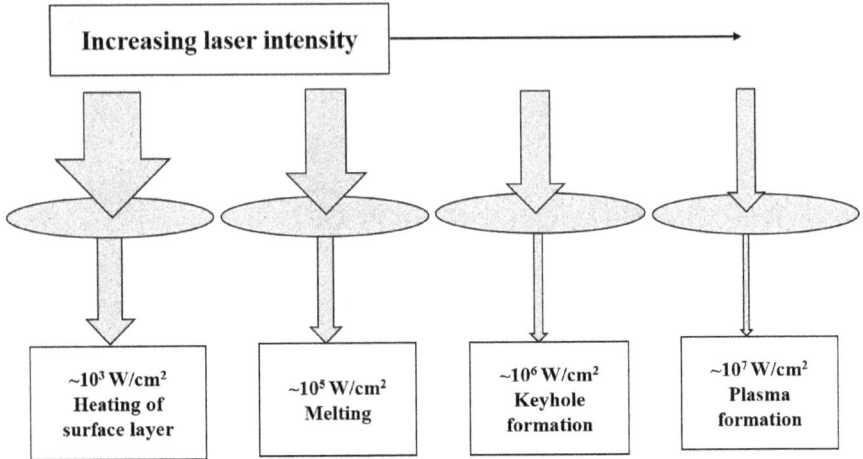

FIGURE 11.2   Physical phenomena at increasing laser intensity.

laser beam is reflected, scattered, absorbed and transmitted in a material. Laser radiation of a certain wavelength is first absorbed by free-electrons in a metal and thus, their energy and temperature increases. Heated electrons share their energy with ions, resulting in lattice vibrations, and hence, the material gets heated up. In most metals, laser radiation is absorbed within a few tens of nanometer depth of the metal surface and heating occurs by thermal diffusion.

The interaction of laser beam depends upon laser wavelength, polarization, intensity, and interaction time. Figure 11.2 shows the sequence of physical phenomena at increasing laser intensity. Laser beam cutting dominants the industrial laser applications and has more than 75% of share of all laser machining applications.

11.4   LASER BEAM CUTTING (LASER DRILLING)

In laser drilling, due to the laser (for example, a Nd:YAG laser), a localized intense heat is produced at the contact point of the workpiece, which increases the temperature of the workpiece, resulting in melting and evaporation of the workpiece material. The evaporated material transforms into plasma and is disposed of from the machining zone. Further, a high pressure fluid jet, usually an oxygen gas jet, is flushed at the zone and assists in the transformation of phase and exit of the removed material from the machining zone. Laser drilling is found to be best suited for the machining of accurate holes in the exotic materials that are hard to drill using other techniques.

11.4.1   Applications of LBM

Laser beam machining is extensively applied for the drilling and cutting of metallic and non-metallic materials. This process can make ultra-precision holes up to approximately 0.005 mm in ceramic materials and other refractory metals, and composites without any warping. LBM is widely used in various industry such as the electronics,

TABLE 11.2
Advantages and limitations of laser cutting

Advantages of laser cutting	Limitations of laser cutting
Flexibility in cutting path as it uses laser	Not economical for high productivity when compared with stamping
Very less support structure requirement	Taper issue leads to thickness limitation
Cuts very hard and abrasive materials	High capital cost
High accuracy parts can be produced	High maintenance cost
Narrow heat affected zone	Inert/Assist gas required
Versatile and multipurpose	

automobile, aircraft, defense and biomedical industries, etc. The advantages and limitations of laser cutting are given in Table 11.2.

11.5 LASER-ASSISTED HYBRID MACHINING PROCESSES

The classification of laser-assisted hybrid machining processes is shown in Figure 11.3. These machining processes are broadly divided into 2 groups, one is traditional hybrid machining and the other is advanced hybrid machining [4]. In traditional hybrid machining, the laser facilitates the cutting tool for material removal through thermally induced softening of the workpiece, whereas in advanced hybrid machining the laser aids in removal of material through melting and evaporation followed by the application of other advanced machining processes. Among these advanced hybrid machining processes, laser-assisted electrical discharge machining (LAEDM) has gained remarkable research and market attention because of its promising industrial applications.

11.5.1 EMERGENCE OF LASER-ASSISTED ELECTRICAL DISCHARGE MACHINING TECHNOLOGY

The rate of production of features using EDM is found to be low owing to the lower material removal rate of the process. To overcome this problem, researchers have made attempts to use laser beam machining instead of EDM. But, the quality of the features produced via laser machining was not able to meet the industrial standard. This is due to the generation of large recast layers and heat affected zones. Hence, a combination of laser machining with EDM could eliminate the problems of the recast layer and heat affected zones usually associated with laser machining, while increasing the material removal rate associated with EDM machining. Therefore, laser-assisted electric discharge machining (LAEDM) has emerged as the recent innovative technique for the enhancement of the capabilities of the EDM process. LAEDM is also known as sequential laser electrical discharge machining, wherein laser beam machining is used sequentially with EDM, and thus decreases the workpiece machining time by assisting in enhancing the MRR, surface finish and improving the product geometry.

```
                    ┌─────────────────────────────────────────┐
                    │  Laser-assisted hybrid machining processes │
                    └─────────────────────────────────────────┘
              ┌──────────────────────┐         ┌──────────────────────┐
              │  Traditional hybrid  │         │   Advanced hybrid    │
              │      machining       │         │      machining       │
              └──────────────────────┘         └──────────────────────┘
                        │    ┌─────────┐                 │    ┌─────────┐
                        ├────│   LAM   │                 ├────│  LAECM  │
                        │    └─────────┘                 │    └─────────┘
                        │    ┌─────────┐                 │    ┌─────────┐
                        ├────│   LAG   │                 ├────│ LAJECM  │
                        │    └─────────┘                 │    └─────────┘
                        │    ┌─────────┐                 │    ┌─────────┐
                        ├────│  LATM   │                 ├────│  LAEDM  │
                        │    └─────────┘                 │    └─────────┘
                        │    ┌─────────┐                 │    ┌─────────┐
                        └────│   LAT   │                 └────│  LAWJM  │
                             └─────────┘                      └─────────┘
```

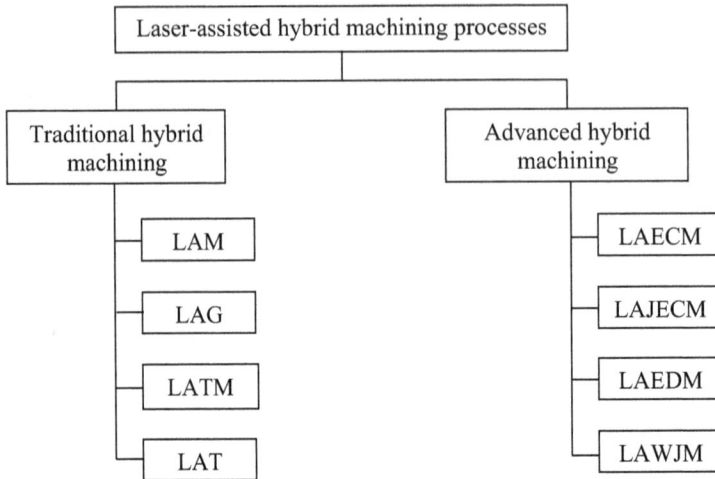

LAM	Laser-assisted milling	LAECM	Laser-assisted electrochemical machining
LAG	Laser-assisted grinding	LAJECM	Laser-assisted jet electrochemical machining
LATM	Laser-assisted turn mill	LAEDM	Laser-assisted electrical discharge machining
LAT	Laser-assisted tuning	LAWJM	Laser-assisted water jet machining

FIGURE 11.3   Classification of laser-assisted hybrid machining processes.

Consequently, the technique becomes more stable and reliable, thereby improving the performance characteristics. Moreover, the machined surface develops high resistance to corrosion and abrasion.

An abstraction of the sequential laser EDM process is shown in Figure 11.4. In Figure 11.4, a short-pulsed laser beam capable of generating a high rate of ablation is utilized to pre-machine the required basic geometries into the workpiece. This indicates that a pilot crater is first machined by the pulsed laser beam, and then the tool is re-positioned followed by the generation of the required feature by the EDM process. Accordingly, the defects on the surface attributed to the heat effects of laser ablation will be removed and the surface finish of the product can be improved. This sequential hybrid process will impart low thermal load to the workpiece and offer precise machining with micron resolution.

Moreover, the process could enable a reduction in total machining time to a great extent and could produce high quality features in comparison with standard EDM (Figures 11.5 and 11.6). Figure 11.6 shows that the size (diameter) of the electrode in EDM can be determined based on the size of the hole machined using the laser, the HAZ and the requirement for a certain final hole size.

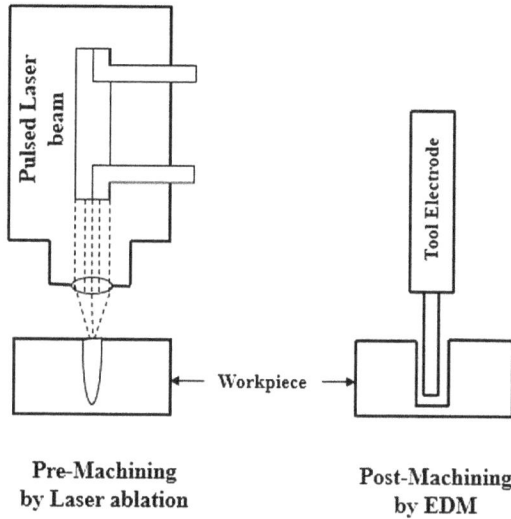

FIGURE 11.4   Sequential laser cum EDM process.

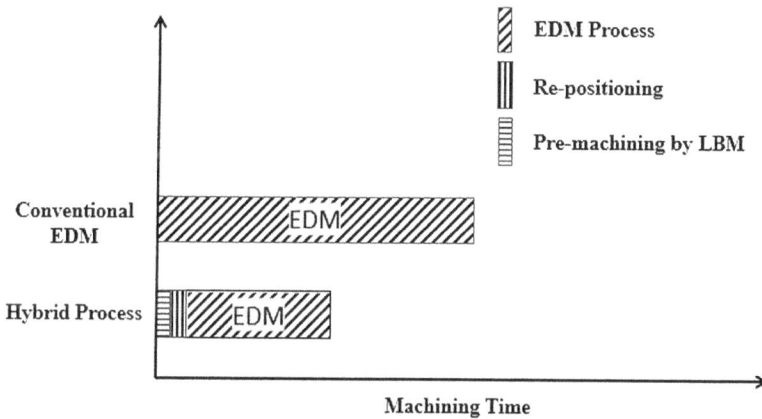

FIGURE 11.5   Process sequence of the hybrid process [8].

11.5.2   EVALUATION OF THE ADVANTAGES AND DISADVANTAGES OF LASER EDM

LAEDM can play a crucial role in overcoming the drawbacks of EDM and LBM. The advantages and disadvantages of conventional machining, EDM, laser beam machining and sequential laser-EDM machining is shown in Table 11.3. Increased productivity, enhanced repeatability, reduction of tool wear, automation flexibility, online process monitoring and control, and increased reliability, are the characteristics associated with the sequential laser-EDM process. Further, in the sequential hybrid process, complex parts can be machined at relatively low costs compared to the respective EDM or laser machining as depicted in Figure 11.7.

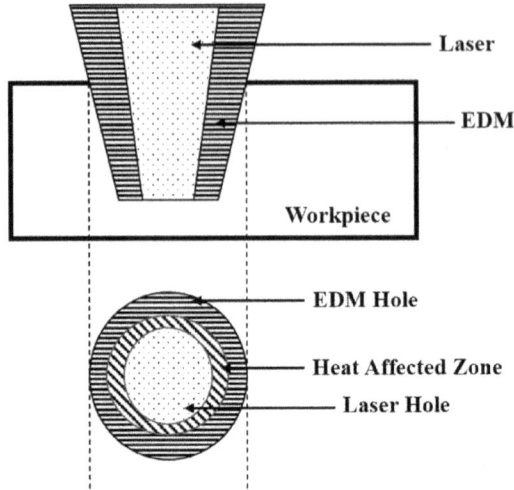

FIGURE 11.6   Schematic of combined laser and EDM drilling LAEDM [4].

TABLE 11.3
Comparison of advantages and disadvantages associated with conventional machining, EDM, laser beam machining and sequential laser-EDM Machining

Comparison	Conventional Machining	EDM Machining	Laser Beam Machining	Sequential laser-EDM Machining
Advantages	Higher Material removal rate	Intricate shapes	Increased MRR	1. Minimalize tool wear 2. Increased MRR and surface integrity 3. Improvement in surface roughness 4. Increased tool life
Disadvantages/ Limitations	1. Tool wear 2. Less feasible for complex shapes	Lower MRR	Lower surface integrity	Higher investment

---

11.6   ILLUSTRATION: STATE-OF-THE-ART SIGNIFICANT FINDINGS

Li et al. [5] were able to produce high accuracy and high-quality holes using a LAEDM drilling technique for application in the manufacture of fuel injection nozzles. The authors observed that the combination of laser and EDM yielded better results such as elimination of the recast layer and of heat affected zone problems commonly associated with the machining of holes using laser. Further, the hybrid process enabled a 70% decrease in total drilling time in comparison to the standard EDM

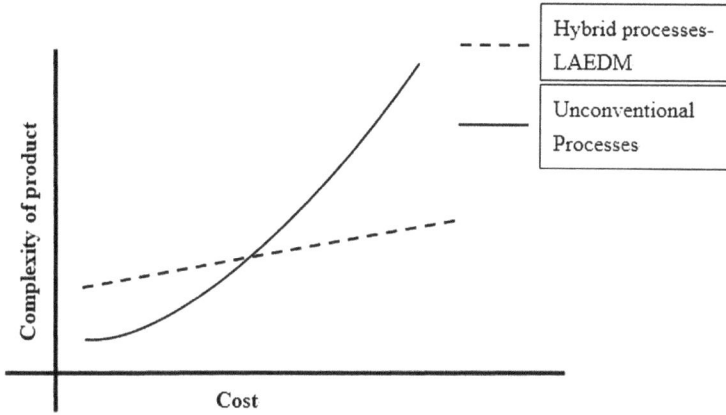

FIGURE 11.7 Comparison of complexity versus cost factor for EDM/Laser Machining and LAEDM.

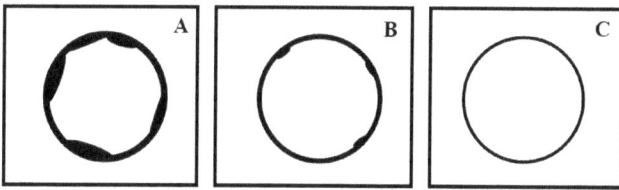

FIGURE 11.8 Production of holes using (a) LBM, (b) EDM and (c) sequential laser cum EDM [5].

drilling operation. The quality of holes drilled by LAEDM was observed to be better than EDM machining. Figure 11.8 shows a schematic of sequential laser EDM holes (size: 140 µm) which is less than the size of the hole (size: 146 µm) obtained from EDM using the same sized EDM electrode. Also, the radial overcut was observed to be small with LAEDM. In this work, self-centering alignment pins were used for the sequential laser and EDM drilling, as these are crucial for the LAEDM process. These pins were designed to be applied on both the fixtures of laser and EDM to minimize deviation and to maintain the position of the hole within ± 20 µm. The study showed that the process of generating the holes using LAEDM was cost-effective, whereby the cost reduced by about 42% [5].

Clerici et al. [9] investigated the modeling of laser guided electric discharge machining. A study of Gaussian, Bessel and Airy beams was carried out to analyze the phenomena of discharge from various configurations of beam shapes. Based on the different laser beams, different discharge shapes were assayed to take corrective actions on the degrees of freedom of the discharges. The research work shows that relevant and suitably shaped laser radiation can be used to operate along a complex path. Al-Ahamari et al. [10] reported the machining of micro holes on Nickel-Titanium based shape memory alloys using laser and the micro-EDM process.

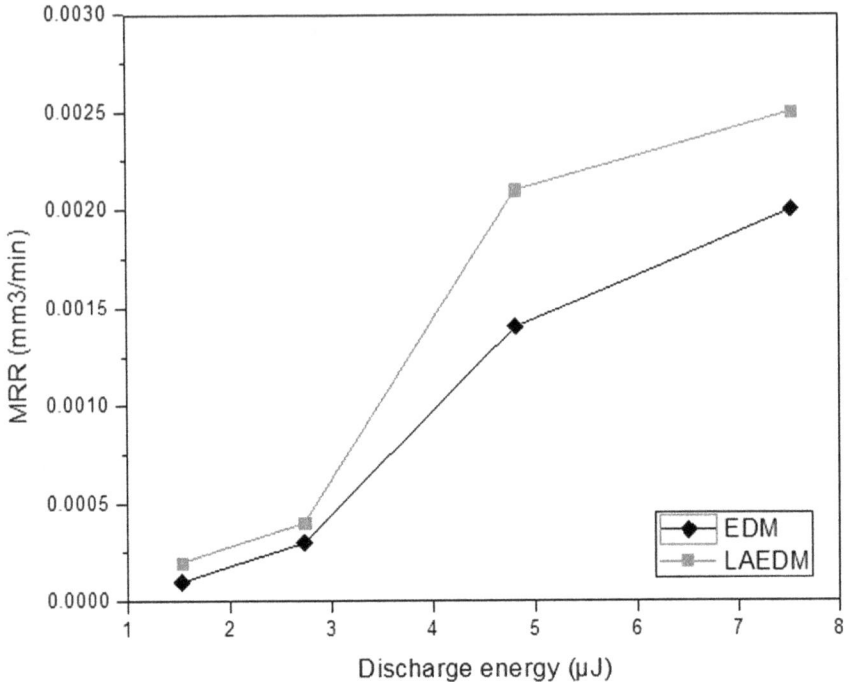

FIGURE 11.9 Effect of discharge energy on the material removal rate in EDM and LAEDM [10].

Dielectric flushing effectiveness had improved considerably due to the influence of pilot laser drilling, thereby carrying the debris from the machining zone, which in turn reduced secondary sparking and reduced tool wear rate. The MRR was found to be improved by the synergic effect of laser pilot drilling followed by EDM drilling. The authors reported that using the proposed LAEDM process, the machining time decreased by 50–65% and the MRR increased by 40–65%. Thus, the performance characteristics such as material removal rate and the geometric tolerance of parts produced using LAEDM was found to be enhanced.

Figure 11.9 shows the influence of discharge energy on the MRR in EDM and LAEDM. It was observed that in case of the EDM process, for a steady pulse duration, the material removal rate was found to have increased because of the increased discharge energy resulting from the high discharge current. But, the stability of the impulse discharge will be vulnerable at a high discharge energy. This will result in initiation of a continuous current flow, namely, arcing or short circuiting. Thus, a fraction of the discharge energy can be utilized for the formation of metallic debris by remelting and evaporation, and most of the discharge energy in the form of heat is lost in a gaseous environment in this process. As a result, there will be inefficient debris removal from the machining zone which leads to a reduction in the material removal rate, unlike in LAEDM, in which the laser will assist the EDM process to

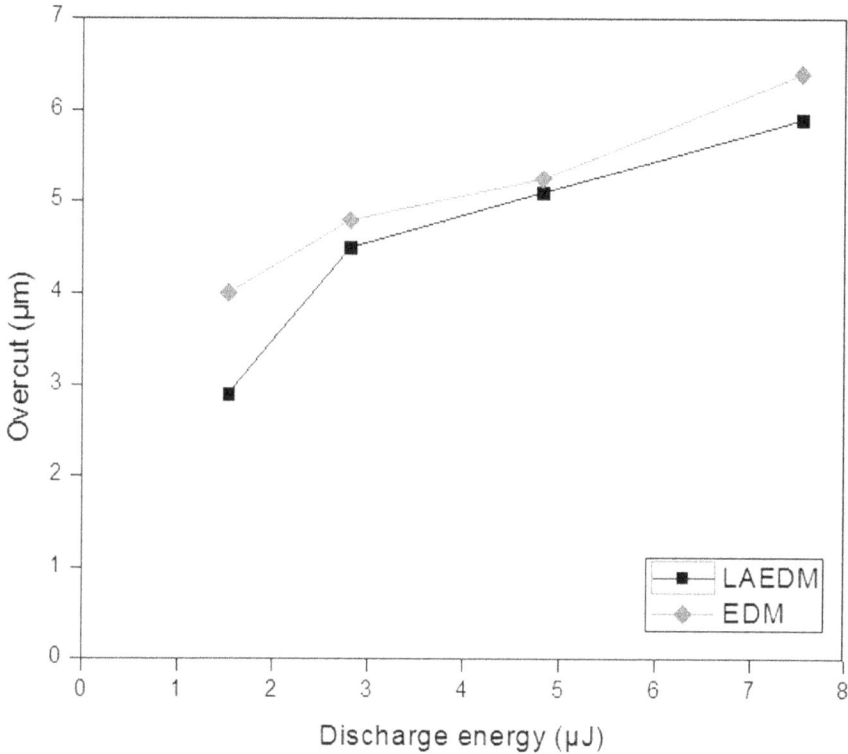

FIGURE 11.10   Effect of discharge energy on the overcut in EDM and LAEDM [10].

avoid the arcing or short circuiting and aid in the removal of debris efficiently with a stable discharge energy and at minimal loss.

Figure 11.10 shows overcut versus discharge energy effect in EDM and LAEDM. Overcut is a major characteristic of EDM drilled holes which can be used to evaluate the quality and accuracy of holes produced. It was measured as the difference between the generated micro-hole diameter and the actual electrode diameter. As the spark discharge energy increases, the dielectric content decomposes to a large extent and thus, a large amount oxygen can be released. Further, this will influence efficiency in machining, and lead to improper flushing of debris, resulting in the production of side sparks. The side sparks impinge on the inner surface of the hole during machining. Thus, overcut was seen to increase with an increase in discharge energy [10].

11.7   CONCLUSION

In this study, at the outset, the need for advanced machining techniques across the global manufacturing industries was briefly detailed. Among the advanced machining techniques, the popularity of respective electrical discharge machining (EDM) and laser beam machining (LBM) along with their conceptual realization and drawbacks were discussed. But, in today's high technology global market, owing to the stringent demand for efficient machining of exotic materials and for the production of intricate geometries to a higher degree of accuracy and favorable cost, the significant importance of sequential laser EDM (one of the hybrid advanced machining techniques to accomplish the above demands) was devoted. Further, state-of-the-art literature findings on the improvement of performance characteristics, using laser-assisted EDM, were illustrated in the study.

Moreover, the performance of hybrid energy source-based sequential laser EDM was found to reliable. The leapfrogging in laser machining and EDM can enhance the capabilities of sequential laser EDM and shows promise for various future breakthroughs that will transform the advancement of technological directions in different potential applications.

REFERENCES

[1] Jain, V.K. Introduction to Micromachining, India: Narosa Publishing House, 2010.

[2] Pankaj K Shrivastava and Avanish K Dubey. Electrical discharge machining–based hybrid machining processes: A review. Proceedings of the Institution of Mechanical Engineers, Part B: Journal of Engineering Manufacture, 228 (6), 799–825, 2013.

[3] Krishna K. Saxena, Mattia Bellotti, Jun Qian, Dominiek Reynaerts, Bert Lauwers and Xichun Luo. Overview of Hybrid Machining Processes, Chapter 2, Hybrid Machining Theory, Methods, and Case Studies, 21–41, 2018.

[4] Choon-Man Lee, Wan-Sik Woo, Dong-Hyeon Kim, Won-Jung Oh and Nam-Seok Oh. Laser-assisted hybrid processes: A review. International Journal of Precision Engineering and Manufacturing, 17, 257–267, 2016.

[5] Lin Li, C. Diver, J. Atkinson, R.Giedl-Wagner and H.J. Helml. Sequential Laser and EDM Micro-drilling for Next Generation Fuel Injection Nozzle Manufacture. CIRP Annals, 55 (1), 179–182, 2006.

[6] Ghosh, A. and Mallick, A.K. Manufacturing Science, 2nd edition, East-West Press (India) Pvt Ltd, 2010.

[7] Payal, H.S. and Sethi, B.L. Non-conventional machining processes as viable alternatives for production with specific reference to electrical discharge machining. Journal of Scientific and Industrial Research, 62, 678–682, 2003.

[8] Sanha Kim, Bo Hyun Kim, Do Kwan Chung, Hong Shik Shinand, Chong Nam Chu" Hybrid micromachining using a nanosecond pulsed laser and micro EDM" Journal of micromechanics and Micro engineering, 20(1), 15–37, 2009.

[9] Clerici, M., Hu, Y., Lassonde, P., Milián, C., Couairon, A., "Laser-Assisted Guiding of Electric Discharges Around Objects," Science Advances, 1(5), 2015.

[10] M. A. Al-Ahmari, M. Sarvar Rasheed, Muneer Khan Mohammed & T. Saleh. A Hybrid Machining Process Combining Micro-EDM and Laser Beam Machining of Nickel–Titanium-Based Shape Memory Alloy, Materials and Manufacturing Processes, 31:4, 447–455, 2016.

12 Powder Mixed Electrical Discharge Machining (PMEDM)

Jibin T. Philip[1,2]* and Basil Kuriachen[2,3]

[1]Amal Jyothi College of Engineering, Kanjirappally, Kerala, India

[2]Department of Mechanical Engineering, National Institute of Technology Mizoram, Aizawl, Mizoram, India

[3]Advanced Manufacturing Centre, Department of Mechanical Engineering, National Institute of Technology Calicut, Kerala, India

*Corresponding author

CONTENTS

DOI: 10.1201/9781003202301-12

12.1 INTRODUCTION

The need for the machining of characteristically superior alloys, namely, Ti6Al4V, Inconel 718, Haynes 25, and the like, for advanced applications has substantially increased the demands on the existing non-conventional machining processes, and the requirement for their modification due to their efficiency in the processing of such materials.

Electrical discharge machining (EDM) is a method that is potentially capable of machining Ti and its alloys, regardless of their production routes, and mechanical and chemical properties. The technique is not only widely utilized for the formation/development of complex shapes and parts exclusively for electrically conductive materials, but is also compatible with a few non-conductive materials subject to tight constraints. The incorporation of suspended particulates/powders in the dielectric fluid for EDM, generally referred to as the powder mixed electrical discharge machining (PMEDM) process, can aid in the homogenization of THE discharge effect throughout the work material surface. Additionally, they assist in the formation of various surface compounds through the phenomenon of material migration [1]. The inclusion of micro/nano powders in the dielectric can help enhance the conductivity of the intermediate fluid and assist in broadening the inter electrode gap (IEG). The suspended particulates can help prevent (through mechanical action) the development of micro cracks (during resolidification) caused by the development of transformational stresses. PMEDM can aid in the formation of surfaces embedded with fine particulates migrated from the dielectric through accelerated penetration of the molten matrix using the velocity gained by the development of negative pressure and the occurrence of electrophoresis at the IEG [2]. The stochasticity associated with the EDM/PMEDM processes makes the modelling and simulation efforts to understand the respective processes to be time consuming and arduous [3].

The potential of EDM/PMEDM to act as a surface modification technique (SMT) is receiving enormous attention to try to improve its mechanical properties, triboresponse, and to improve the bio-degradability of various materials. The approach has advantage over other similar SMTs due to its capabilities such as (i) the nonrequirement/non-essentiality of the pre/post-processing of material surfaces; (ii) the development of various surface compounds (oxides/carbides) imparting unique properties, namely, biocompatibility and hydrophilicity; (iii) the formation of nanoporous structures; (iv) the accretion of hard layers/coatings; (v) the development of wear and corrosion-resistant surfaces. The surface so formed is composed of many randomly distributed overlapping craters. Besides, a positive skewness is imparted to material surfaces by the EDM process through the existence of interspersed peaks and valleys. During tribo-interactions, the vacant space within such craters may entrap the debris particles emanating from the mating surfaces (during dry surface interactions) and entrain lubricants (during wet surface interactions), thereby the effect of friction/wear gets subdued [1].

Figure 12.1 shows the schematic diagram of the PMEDM setup. The evident/obvious distinction between the EDM and PMEDM setups is that the latter has an external attachment for mixing and circulation of the dielectric including its powder

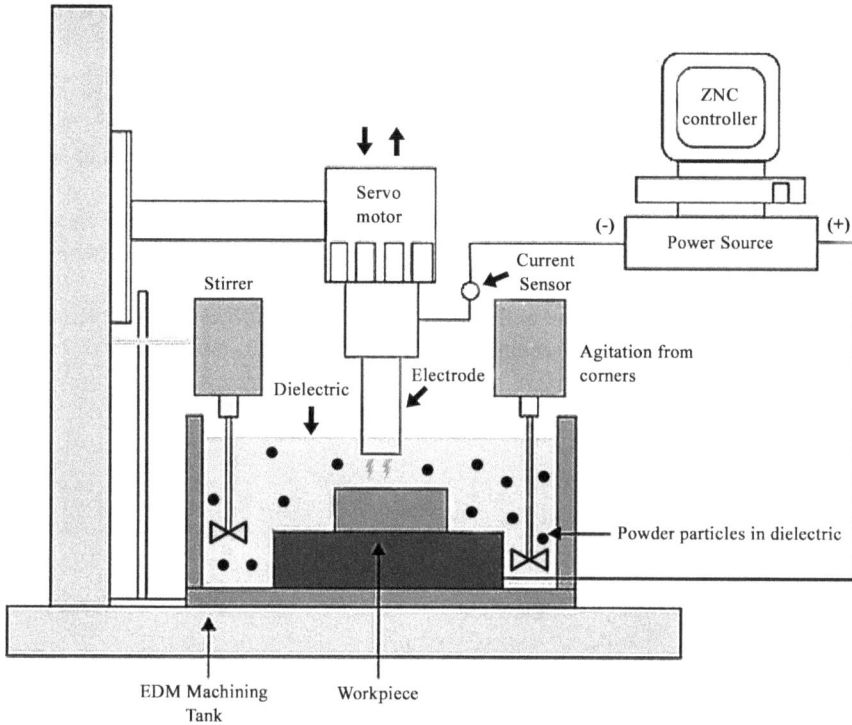

FIGURE 12.1 Schematic diagram of the PMEDM setup [4]. Reproduced in accordance with Creative Commons License from Ref. [4]. Copyright MDPI (2020).

additives. Generally, the standard EDM machine forms the control equipment/unit and the addition of a portable-fabricated setup for supplying and monitoring the powder assisted machining aids in carrying out the PMEDM process.

The modification of the EDM mechanism to implement its advanced variant, namely, PMEDM to improve the process dynamics and responses of the former may seem like a minor modification; nevertheless (in reality), the approach is highly complex. The various process parameters associated with the type of dielectric, type of powder, type of tool/workpiece electrodes, and process circuit parameters influence the PMEDM process. The complexities of the process have led to the technique being prevented from entering the mainstream manufacturing/machining process chains for industrial/commercial applications. There is a need to present a comprehensive review on the most current information on the PMEDM method, so that the researchers can capitalize on the available knowledge and develop it. Besides, the utilization of powders (suspended in the dielectric) with unique characteristics can impart superior properties to the machined surfaces processed through the PMEDM method. Hence, this manuscript also summarizes the advanced properties that can be developed on the various material surfaces through this method by utilizing different powder additives in the dielectric during processing.

12.2   MECHANISM OF POWDER MIXED ELECTRICAL DISCHARGE MACHINING (PMEDM)

The concept of PMEDM came into existence with the basic need to improve the efficiency of its parent counterpart, namely EDM. Initially, the effort primarily focused on the improvement of the process responses, namely, the MRR, TWR, and surface finish. When a voltage is applied across the electrodes (tool and the workpiece), the conductive powder additives added in the dielectric get energized and move in a criss-cross pattern [5]. Figure 12.2 shows a schematic representation of the mechanisms of the EDM and PMEDM processes.

The EDM process can assist in the formation of a resolidified layer formed by the impact of primary discharges, generally consisting of surface defects such as cracks.

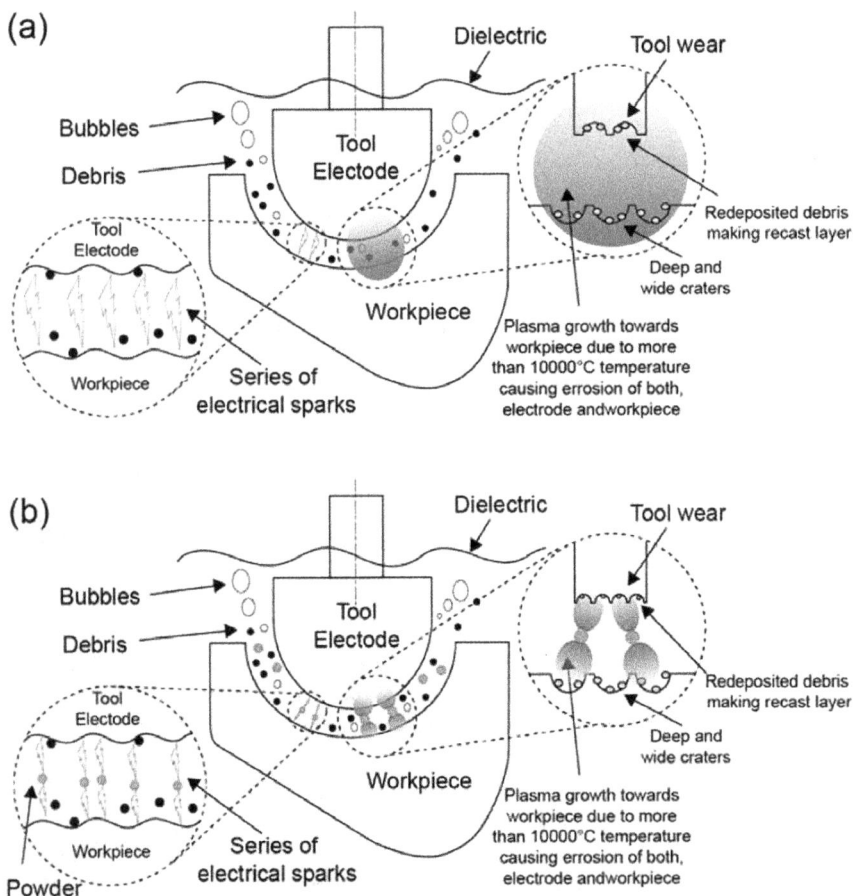

FIGURE 12.2 Schematic representation of the mechanisms of EDM and PMEDM processes (a) EDM and (b) PMEDM [4]. Reproduced in accordance with Creative Commons License from Ref. [4]. Copyright MDPI (2020).

These cracks propagate through the appendage layer and usually terminate at the interface between the deposited matrix and the subsurface heat affected zone (HAZ). At the end of each pulse cycle, the discharge ceases and the dielectric rushes to fill the developed void. Due to the dielectric cooling activity, the surface of molten material cools more rapidly than the sub-surface materials. The re-solidification process induces transformational stresses which can have detrimental effect on the deposited layer. Crack initiation and propagation occurs if these stresses exceed the ultimate strength of the work material [6].

During PMEDM, the main discharge channel is sub-divided into several secondary discharges through the interaction of the added powders in the dielectric. These distributed discharges can heat the powder particles, and the activity of the dielectric can assist in their settling in the distributed cavities formed over the work material [6]. Kunieda and Yanatori [7] (based on their experimental findings) suggested that during initial discharging, the powders align themselves in the form of chains in the primary discharge channel. Progressively, as the plasma expands, the channel is disrupted, leading to the breakage and separation of the chain formations. This leads to a phenomenon called, the discharge separation effect (DSE), due to which the primary discharge channel is scattered and distributed in the form of multiple/multi-level secondary discharges. At the end of each pulse discharge, a negative pressure develops in the IEG due to the instantaneous collapsing of the secondary discharges. Hence, a sudden progression of the stacked particles (powders) assisting the secondary discharges along which the molten material happens at infinitesimally small-time intervals, assisting the penetration of the powder additives into the molten mass. Consequently, an evolution of a powder-additive rich layer on the work material occurs (which is generally referred to as the coated layer), when the PMEDM process is used as a surface modification technique. Reportedly, adjusting the pulse current and the pulse time can efficiently aid in controlling/monitoring the build-up of the deposited matrix. Processing at high pulse current and low pulse time can lead to the development of high temperatures substantially favoring the formation of the deposited layer [6].

There are several reasons for the implementation/development of the PMEDM process. One of these primary factors is the need for optimum utilization of the discharge energy. During transition from EDM to PMEDM, there are specific variations associated with the process which are as follows:

(i)   Formation of Secondary Discharges: The inclusion of powder additives can result in a considerable decline in the insulating strength of the dielectric. This is due to the random movement of the particles, in other words, Brownian motion, and their scattering achieved by the proper mixing of the powder in the dielectric using suitable mechanisms.

(ii)   Widening/broadening of the IEG: The existence of conductive/semi-conductive particles in the dielectric can lead to electric field aberrations. Dielectric breakdown can easily occur due to random short circuiting between the powder particles, leading to DSE.

(iii)   Multiple Discharges: The rapid zig-zag movement of the powder additives in the dielectric due to Brownian motion and enhanced dispersibility achieved through effective techniques such as ultrasonication, stirring, and surfactant modification (of the powders/dielectric) can lead to the formation of multiple/multi-level discharge paths assisting the near uniform distribution of the discharge energy and formation of multi-overlapping craters. In contrast to the discharge waveforms (in terms of current and voltage signals) of the conventional EDM, its modified counterpart, namely, PMEDM is associated with disturbed/uneven waveforms, due to the occurrence of multiple discharges [8].

Ekmekci et al. [9] proposed a discharge separation model which clearly demonstrates the various possibilities (types/forms) for the occurrence of discharges during the PMEDM process (Figure 12.3). Reportedly, altering the IEG through the incorporation of specific powders in the dielectric can aid in the development/formation of specific textures and aid in controlling the HAZ. Partial or complete decomposition of the powders is possible through the confrontation of the main/primary channel with the dispersed particles. This can lead to the formation of superior hard surfaces on the work material through material migration/transfer. Besides, reducing the pulse on time (slightly) and allowing the penetration of the powder particles into the molten matrix (at the end of discharge time), can substantially enhance the material migration rate.

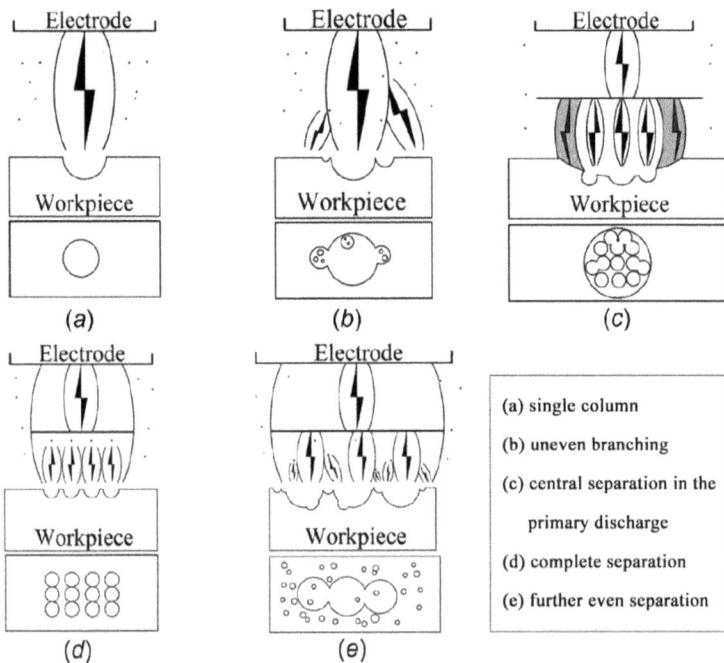

FIGURE 12.3 Types/forms of electric discharge occurrences at the IEG [9]. Reproduced with permission from Ref. [9]. Copyright ASME (2016).

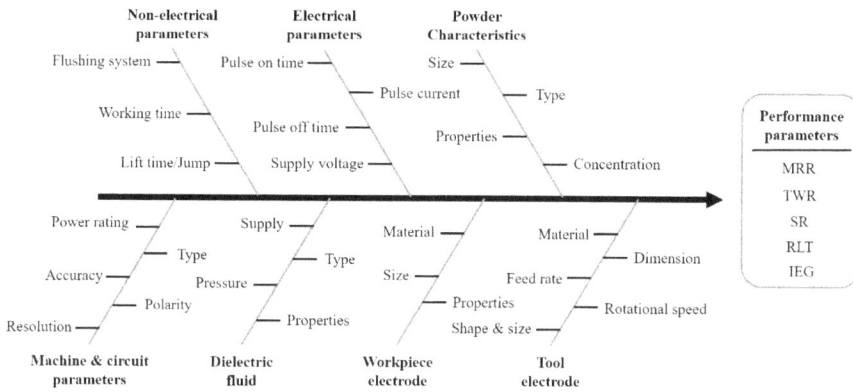

FIGURE 12.4 Cause and effect diagram of the PMEDM process featuring the control parameters and the output responses [1]. Reproduced with permission from Ref. [1]. Copyright Elsevier (2021).

12.3 PROCESS CONTROL PARAMETERS

There are several input parameters that control the EDM and PMEDM processes, respectively. Figure 12.4 shows the cause-and-effect diagram of the latter. Table 12.1 discusses the primary process parameters of the PMEDM, and their significance in governing the performance and efficiency of the technique.

12.4 TYPES OF DIELECTRICS AND THEIR ROLE

12.4.1 SELECTION CRITERIA

The role of the dielectric is to fill the IEG, act as a conductive medium and control the discharges, maintain the stability of the process, cool down the electrodes (tool and the workpiece), and flush the machining zone to remove the debris particles. There are certain primary characteristics that distinguish a good dielectric, such as efficiency in enabling and controlling electrical discharges, potential to act as a heat sink, moderate viscosity, permissible transparency, acceptable odor, and low cost. In order to use the powder additive in the dielectric to assist the PMEDM process, there should be a mutual compatibility between the two.

12.4.2 TYPE AND ROLE OF DISTINCT DIELECTRICS

12.4.2.1 Oil-based

The oil-based dielectrics are the most commonly and widely used discharge media in die-sink EDM machining and/or processing of materials. They include kerosene, mineral oils, vegetable oils, transformer oil, and mineral seals [15]. Singaravel et al. [16] conducted experimental investigations using sunflower oil as a dielectric for the machining of Inconel 800. Trial tests were carried out using distinct tool materials,

TABLE 12.1
Major control parameters that govern the PMEDM process

Sl. No.	Control Parameters	Specific categories	Significance	Ref.
1	Powder characteristics	Powder type	• The type of powder and its specific properties will govern the efficiency of the modified process and the unique properties that could be imparted to the machined surface.	[10–14]
		Powder size	• Larger sized powders can increase the IEG, but can cause contamination and also lead to short circuiting.	
		Powder concentration	• There is an optimum powder concentration above and below which the process responses will fail to demonstrate superior performance.	
2	Electrical parameters	Peak current	• The maximum amount of current the power source can supply within a limited period of time (for each pulse).	
			• Peak current can significantly affect MRR and SR.	
		Average current	• For a complete cycle, the average of the amperage at the IEG denotes the average current.	
			• Product of duty cycle and peak current.	
			• Demonstrate the machining operational efficiency in terms of MRR.	
		Pulse on time	• There exists a near proportionality with the MRR.	
			• Rough surfaces can be created through the formation of wide and deep craters by machining at high pulse on time.	
			• Reportedly, TWR increases with T_{on}.	
		Pulse off time	• Governs the frequency and stability of the material removal.	
			• TWR is expected to increase with T_{off}.	
			• Short T_{off} cannot assist the removed material from getting evacuated (by the dielectric) from the IEG, thereby unstable discharging can occur. This condition	
		Duty cycle	• It represents the percentage of on-time relative to the total cycle time. Increased duty cycle can improve the machining efficiency up to admissible limits.	
		Gap voltage	• The discharge energy is determined by the total applied voltage. The IEG is controlled by the servo, based on the gap voltage and larger IEG can help improve the flushing efficiency and hence the MRR.	
		Polarity	• Polarity defines the direction of current flow relative to the tool and the work material (electrodes). Based on the application requirement, the polarity is reversed.	

3	Non-electrical parameters	Pulse frequency	• Number of discharge cycles produced across the IEG per second. Depending upon the pulse frequency the value of T_{on} increases and decreases, relative to T_{off}. • At high pulse frequency wide and deep craters are formed with greater heat impact on the work material, leading to the formation of rough surfaces and vice-versa. The RLT and HAZ will be considerably large during machining under such circumstances.
		IEG	• It denotes the gap between the two operational electrodes (tool and the workpiece). Based on the average gap voltage, hydraulic and electro-mechanical systems aid in maintaining the IEG. • In order to maintain maximum flushing efficiency, optimum machining performance and high MRR, adequate IEG is required.
		Flushing pressure	• Pressure of the dielectric supplied at the cutting/machining zone, significantly monitors the removal of the debris, availability of pure dielectric, and adequate cooling of the electrodes. • Improper flushing can lead to short circuiting and tendency for erratic machining. • Forced/injection flushing is highly favored due to its efficiency in performing the intended functions.
		Type of dielectric	• The dielectric fluid acts as an insulating medium and when voltage is applied across the electrodes, it breaks down (partially) allowing stable and controlled discharging, leading to material removal. • A wide range of dielectrics are experimented to evaluate their suitability in *in-situ* machining conditions ranging from oil-based dielectrics to gaseous dielectrics. • The demand and usability of a dielectric depends upon their ability of properly control the machining process, aid in achieving superior surface finish and perform various activities, viz. cooling of the electrodes, better evacuation of the debris, ability to prevent the formation/release of unwanted elements/compounds through decomposition, etc.
4	Electrode characteristics	Electrode material	• The materials chosen as the tool and workpiece material governs the MRR, TWR, SR, IEG, and RLT. Although the latter cannot be chanced in many circumstances, the former must be chosen based on specific material properties such as thermal and electrical conductivity (high), melting point (high), wear rate (low), and resistance to deformation (high). • Commonly used tool materials include: metals, alloys, and composites.
		Electrode size	• Size of the electrode should be controlled/chosen based on the machining requirement. • With increase in surface area available for discharging, available energy will be accordingly distributed.
		Electrode shape	• The shape/cavity/surface formed on the work material will be a negative imprint of the shape of the tool electrode.

namely, copper, brass, and tungsten-copper. Reportedly, there was a considerable increase in the MRR, TWR, and SR through the utilization of vegetable oil dielectrics.

12.4.2.2 Water-based

Erdel and Temel [17] evaluated the efficiency of deionized water as a dielectric for EDM, and based on the results, the MRR substantially improved with a significant reduction in the TWR. Although the utilization of plain water has many advantages over hydrocarbon dielectric for specific conditions, the latter is predominantly superior due to the low viscosity of water, limiting the restrictions on the discharge channel and leading to a significant decline in the energy density. Consequently, the MRR achievable through EDM using plain water is considerably lower in comparison to hydrocarbon oils [18]. Besides, a huge amount of energy is required to heat and vaporize water and therefore, comparatively lower vapor pressure is developed through the discharge at the IEG, which cannot aid in the proper removal of the molten metal [19]. Contrarily, the reports suggest that distilled water can demonstrate better performance than hydrocarbon oils. When machining Ti6Al4V (with copper as the tool electrode) the hydrocarbon oil assisted the deposition of a large amount of decomposed carbon on the work material surface together with the formation of hard compounds such as titanium carbide (TiC). The high MP of the TiC together with the hardness imparted by its presence on the surface considerably reduced the MRR, caused by a stark decline in the impact of the impulse forces on the work material surface. Nevertheless, the incorporation of distilled water led to the development of oxides such as TiO_2 (having lower MP) assisting in better removal of material from the workpiece surface [20]. Generally, water-based dielectrics are regarded as highly effective and are expected to replace hydrocarbon-based dielectrics, particularly due to its environmental friendly nature and ability to enable superior machining performance [21].

12.4.2.3 Emulsions

Liu et al. [22] tested the efficiency of water-in-oil emulsions as dielectrics for EDM and compared its efficiency in terms of the MRR, SR, and IEG with the case of when kerosene dielectric was used. Reportedly, high a MRR, low SR, large IEG, and compatible work environment is achievable through the utilization of water-in-oil emulsions. Surfaces machined with the assistance of emulsions are free from micro-cracks and the considerably shallower craters developed have smooth edges. With the percentage variation in water and oil in the emulsion, globules may form on the machined surface, and there could be an increased electrode wear ratio over that of kerosene. Zhang et al. [23] also used water-in-oil emulsion for the EDM of mild steel with a copper electrode, and compared its efficiency to that of kerosene. In cases when the peak current and pulse duration are significantly large, water-in-oil emulsions can assist in producing a better MRR. Superior and stable machining is achievable through the utilization of emulsions. Its relative electrode wear rate (REWR) is higher as there is low carbon adherence to the electrodes. The economic and environmental friendliness adds to its advantages for consideration as a replacement dielectric for kerosene and other hydrocarbon-based oils.

12.4.2.4 Gaseous

Under specific machining conditions, gaseous dielectrics, namely, air and/or oxygen can assist in achieving a higher MRR than when hydrocarbon dielectrics are used. From a commercial point of view, this argument needs the back up of a proven technology and deeper research in the subject context [21]. Kunieda et al. suggested a new concept of supplying oxygen when water-based dielectrics are used to assist the EDM process [24]. Gradually, the concept of dry EDM (gas assisted machining) came into existence [25]. Reportedly, a superior MRR and a low TWR is achievable through an increment in the oxygen content in the air when 3D cavities are fabricated through a high speed EDM milling process [26]. Economies in terms of time and cost are attainable during the dry (EDM) milling of cemented carbide in contrast to oil assisted EDM milling [27].

12.4.3 Summary

Strong arguments (based on valid results) exist regarding the superiority of mineral oils over the water-based oils, namely, deionized water, distilled water, and so forth.. [21]. On the flip side, researchers have also found out that EDM (in high pulse ranges) using water-based dielectrics such as deionized water can aid in achieving a peak MRR, better surface finish, and a low TWR, nevertheless with a short drop in the machining accuracy [28].

Despite many advantages of machining through dry EDM, the technique suffers from process instability and arcing [29]. This is because of the short IEGs that are frequently prevalent in dry EDM. Kunieda et al. recommended a solution through the utilization of a piezo electric actuator to control the IEG and thereby prevent the short circuiting [30]. Another proposed approach was the implementation of a magnetic field to further facilitate the electron movement and the extension of ionization in the plasma channel, by applying the magnetic field at a tangent to the existent electric field [31].

12.5 THE IMPACT OF POWDER ADDITIVES

12.5.1 Selection Criteria

The inclusion of powder additives in the dielectric can positively influence the efficiency of the PMEDM process, if specific considerations in the context of powder material, powder size, and powder concentration is given careful attention. Reviews of the literature, in-depth analysis through *in-situ* experimentation, trial tests, and specialized optimization methodologies are required in order to produce useful results.

A wide range of powder additives, namely, metallics, non-metallics, ceramics, and solid lubes have been tested for their respective efficacy in improving machining efficiency in terms of process response parameters (MRR, MDR, TWR, RLT, IEG) and machined surface parameters (SR, SF, SD, WR, CR, Hardness, Hpty, Bcty). The incorporation of specific powders with unique properties can aid in imparting advanced physical and chemical properties to material surfaces, regardless of their production route, type, size, and properties (provided they meet all the basic criteria for being processed on an EDM machine).

Powder size is another major parameter that governs dynamics and response of the PMEDM process. Based on the utilization of large and small particles in the dielectric, the IEG can broaden/widen and enable discharges at close proximity to the working electrodes, respectively. At high values of T_{on}, utilization of large sized powders can expand the IEG, caused by extensive contamination and deionization between the tool and the work material [32]. Reports suggest that the utilization of smaller sized particles (70-80 nm) is advantageous in terms of achieving a high MRR, low TWR, and superior surface finish, despite the fact that the increment in the IEG is gradual and minimal [33].

Powder concentration (PConc) plays a significant role in determining whether or not the powder addition can be advantageous or disadvantageous, regardless of the type and size of the powder material. An optimum PConc should be maintained in the dielectric, an increment beyond which can lead to abnormal conditions such as arcing, short circuiting, and the generation of abnormal discharges [32]. Excessive powder addition can create a condition at which spark/discharge concentration can occur which is generally caused by the production of an excessive amount of debris at the IEG (due to low/poor flushing efficiency). It can also affect machining efficiency due to a negative bridging effect and problems caused by excessive settling of added powders (in the dielectric), which can substantially deteriorate the machined surface [34].

12.5.2 Types and Role of Distinct Powder Additives

A wide range of powder additives (metallics/non-metallics/ceramics/solid lubes) have been tried frequently in the past to improve the process efficiency of the EDM/PMEDM technique, and/or to impart specific/advanced characteristics to the machined surfaces. The suspended powders in the dielectric have a dominant influence/impact on the process responses, namely, the MRR, TWR, SR, RLT, and SD. Sub-sections (12.5.2.1–12.5.2.4) discuss a few major investigations carried out using metallic, non-metallic, ceramic, and solid lube powder additives added to the dielectric for PMEDM on process efficiency and the characteristics of the machined surface. For brevity and for effective utilization of the available data, Table 12.2 and Table 12.3 extensively cover the impact of a wide range of distinct powder additives on the process characteristics and responses of the PMEDM technique in a concise and efficient manner.

12.5.2.1 Metallics

In the context of chromium powder, the size of the added particles, plays a major role in controlling the process parameters/responses. The respective increment and decrement in the MRR and TWR are achievable when nano-chromium powder is added to the dielectric for EDM. Reportedly, a reduced powder consumption (by a factor of 6) is achievable when nano-powders are used instead of micro-powders as additives. Although, the purchase of nano-chromium powder is comparatively

TABLE 12.2
Consolidated list of PMEDM experiments assisted by dielectric added with distinct powders/particulates

Sl. No.	Additive	Process Conditions	Dielectric Fluid	Additive Size & Concentration	Tool Material	Workpiece Material	Major findings	Ref.
1	Silicon (Si)	SV: 180 V I: 3, 9, 17 A T_{on}: 82, 145, 256 µs T_{off}: 55, 110, 183 µs	Kerosene	Size: < 20 µm PConc.: 0–4 g/l	Copper (Cu)	Ti64	• PConc. has dominant influence on the MRR and R_a. • TWR is governed by the variation in I.	[41]
		SV: 200 V I = 10, 65 mA T_{on} = 10 µs τ: 80% FP: 1 bar	Oel Held IME 63	Size: < 10 µm PConc.: 2–10 g/l	Tungsten (W)	Inconel 718	• Comparatively, Si powder dispersed in the dielectric produced thicker rim zone and higher MRR than Al. • Due to the high heat of fusion of Si, it can store large energy which opposes the cooling action leading to the formation of a gradient grey rim zone.	[42]
		I: 3, 12 A T_{on}: 50, 150 µs τ: 0.7, 0.9	Kerosene	Size: 20–30 µm PConc.: 0, 2 g/l	Copper (Cu)	EN31	• Maximum MRR got obtained at I = 12 A, T_{on} = 150 µs, and PConc. = 2 g/l. • Enhanced surface finish (low SR) of the work material occurs at I = 3 A, T_{on} = 50 µs, and PConc. = 2 g/l.	[43]
		I: 3, 6, 10 A T_{on}: 50, 100, 150 µs T_{off}: 15, 20, 25 µs τ: 0.7, 0.9 DFR: 2 l/m PR: 0.3 hp	Kerosene	Size: 30 µm PConc.: 0, 2, 4 g/l	Copper (Cu)	AISI D2 Die Steel (59 HRC)	• I and PConc. are the primary significant input parameters governing/monitoring the MR. • Optimum condition for MR got achieved at I = 16 A, T_{on} = 100 µs, T_{off} = 15 µs, PConc. = 4 g/l, and gain = 0.83 mm/s.	[44]

(continued)

TABLE 12.2 (Continued)
Consolidated list of PMEDM experimentations assisted by dielectric added with distinct powders/particulates

Sl. No.	Additive	Process Conditions	Dielectric Fluid	Additive Size & Concentration	Tool Material	Workpiece Material	Major findings	Ref.
		Dia. of ejector nozzle: 4 mm; FP: 0.5 Kg/cm^2; Gain: 0.83, 0.84, 0.85 mm/s						
		I: 20 A; T_{on}: 2 μs; T_{off}: 2 μs	Kerosene	Size: < 20 μm; PConc.: 20 g/l	Sintered tungsten carbide (WC)	H13 Steel and Copper, Anode	• Mirror-like surfaces (free form) were generated.	[45]
		DFR: 0.25, 0.5, 1, 1.5, 2, 2.5 l/min	Kerosene-Castrol SE Fluid 180	Size: 10–15 μm; PConc.: 0, 1, 2, 3, 5, 10, 20 g/l; Optimum 2–3 g/l	Copper (Cu)	AISI H13 (54 HRC)	• Size of the developed craters, and RLT gets reduced through the incorporation of Si powder in the dielectric. • For rough and finish machining, smaller DFR can generate abnormal discharges assisting in the development of surface defects.	[46]
		SV: 200 V; I: 0.5–8 A; τ = 50%	Castrol SE Fluid 180	Size: 10–15 μm; PConc.: 2 g/l	Electrolytic Copper (Cu)	AISI H13	• Fine-polished surfaces can be realized through the inclusion of silicon powders in the dielectric for EDM. • Extended polishing time can aid in reducing the SR.	[47]
2	Silicon carbide (SiC)	I: 2, 7, 12, 22, 42, 72 A; PD: 2, 6, 25, 100, 400, 1600 μs	De-ionized water	PConc.: 20 g/l	Aluminium (rod)	Ti64	• Dispersed particles get decomposed (partially/fully) on confrontation with the main discharge channel.	[48]

Parameters	Dielectric	Materials	Powder	Findings	Ref.
I: 7, 14, 22 A T_{on}: 25, 100, 400 μs	Tap water and oil	Copper (Cu, Pure), Anode Steel (Interstitial free), Cathode	Size: 500 mesh PConc.: 5, 10, 15 g/l	• Extremely hard layers get developed over the substrate surface due to DSE. • Machined surfaces get engraved with fine particles. • The dispersed particulates gain momentum through electrophoresis and negative pressure developed ensuing electric discharges.	[2]
I: 8, 22 A V: 140 V T_{on}: 40, 120 μs τ/η: 0.75 FP: 0.73 bar	Kerosene	Copper (Cu) and graphite AISI D2 die steel	Size: 95.502 μm PConc.: 0, 5 g/l	• The total heat flux developed by the graphite electrode was much higher than copper electrode when used with pure kerosene (82.4%) and kerosene added with SiC as dielectric fluid (91.5%). • The use of graphite electrodes with SiC dispersed dielectric produced higher fatigue stresses (7%) compared to copper but has improved fatigue life than when copper (14.58%) and graphite (18.54%) electrodes are used with dielectric without additives.	[37]
I: 2, 7 A T_{on}: 25, 50, 100 μs	De-ionized water	Aluminum 6081 T6 Ti64	Size: 800 mesh PConc.: 5, 10, 20 g/l	• High PConc. in the dielectric, together with a low pulse current, can enhance material migration in the form of particulate matter. • The hardness and thickness of the RL are directly and inversely proportional, respectively, to the PConc. in the dielectric.	[49]

(continued)

TABLE 12.2 (Continued)
Consolidated list of PMEDM experimentations assisted by dielectric added with distinct powders/particulates

Sl. No.	Additive	Process Conditions	Dielectric Fluid	Additive Size & Concentration	Tool Material	Workpiece Material	Major findings	Ref.
		I: 7, 12, 22 A T_{on}: 3, 6, 12, 25, 50, 100 μs	De-ionized water and oil	Size: 500, 800 mesh PConc.: 25 g/l	Aluminium (Al), Titanium (Ti), and Copper (Cu)	Ti64	• Material migration from the dielectric fluid depends on the process parametric conditions. • The possibility to control the transfer of material, assist the process to function as a surface alloying technique.	[50]
		SV: 250 V I: 0.1, 1 A V: 25 V PD: 2, 5, 10, 20 μs τ: 0.8 DFR: 60 l/h RoE: 200 rpm	De-ionized water	Size: 3-5 μm PConc.: 25 g/l	Copper (Cu)	Ti64	• Improved MRR, R_a and surface textures compared to the usage of dielectric without additives.	[51]
		–	Kerosene	PConc.: 2-6 g/l	Copper-tungsten	Ti64	• At high current discharge, higher MRR and R_a was observed.	[52]
		SV: 250 V I: 0.1, 0.5, 1 A V: 25 V PD: 2, 5, 10, 20 μs τ: 0.8 DFR: 60 l/h MT: 3 min RoE: 0-150 rpm	Kerosene	Size: 1 μm PConc.: 0–75 g/l	Copper (Cu)	Ti64	• SiC added dielectric assisted ED machining resulted in increased MRR and R_a, in comparison to the case when aluminium (Al) was used as additive.	[53]

#	Powder material	Process parameters	Dielectric	Powder size/concentration	Tool	Workpiece	Observations	Ref.
3	Boron carbide (B$_4$C)	Combined EDM-USM technique; I: 3, 6, 12, 15 A; V: 60 V; PD: 100, 200, 400, 500 μs; Amplitude of ultrasonic: 30 μm	Kerosene and distilled water	Size: 3, 9 μm; PConc.: 60, 90, 120 g/l	Electrolytic Copper (Cu)	Ti64	• The coupled EDM-USM process produced better MRR. • Low SR was observed with coupled EDM-USM process assisted by kerosene dielectric. • RLT on surfaces ED machined using distilled water is comparatively lower	[54]
		I: 4, 8, 12 A; V: 20, 30, 40 V; T$_{on}$: 90, 150, 200 μs; T$_{off}$: 30, 45, 60 μs	Kerosene	Size: 100 μm; PConc.: 0, 5, 10 g/l; Optimum PConc.: 10g/l	Copper (Cu)	Ti64	• Optimal increment in MRR was achieved at a peak I value of 8 A. • High MRR and low TWR is achievable at a PConc. of 10 g/l.	[55]
		SV: 110 V; I: 20A; T$_{on}$: 65 μs; T$_{off}$: 45 μs; DFP: 0.5 MPa; MT: 30 min	SAE 450 (EDM Oil)	Size: 20-30 μm; PConc.: 1-20 g/l; Optimum PConc.: 15g/l	Copper (Cu)	Ti64	• The Ti64 work material was associated with adhering effect at high concentration of the added B$_4$C. • High electric field density and increment in IEG.	[56]
4	Carbon (C)	SV: 90 V; I: 3 A; PD: 2-200 μs; T: 10%	Kerosene	Size: 30 μm; PConc.: 0-15 g/l	Titanium (Ti)	SDD 61 (Tool steel)	• Formation of TiC (hard) layer on the workpiece.	[57]
5	Carbon nanotubes (CNTs)	I: 5, 12, 24, 48 A; T$_{on}$: 10, 25, 100, 400 μs	Oil Flux ELF2	PConc.: 2 g/l	Copper (Cu)	Ti64	• Enhanced machining stability through decrement in number of inadequate discharges. • Enhanced MRR (at long PD combined with low values of I) and reduced TWR. • Size and length of microcracks are minimized.	[58]

(continued)

TABLE 12.2 (Continued)
Consolidated list of PMEDM experimentations assisted by dielectric added with distinct powders/particulates

Sl. No.	Additive	Process Conditions	Dielectric Fluid	Additive Size & Concentration	Tool Material	Workpiece Material	Major findings	Ref.
6	Hydroxyapatite powder	I: 22 A T_{on}: 12 µs	Distilled water	Size: 30 µm PConc.: 0, 5, 10, 15 g/l	Ti (Pure)	Ti64	• High SR was produced, compared to the conventional EDM with deionized water and abrasive polished surfaces. • PConc. was the major factor that governs bio-compatibility, cellular activity, and surface wettability.	[59]
7	Graphite	SV: 150 V PD: 5-40 µs Pulse energy (PE): 1-500 mJ IEG: 25, 50, 75 µm	Kerosene	Size: 10 µm PConc.: 0.25-6 g/l Optimum PConc.: 4g/l	Copper (Cu), Cathode	Mild steel (MS), Anode	• At optimum PConc. and for a spark gap of 50 µm, the breakdown voltage of the dielectric was reduced by 30%. • A photocell or light detecting resistance (LDR) setup was suggested to be placed to monitor the PConc. of graphite in the dielectric.	[60]
		I: 12 A PD: 25, 100 µs T_{off}: 12.5 µs MT: 13 s	Kerosene	Size: 40.2 µm PConc.: 5, 10, 15 g/l	Copper (Cu)	Prismatic steel	• The powder concentration and T_{on} were the major factors that govern the process characteristics.	[61]
		SV: 110 V C*: 3300 pF VF: 1000 Hz VA: 1.5 µm	Kerosene	Size: 55 µm PConc.: 0, 2, 5, 7, 10, 15 g/l	Tungsten (W)	Silver-tungsten	• The MRR was improved quintuple times compared to the ED machining in pure kerosene.	[62]

S. No.	Powder	Parameters	Dielectric	Powder/Surfactant details	Tool electrode	Workpiece	Observations	References
		I: 6, 12 A T_{on}: 50, 100 μs T_{off}: 100 μs DFP: 0.1, 0.4 bar	Kerosene	Powder Size: < 37 μm PConc.: 1.3, 5, 10, 15 g/l	Electrolytic copper (Cu)	Ti64, AISI 1040 steel	• The addition of graphite powder aided in stabilizing the process and reduced the carbon deposition on the machined surface. • The MRR, RW and R_a got improved substantially as compared to machining with kerosene dielectric.	[63]
		SV: 110 V I: 10, 15, 20 A V: 65 V	EDM Oil (spark erosion 450)	Powder Size: 20 μm PConc.: 4.5, 9, 13.5 g/l Surfactant Span 20 SConc.: 4, 6, 8 g/l	Electrolytic Copper (Cu), Cathode	Ti6Al4V, Anode	• A proportionate increment in RLT got observed with the rise in I and PConc. • I and surfactant concentration have primary control over MMR and TWR whereas, I and PConc. possesses more monitors the SR and RLT.	[64]
			EDM Oil	Powder Size: 20-30 μm PConc.: 4.5, 9, 13.5 g/l Surfactant Non-ionic SPAN 20 SConc.: 4, 6, 8 g/l	Copper (Cu)	Stainless steel (PH17-4)	• The increment in surfactant conc. improved the electrical conductivity of the dielectric.	[13]
8	Molybdenum (Mo)	SV: 200 V I: 1, 3, 6, 12 A T_{on}: 12.8, 25, 50 μs T_{off}: 50 μs	Hydro-carbon fluid	Size: < 5 and <15 μm PConc.: 3 g/l	Electrolytic copper (Cu)	AISI H13 tool steel	• The infusion of Mo into the surface layers enhanced the hardness by quadruple times, as compared to the base material. • The formation of various phases viz., Mo_xC, Fe-Mo, and other crystalline phases of Mo was identified.	[65]

(continued)

TABLE 12.2 (Continued)
Consolidated list of PMEDM experimentations assisted by dielectric added with distinct powders/particulates

Sl. No.	Additive	Process Conditions	Dielectric Fluid	Additive Size & Concentration	Tool Material	Workpiece Material	Major findings	Ref.
9	Molybdenum Di-sulphide (MoS_2)	SV: 110 V C: 330 pF Feed: μm/s	Kerosene	Size: < 2 μm PConc.: 0, 2, 5 g/l	Copper (Cu), Copper-tungsten (Cu-W), and Silver-tungsten (Ag-W)	Copper (Cu), Copper-tungsten (Cu-W), and Brass	• Improved MRR and surface quality. • Generation of flat surface, devoid of black spots.	[66]
10	Chromium (Cr)	I: 0.4, 1.2, 1.6 A V: 90 V PD: 2 μs Electrode oscillation (radius: 250 μm)	Chromium powder mixed fluid (CrPMF)	Size: 5 μm PConc.: 1-5 g/l	Copper (Cu)	SKD 11 (alloy tool steel)	• Considerable reduction in SR can be achieved through stirring effect. • The presence of Cr improves the hardness, corrosion resistance and water repellency.	[67]
		SV: 415 V I: 4, 6, 8 A τ: 54, 63, 72 % TED: 8, 12, 16 mm DFR: 50 l/h PF: 0.3 MT: 15 min	Kerosene	Size: 45-55 μm PConc.: 2, 4, 6 g/l	Copper (Cu)	EN8 steel	• I, PConc., and TED are the primary factors that govern the MRR and TWR. • For the chosen range, MRR increases with increment in PConc.	[68]
		SV: 415 V I: 4, 6, 8 A τ: 54, 63, 72 % DFR: 500 l/h PF: 0.3 MT: 15 min AoE: 50, 90, 130°	Kerosene	Size: 45-55 μm PConc.: 2, 4, 6 g/l	Copper (Cu)	EN8 steel	• Maximum MRR got achieved at I = 8 A, τ = 72%, PConc. = 6 g/l, and AoE = 130°. • For minimum TWR, the parametric ranges are, viz. I = 4 A, τ = 54%, PConc. = 2 g/l, and AoE = 50°.	[69]

#	Powder	Dielectric	Powder (size/conc.)	Process parameters	Electrode	Work material	Observations	Ref.
11	Titanium (Ti)	Kerosene	Size: 45-55 µm; PConc.: 2, 4, 6 g/l	I: 4, 6, 8 A; τ: 54, 63, 72 %; TED: 8, 12, 16 mm; MT: 15 min	Copper (Cu)	EN8 steel	• SR is governed by the variation in I (avg) and PConc. Besides, the impact of τ got suggested to as insignificant.	[70]
		Hydrocarbon oil	Size: 40-60 nm; PConc.: 2 g/l	SV: 120 V; I (peak): 6, 12, 20 A; T_{on} + T_{off}: 120+80, 210+140, 340+225 µs; τ: 60%; MT: 10 min	Copper (Cu)	Die steel D2	• Notable improvement in MRR and SR. • Limited/negligible material migration of Ti to the work material. • Ti powder has the capacity/potential to reduce the surface cracks and thereby enhance the surface finish	[71]
		Shell EDM Fluid 2A	Size: <36 µm; PConc.: 50 g/l	SV: 150 V; T_{on}: 510 µs; τ: 20, 40, 50, 80 %; MT: 15, 30, 60 min; JT: 0.5 s; WT: 2 s; DFP: 1 kg/cm²; DFR: 12 l/min; ERS: 100 rpm	Copper (Cu)	Tungsten carbide (WC90–Co10)	• At I = 20 A and τ = 50 % a high hardness (990-1750 HV) layer got formed on the work material composed of Ti and C. • The formation of TiC assisted in achieving strong bonding with the substratum. • Ti powder inclusion in the dielectric can superiorly decrease the micro-crack formation and improve micro-hardness of the layer.	[72]
		EDF-K (Mitsubishi oil)	Size: <36 µm; PConc.: 50 g/l	SV: 80-320 V; I: 1-20 A; T_{on}: 2-1024 µs; T_{off}: 2-4096 µs; τ: 0.05-50 %	Copper (Cu)	Carbon steel (AISI-1049)	• Hard TiC coatings over the substrate is possible to be developed with high PConc. • The thickest layer got achieved at I = 3 A, T_{on} = 2 µs, and Toff = 1024 µs. • The hardness of the accreted layer reaches up to 1600 HV.	[73]

(continued)

TABLE 12.2 (Continued)
Consolidated list of PMEDM experimentations assisted by dielectric added with distinct powders/particulates

Sl. No.	Additive	Process Conditions	Dielectric Fluid	Additive Size & Concentration	Tool Material	Workpiece Material	Major findings	Ref.
		I: 0.1-0.8 A V: 150 V T_{on}: 10, 30, 50, 80 μs T_{off}: 30 μs	Deionized water	Size: 35-45 μm PConc.: 3, 6 g/l	Titanium (Ti, grade IV)	Titanium (Ti, grade IV)	• Thickness of the RL increases with I, T_{on} and PConc. • Ti surface was devoid/free of micro-cracks ensuing processing at I = 0.1 A, $T_{on} \leq$ 50 μs, and PConc. = 6 g/l.	[74]
12	Titanium dioxide (TiO_2)	I: 2, 5, 6, 9 A V: 80 V T_{on}: 35, 50, 100 μs T_{off}: 25 μs ERS: 0, 200, 400, 600 rpm MT: 15 min MFI: 0.38 T	Kerosene	Size: 20 nm PConc.: 0, 1, 2, 3 g/l Optimum PConc.: 1 g/l	Copper (Cu), Anode	H13 steel, Cathode	• The coupled influence of the added particulates and the rotational tool can improve the efficacy of the process. • The optimum PConc. for enhanced performance using TiO_2 nanoparticles is 1 g/l, above and below which MRR decreases and increases, respectively.	[75]
13	Titanium Carbide	I: 3, 6, 12, 15 A PD: 50, 100, 200, 400, 500 μs τ: 0.71 UA: 30 μm	Kerosene	Size: 10 μm PConc.: 60 g/l	Electrolytic Copper (Cu)	Ti-Zn-Mg alloy	• An alloyed layer got formed over the surface through combined EDM-USM technique with superior hardness and wear resistance. • TiC particles migrated/penetrated the machined surface assisting in dispersion hardening.	[76]
		I: 2.5, 4.5, 6.5 A T_{on}: 10 μs T_{off}: 9 μs IEG: 10 mm Jump: 1	Kerosene	–	Copper-tungsten (Cu (30%)-W (70%)) (90 HRB)	Mild steel (45.6 HRB)	• Low current (I = 2.5 A) and power addition in the dielectric can develop smooth surfaces.	[77]

No.	Powder	Electrode	Dielectric	Powder properties	Machining parameters	Workpiece	Observations	Ref.
14	Tungsten (W)	Electrolytic Copper (Cu)	Kerosene	Size: 10–15 μm PConc.: 4 g/l	I: 1, 3, 6, 9, 12 A V: 30, 40, 50, 60, 70 V T_{on}: 30, 60, 90, 120, 150 μs T_{off}: 8, 56, 104, 152, 200 μs	Al-alloy 6061-10% SiC	• Addition of TiC in the dielectric for PMEDM produce improved surfaces (low SR) compared to the case when Al_2O_3 is used as additive and conventional EDM. • Peak MRR achieved at high values of I and T_{on} at optimum conditions of T_{off} and V. • The MRR improved by 48.43% and RLT got reduced by 42.85%. • Transfer of W, aided in improving the quality of the work surface.	[78]
		Electrolytic Copper (Cu), Cathode	Kerosene (commercial grade)	Size: 30–40 μm PConc.: 15 g/l	V: 135 ± 5% V I: 2, 4, 6 A T_{on}: 5, 10, 20 μs T_{off}: 38, 57, 85 μs MT: 10 min	O2 (OHNS), D2 (HC-HCr), and H13 die steel, Anode	• Phenomenal increase (100%) in micro-hardness of all the specimens processed through PMEDM got reported. • Abrasion resistance of materials can be substantially improved by PMEDM assisted by dielectric added with W powders.	[79]
15	Tungsten di-sulphide (WS₂)	Brass (rod)	De-ionized water	Size: ≈ 15 μm PConc.: 6, 8, 10, 12 g/l	V: 20, 40, 60 V τ: 30, 50, 70% MT: 3 min	Ti64	• The formation of various oxides of Ti, Al, and Zn resulted in improved hardness of the coated layer. • The surface exhibit lubricating behavior due to the presence of WS_2, oxides of W and Zn leading to low SWR.	[80]

(continued)

TABLE 12.2 (Continued)
Consolidated list of PMEDM experimentations assisted by dielectric added with distinct powders/particulates

Sl. No.	Additive	Process Conditions	Dielectric Fluid	Additive Size & Concentration	Tool Material	Workpiece Material	Major findings	Ref.
		V: 20, 40, 60 V τ: 40, 50, 60% MT: 20 min	De-ionized water	PConc.: 6, 8, 10 g/l	Tungsten	Ti64	• Peak hardness value of the deposited layer is achieved at τ = 50% and PConc. = 10 g/l. • Self-lubricating characteristics can be imparted to surfaces through alloying using PMEDM technique assisted by appropriate electrode (tool) and additive materials (in dielectric fluid).	[39]
16	Copper (Cu)	SV: 200 V I: 1.5, 4 A V: 120 V T$_{on}$: 6, 25, 75 µs τ: 1/2, 2/3	Kerosene	Size: 0.07–0.08, 10–15, 100 µm PConc.: 0.25, 0.5, 1.0 g/l	Copper (Cu)	SKD-11 (mould steel)	• Negligible influence on the performance of the process.	[32]
17	Nickel (Ni)	I: 4, 6, 8 A τ: 54, 63, 72% AoE (triangular): 8, 12, 16° DFR: 500 l/h PF: 0.3 MT: 15 min	Kerosene (commercial grade)	Size: 1-2 µm PConc.: 2, 4, 6 g/l	Copper (Cu)	EN19 steel (AISI 4140)	• Maximum MRR got achieved at I = 8 A and PConc. = 6 g/l. • MRR and TWR increases with PConc. • AoE and τ has limited effect on the output parameters for the chosen range.	[81]
		I = 3 A PD: 2 µs τ: 10%	Kerosene	Size: 1-2 µm PConc.: 0, 5, 10, 40 g/l	Copper (Cu), Anode	Aluminium bronze	• The developed surface possesses superior hardness and can resist damage due to sand abrasion. • EDM of aluminium-bronze results in low electrode wear.	[82]

No.	Powder	Process parameters	Dielectric	Powder details	Tool	Workpiece	Observations	Ref.
		SV: 200 V I: 3 A V: 90 V T_{on}: 2 μs τ: 0.1	Kerosene	Size: < 5 μm PConc.: 0, 5, 10, 40 g/l	Copper (Cu)	AlBC3Al-bronze	• Reduced SR. • Increment in PConc. led to a corresponding increase in RLT. • High resistance to sand abrasion.	[57]
18	Aluminium (Al)	I: 15, 19, 23 A PD: 10, 60 100, 150 μs	EDM Oil	Size: 10 μm PConc.: 40 g/l	Red Copper (Cu)	45# Steel	• Low machining efficiency and high surface finish.	[83]
		I: 27-51 A V: 80-150 V PD: 1.5, 3, 9, 18, 25, 50, 100, 300 μs	–	Size: 40 nm PConc.: 1-3 g/l	Copper-tungsten	Ti64	• Material migration from the tool, dielectric, and the added powders can enhance the fatigue performance. • PMEDM using Al powder can infuse corrosion resistant characteristics on the Ti64 alloy surface.	[84]
		SV: 240 V I: 1, 2, 5 A V: 80-150 V T_{on}: 64-128 μs T_{off}: 128 μs τ: 0.5	Kerosene	Grain size: 1, 10, 50 μm PConc.: 2, 6 g/l	Copper (Cu)	SKD 61	• Improvement in hardness and corrosion resistance of surfaces are achievable by the utilization of a mixture of Al and Cr particulates in the dielectric for PMEDM • Use of smaller grain size particulates can assist the formation of smoother surfaces.	[85]
19	Manganese (Mn)	I: 2, 6 A T_{on}: 6.4 μs T_{off}: 100 μs	–	Size: <10 μm PConc.: 5, 10 g/l	Electrolytic Copper (Cu), Cathode	AISI H13 (tool steel)	• The PMEDM using Mn and Si powders assisted in the development of RL with high hardness value and free of defects (pores and cavities). • Various carbides (Mn_4C, Mn_4C_2 and SiC) got developed over the surfaces machined using distinct powders.	[86]

(continued)

TABLE 12.2 (Continued)
Consolidated list of PMEDM experimentations assisted by dielectric added with distinct powders/particulates

Sl. No.	Additive	Process Conditions	Dielectric Fluid	Tool Material	Additive Size & Concentration	Workpiece Material	Major findings	Ref.
		SV: 135 ± 5 V I: 4, 8, 12 A T_{on}: 100, 150, 200 μs T_{off}: 75 μs MT: 30 min	Kerosene	Copper (Cu)	PConc.: 0, 2, 4 g/l	AISID3, AISID6, H13 (die steel)	• Smooth surfaces (low SR) got realized at parametric conditions, viz. I = 4 A, T_{on} = 150 μs, and PConc. = 2 g/l, with the utilization of Mn powder dispersed in the dielectric for PMEDM.	[87]
20	Zinc (Zn)	I: 38, 47, 55 A T_{on}: 16, 32, 64 μs T_{off}: 128, 256, 512 μs	–	Copper (Cu)	Size: 80 nm PConc.: 1, 2, 3 g/l	AZ31, Magnesium (Mg) alloy	• There was significant interaction between the input parameters governing/monitoring the performance measure. • The lowest corrosion rate got achieved at PConc.: 2 g/l, I: 38 A, T_{on}: 16 μs, and T_{off}: 512 μs	[88]
		I: 170, 190, 210 A T_{off}: 48, 50, 52 FP: 0.5, 1, 1.5 Kgf/cm² TED: 8, 10, 12 mm	Kerosene	Copper (Cu, 40 BHN)	Size: 100 mesh PConc.: 2, 4, 6 g/l	EN8 steel (255 BHN)	• The MRR is primarily governed by the I, PConc. and the interaction between them. • T_{off} and TED got considered to have least impact on the MRR.	[89]

TABLE 12.3

Effect of distinct powders suspended in the dielectric on the PMEDM process responses

Sl. No.	Powder Additive	Dielectric	Work Material	Tool Material	Process Responses					
					MRR	TWR	SR	RLT	SD	Ref.
1	Si	Kerosene	Ti6Al4V	Copper	↑	↑	→	–	–	[41]
		Kerosene	Ti6Al4V (ELI)	Copper	–	–	→	←	–	[4]
2	SiC	Deionized water	Ti6Al4V	Al6081 T6	↑	–	→	↑	–	[49]
		Hydrocarbon-based dielectric	Interstitial Free Steel (IF Steel)	Copper	↑	–	←	–	→	[2]
3	C	Kerosene	SKD61	Titanium	–	–	→	–	→	[36]
4	CNTs	Oil Flux ELF2	Ti6Al4V	Copper	→	→	→	–	–	[58]
5	Graphene	Deionized water	Inconel 718	Copper	→	→	←	–	–	[90]
6	Graphite	Paraffin	EN19	–	←	–	←	→	–	[91]
7	Graphite + Span 20	EDM oil	Ti6Al4V	Copper	↑	→	→	←	–	[64]
8	HAp	Deionized water	Ti6Al4V	Titanium	–	–	→	–	–	[6]
9	Nano HAp	EDM oil	SS316 L	Copper	←	–	→	←	↓	[92]
10	Cu	Transformer oil	Inconel 718	–	↑	←	→	–	–	[14]
11	Zn	Deionized water	Ti6Al4V	Brass	←	–	→	–	–	[93]
		Compressed air + oil + metallic powder	EN31	Copper	–	–	→	–	–	[94]
12	Mn	Hydrocarbon-based oil	AISI H13 tool steel	Copper	–	–	→	←	→	[86]
13	Al	Kerosene	EN19	Brass	–	–	←	–	–	[95]
		Electra oil	AISI D3 Die Steel	Copper	←	–	–	–	–	[96]
14	Ti	Kerosene	AISI P20	W	↑	→	–	–	–	[97]
		Shell EDM Fluid 2A	WC	Copper (–)	–	–	–	–	→	[72]
15	Nano-Ti	Hydrocarbon oil	AISI D2 Steel	Copper	←	–	→	–	→	[71]
16	TiC	Kerosene	Mild Steel	Copper-Tungsten	–	–	→	↑	–	[77]
17	TiO$_2$ (US3500)	Kerosene	H13 Steel	Copper	→	→	→	–	–	[75]

(continued)

TABLE 12.3 (Continued)
Effect of distinct powders suspended in the dielectric on the PMEDM process responses

Sl. No.	Powder Additive	Dielectric	Work Material	Tool Material	Process Responses					Ref.
					MRR	TWR	SR	RLT	SD	
18	B₄C	Kerosene	Ti6Al4V	Copper	↑	←	→	-	-	[55]
19	Al₂O₃	Kerosene	A413	Copper	↑	→	→	-	-	[98]
		EDM Oil	Inconel 825	WC	↑	-	→	-	-	[99]
		Kerosene	Mild Steel	Copper-Tungsten	-	-	→	←	→	[77]
20	Nano-Al₂O₃	EDM Oil	Inconel 825	Copper	↑	-	→	←	→	[100]
21	W	Kerosene	AA6061/10%SiC	Electrolytic Copper (+)	↑	-	→	→	-	[78]
22	WC	Kerosene	SKD 61 Steel	Copper (−)	-	-	→	-	-	[101]
23	WS₂	Deionized water	Ti6Al4V	Brass	-	-	→	←	→	[80]
24	Cr	-	H11 Tool Steel	Copper	-	-	-	→	→	[12]
25	Ni	Kerosene	Aluminum Bronze, (AlBC3)	Copper	-	-	→	←	-	[36]
			EN-19 steel (AISI-4140)		↑	←	-	-	-	[81]
26	Mo	Hydrocarbon fluid	AISI H13 Tool Steel	Electrolytic Copper	-	-	→	←	→	[65]
27	MoS₂	Deionized water	Ti6Al4V	Brass	-	-	-	←	→	[102]
28	MWCNTs	-	Al–30SiCp	Copper	↑	←	→	-	-	[103]
29	Nano-Chromium + Span 20	Kerosene	AISI D2 Steel	Copper	↑	→	→	-	-	[104]
				Electrolytic Copper	-	-	-	-	-	[35]
30	Zn	-	AZ31 (Mg alloy)	Copper	-	-	→	-	→	[88]

expensive, the reduced powder consumption during processing aids in making the process more (four times than when micro-powders are used) economical. The specific process characteristic is the formation of smaller craters, but at higher spark frequency [35].

12.5.2.2   Non-metallics

Uno et al. [36] used carbon powder dispersed dielectric (kerosene) for machining of alloy tool steel (SKD61) by PMEDM technique. Titanium was used as the tool electrode. Based on the results, it was suggested that a superior hard surface can be developed on steel material through carbon powder assisted PMEDM process. The hardness of the modified surface was due to the formation of TiC. The SR was also substantially reduced as compared to the normal EDM using kerosene dielectric.

12.5.2.3   Ceramics

Al-Khazraji et al. [37] carried out the EDM of die steel (AISI D2) with SiC dispersed in the dielectric during processing and evaluated the RLT, fatigue life, and heat flux. Graphite and copper were used as tool electrodes. Reportedly, the utilization of a graphite electrode can aid in achieving higher heat flux, with and without powder additives in the dielectric. A phenomenal increment in RLT was also observed through the incorporation of SiC powder additives in the dielectric. The utilization of a combination of graphite (as tool electrode) and SiC (as dielectric additive) improve the safety factor by 7.30%. This specific observation is in contrast to that of the case when copper was used as the electrode (with and without powder additives in the dielectric) and when graphite was used as tool for machining without powder additives. Kibria et al. [38] compared the effectiveness of dielectric with B_4C added to that of pure kerosene and deionized water for micro-hole machining using the PMEDM technique. When deionized water was used as dielectric MRR and TWR significantly increased to that of kerosene. MRR increases and TWR decreases, when B_4C was suspended in base fluid for machining using deionized water and kerosene, respectively. Utilization of deionized water reduces the RLT, in contrast to kerosene. The presence of B4C in the dielectric can assist in reduced formation of deposited layer and also increases the machining time.

12.5.2.4   Solid-lubes

Mohanty et al. [39] evaluated the efficacy of EDM as a surface alloying technique through the incorporation of WS_2 powders in the dielectric (deionized water) for the processing of Ti6Al4V alloy. Tungsten was used as the tool electrode. Based on the observations, it was noticed that significant migration of WS_2 powders occurs from the dielectric to the work material surface occurs during the process and becomes a part of the modified surface. Arguably, self-lubricating properties can be imparted on distinct material surfaces through the utilization of this technique and incorporation of solid lubes to assist the PMEDM process. Tyagi et al. [40] utilized EDM as a coating method by suspending a combination of WS_2 and Cu nano-powders (30:70 and 50:50 wt.%) in the dielectric for processing of steel surfaces (SS304). The incorporation of such additives can increase the coating thickness and aid in the evolution of soft and lamellar structures. The wear behavior of the modified surface was substantially improved.

12.5.2.5 Summary

The additives in the dielectric can govern the PMEDM process and the characteristics/ features/properties of the developed surfaces. Table 12.2 shows a consolidated list of PMEDM experimentations carried out using dielectrics with distinct powder additives.

12.6 SURFACE MODIFICATION THROUGH THE EDM/PMEDM ROUTE

The effectiveness of EDM/PMEDM techniques in acting as surface modification/ coating techniques was under investigation from the late 1990s. Shunmugam et al. [105] used a powder metallurgy electrode (40%WC + 60% Fe) as a tool for improving the efficiency of the cutting tool material (workpiece, high-speed steel (HSS)). Compositional analysis and metallurgical studies confirm the transfer of anodic tool material (namely WC) to the work surface. WC-coated HSS tools exhibit improved wear resistance even under extreme pressure and temperature conditions encountered in metal cutting. Reportedly, the improvement in abrasive wear resistance and reduction in cutting forces are 25–60% and 20~50%, respectively. Tyagi et al. [106] tried to improve the tribological behavior of mild steel using an EDM technique. A powder metallurgy material (WS_2 + Cu (40:60, 50:50 and 60:40)) was used as the tool electrode. The pure WS_2 coating was not possible through this method as the powder is very dry, a solid electrode was not possible to prepare. So, a second conductive metal powder was mixed to make the green compact. Significant reduction in wear occurred from 95.75 μm (parent surface) to the least value of 6.71 μm (coated surface). The XRD profile of coating showed the various peaks of W, WS_2, Cu and Cu_2S. Another work by Wang et al. [107] incorporated a Ti powder based green compact electrode for surface modification of carbon steel through the EDM process. A compact TiC layer can be developed over the work surface. The hardness of the ceramic layer was three times more than that of the base material. It varied gradually from the surface to the base material. Reportedly, the method can find wide applicability in the field of surface modification, such as surface repairing and strengthening of cutting tools and molds. Murray et al. [108] compared the competency of electrical discharge coatings (EDCs) and laser claddings (Co-based) in imparting high hardness and wear resistant characteristics to AISI 304 (stainless steel). Laser coatings against all counter-bodies show wear tracks characterized by sharp abrasive grooving. Laser coatings show distinctive abrasive grooves typical of a softer material. EDCs show less extensive abrasion, explained by generally greater resistance to abrasive wear. The superior wear performance of the EDCs seen in most scenarios can be explained by the higher hardness of dense regions of the EDCs in comparison to those of the laser coatings. Electron backscatter diffraction (EBSD) analyses of both coating types revealed a significantly lower average grain size in the EDCs compared to in the laser coatings, which explains the higher hardness in these dense regions. In the recent times, investigations evaluating the efficiency of EDM technique to act as a coupled machining and surface enhancement technique is getting superior attention [109,110].

Mohanty et al. [102] used the μ-EDM process for developing self-lubricating surfaces on Ti6Al4V surfaces through the incorporation of MoS_2 powder (6, 8, 10, 12 g/L) in the dielectric (deionized water). The coefficient of friction decreased from 0.47 (base material) to 0.13 (sample prepared at 70% duty factor). Hence, a hard, self-lubricating, and wear-resistant coated layer is deposited over the Ti alloy surface. The average micro-hardness values (426.434 HV to 714.21 HV) increased significantly up to twice the micro-hardness of the substrate or base metal (417.61 HV) which is due to the formation of intermediate phases of titanium with molybdenum disulfide and oxides. Mohanty et al. [80] carried out the surface modification of Ti6Al4V μ-EDM process using dielectric (deionized water) added with WS_2 powder additives (6, 8, 10, 12 g/L). Reportedly, there was an improvement in micro-hardness from 421.38 HV to 881.34 HV when compared to that of the base material (417.61 HV). Specific wear rate (SWR) decreased as compared to that of the base material. Philip et al. [111] used SiC as an additive in the dielectric (deionized water) and evaluated the wear and frictional characteristics of the modified surfaces. The formation of appendage layers with ceramic characteristics has improved the wear resistance of the modified Ti6Al4V surfaces. Several researchers around the globe have tried numerous powder additives (in the dielectric) for improving the surface properties of various materials and/or for the development of functional surfaces for specific applications. Table 12.4 presents a compiled list of such research initiatives and the specific properties that can be developed on distinct material surfaces through the utilization of different powder additives for PMEDM.

12.7   LIMITATIONS/CHALLENGES

There are several drawbacks to the PMEDM process that hinder the progression of the technique to establish itself as an advanced/modern technique and a surface modification method for practical/industrial applications. A few challenging areas are as follows:

- Fabrication/development of an exclusive setup for pumping and circulation of the dielectric added with suspended particulates/powders is needed.
- There exists a critical need to ensure that the powder particles are properly dispersed in the dielectric fluid. Measures have to be taken to ensure that a uniform distribution is achieved throughout the dielectric medium.
- The type, size, shape, and properties of the additives should be chosen in such a way that it can positively complement the improvement in process dynamics and responses of the technique, and can impart unique and competent properties to the machined surface.
- The PConc has to be fixed for various types of powder additives. As the required concentration can vary substantially based on the discrete characteristics of the added powders (metallic/non-metallic/ceramic/solid lubes).

TABLE 12.4
Impact of distinct powders used to assist the PMEDM process on the properties of the machined surface

Sl. No.	Powder Additive	Dielectric	Work Material	Tool Material	WR	CR	Hardness	Hpty	Bcty	Ref.
1	SiC	Deionized water	Ti6Al4V	Copper	↑	-	↑	-	-	[111]
				Al6081 T6	-	-	↑	-	-	[49]
				Aluminium	-	-	↑	-	-	[9]
2	C	Kerosene	SKD61	Titanium	↑	-	↑	-	-	[36]
3	Mg	Oil	AISI D3 Die Steel	Copper, Graphite	-	-	↑	-	-	[112]
4	HAp	Deionized water	Ti6Al4V	Titanium	-	-	-	-	↑	[6]
5	Nano HAp	EDM oil	SS316 L	Copper	-	-	-	-	↑	[92]
6	Mo	Hydrocarbon fluid	AISI H13 Tool Steel	Electrolytic Copper	-	-	↑	-	-	[65]
7	MoS$_2$	Deionized water	Ti6Al4V	Brass	↑	-	↑	-	-	[102]
8	Ti	Shell EDM Fluid 2A	WC	Copper (-)	-	-	↑	-	-	[72]
9	Mn	Hydrocarbon-based oil	AISI H13 tool steel	Copper	-	-	↑	-	-	[86]
10	Ni	Kerosene	Aluminum Bronze, (AIBC3)	Copper	↑	-	↑	-	-	[36]
11	Zn	-	AZ31 (Mg alloy)	Copper	-	↑	-	-	-	[88]
12	WC	Kerosene	SKD 61 Steel	Copper (-)	-	-	↑	-	-	[101]
13	WS$_2$	Deionized water	Ti6Al4V	Brass	↑	-	↑	-	-	[80]
14	TiC	Kerosene	Mild Steel	Copper-Tungsten	-	-	↑	-	-	[77]
15	Nil	EDM3 Oil	Ti6Al4V	Copper	↑	-	↑	-	-	[113–115]
16	Nil	Oil	SS304 (Steel)	Co-based electrodes (includes Stellite 31)	↑	-	↑	-	-	[108]
17	Nil	-	SS304 (Steel)	-	-	-	-	↑	-	[116]

- Apart from maintaining one of the additives at constant PConc, care should be taken to fix the duration of machining/processing and ensure that there is no drastic impact on the PConc during the time of processing. As significant variations in the PConc (of the powder additives) can affect the performance of the PMEDM process.

- Powder aggregation/agglomeration can lead to excessive settling at the bottom of the machining tank. There are several approaches to tackle this situation; such as: (i) utilization of suitable surfactants (for example,: tween 20, tween 40, span 20, span 40, etc.); (ii) incorporation of a stirrer arrangement inside the machining tank; (iii) circulation of the dielectric-additive blend through the machining tank after proper mixing outside the system; (iv) utilization of surfactant modified powder additives or addition of compatible surfactants to the dielectric which can assist the proper separation of the powders; (v) ultrasonic vibration assisted machining; (vi) use of rotary electrodes; (vii) magnetic field assisted machining; and (viii) establishment of a closed setup in which the pump and stirrer is integrated together in the machining tank.

- Depending upon the type of powder material chosen for machining, disposal of the dielectric (added with additives) is an issue. Harmful chemicals which can adversely affect the ecosystem, cannot be carelessly dumped in the environment without proper pre-treatment. Filtering methods are widely employed for magnetic materials (additives). So, advanced filtering techniques are essential to separate the dielectric and the powders after processing. The by-products formed through decomposition of the dielectric and/or the powders can cause negative effects.

- There also exists the possibility for the formation of undesirable compounds on the machined surface, powders can get deposited on the tool surface and block the initiation of the progressive discharges, and excessive deposition in the tank can degrade the transparency of the dielectric and prevent the operator from being able to monitor the process. Hence, adequate attention in this regard should be given during the selection of dielectric and the additive.

- The cost of the powder can be another factor that can create a problem in the selection of suitable powders for specialized applications. As per the market demand, the cost of nano-sized particles is significantly higher than the micro-sized particles, although they have an upper hand in producing better results in terms of achieving stable machining and production of finished surfaces.

- Selection of dielectric is another concern. Their choice should be made in such as way that it can help the added additives to achieve the desired/expected results.

- Risk to the operator should also be prevented. The gases emitting from the EDM machine are highly toxic and can lead to serious/severe health problems. Preventive measures should be taken to ensure that the gases emitted during decomposition of the dielectric/additives are not inhaled by the operating personals.

12.8   CONCLUSION

Based on an in-depth review of the PMEDM process, through a multi-level/multi-dimensional approach, the following conclusions are derived. This work can be extremely useful in selection of specific powders (as additives), dielectric (as discharge medium), work material, and tool electrodes to meet various machining/processing requirements for distinct basic and advanced applications. The following points should be noted:

- EDM with distinct dielectrics have a significant impact on the surface roughness and/or the surface integrity of the work material. Utilization of water-based dielectrics can produce surfaces with lower roughness; nonetheless, the width of the HAZ and the propensity for the formation of microcracks and other surface defects are comparatively higher than in the case when hydrocarbon (oil-based) dielectrics are used.
- In PMEDM, the suspended particles/powders in the dielectric can divide the main discharge channel into multiple (secondary) sub-discharges, generally referred to as the DSE. The particles involved in activating the secondary discharges have a higher probability of penetrating the molten metal pool and forming a part of the machined surface. This mechanism assists the migration of material from the powder particles dispersed in the dielectric, which is largely governed by the concentration of such particulates which help in the formation of near-uniformly distributed sub-discharges. Besides, the bridging effect enabled/caused by the particles, leads to the generation of several small and shallow craters, which is aided by the increased propensity for distributed discharges or the DSE.
- The utilization of micro and/or nano powder additives in the processing of various materials can improve the efficacy of the PMEDM process. Considering the efficiency and the ability to develop surfaces with superior surface integrity, the latter has an upper hand. Although, nano-powders can be expensive, if they are capable enough to reduce the powder consumption, an economic status is not far from reachable.
- A general observation regarding the impact of the powder addition in the dielectric for PMEDM on the process responses is the increment in MRR and decrement in SR (improved surface integrity). Nevertheless, the trends may vary substantially if the process input parameters and the powder characteristics (such as PConc) are not optimized using standard methods. A negligible number of investigations have focused on the impact of powder inclusion on the RLT and SD, hence it remains an open arena for advanced research.
- The specific properties that can be imparted on material surfaces through PMEDM includes WR, CR, hardness, Hpty, and Bcty. Inclusion of majority of the powder additives (regardless of their type and properties) can substantially improve the hardness of the modified parent material surfaces. Specific powder additives should be chosen to improve the WR (SiC, C, MoS_2, Ni, WS_2), CR (Zn, Cr), Hpty, and Bcty (HAp, Nano-HAp).

Abbreviations

Bcty – Biocompatibility
CR – Corrosion resistance
CNTs – Carbon nano-tubes
DFP – Dielectric flushing pressure

DFR – Dielectric flushing rate
DSE – Discharge separation effect
EBSD – Electron beam secondary diffraction
EDM – Electrical discharge machining
ERS – Electrode rotational speed
FP – Flushing pressure
HAp – Hydroxyapatite (Powder)
HAZ – Heat affected zone
Hpty – Hydrophobicity
IEG – Inter electrode gap
JT – Jump time
MDR – Material deposition rate
MFI – Magnetic field intensity
MP – Melting point
MR – Machining rate
MRR – Material removal rate
MT – Machining time
MWCNTs – Multi-walled carbon nano-tubes
PConc – Powder Concentration
PD – Pulse duration
PE – Pulse energy
PMEDM – Powder mixed electrical discharge machining
PR – Power rating
REWR – Relative electrode wear rate
RLT – Recast layer thickness
RoE – Rotation of the electrode
SD – Surface defects
SF – Surface finish
SMT – Surface modification technique
SR – Surface roughness
SV – Servo voltage
SWR – Specific wear rate
TED – Tool electrode diameter
TWR – Tool wear rate
UA – Ultrasonic amplitude
USM – Ultrasound machining
VA – Vibration amplitude
VF – Vibration frequency
WR – Wear resistance
WT – Working time

Notations
Al_2O_3 – Aluminium oxide (Alumina)
B_4C – Boron carbide
BHN – Brinell harness number
C – Carbon

C* - Capacitance
Co - Cobalt
Cu – Copper
HRB – Rockwell hardness (B)
HRC – Rockwell hardness (C)
I - Current
Inconel 718 – Nickel alloy
Mg - Magnesium
Mn – Manganese
Mo – Molybdenum
MoS_2 – Molybdenum disulphide
Ni – Nickel
Si – Silicon
Si – Silicon carbide
τ – Duty factor
Ti – Titanium
TiC – Titanium carbide
TiO_2 – Titanium dioxide
Ti6Al4V – Titanium alloy (Grade V)
T_{on} – Pulse on time
T_{off} – Pulse off time
V - Voltage
W – Tungsten
WC – Tungsten carbide
WS_2 – Tungsten disulphide
Zn – Zinc

REFERENCES

[1] J.T. Philip, J. Mathew, B. Kuriachen, Transition from EDM to PMEDM – Impact of suspended particulates in the dielectric on Ti6Al4V and other distinct material surfaces: A review, J. Manuf. Process. 64 (2021) 1105–1142. https://doi.org/10.1016/j.jmapro.2021.01.056.

[2] B. Ekmekci, Y. Ersöz, How Suspended Particles Affect Surface Morphology in Powder Mixed Electrical Discharge Machining (PMEDM), Metall. Mater. Trans. B. 43 (2012) 1138–1148. https://doi.org/10.1007/s11663-012-9700-0.

[3] J.T. Philip, J. Mathew, B. Kuriachen, Numerical simulation of the effect of crater morphology for the prediction of surface roughness on electrical discharge textured Ti6Al4V, J. Brazilian Soc. Mech. Sci. Eng. 42 (2020) 248. https://doi.org/10.1007/s40430-020-02321-6.

[4] M. Umar Farooq, M. Pervez Mughal, N. Ahmed, N. Ahmad Mufti, A.M. Al-Ahmari, Y. He, On the Investigation of Surface Integrity of Ti6Al4V ELI Using Si-Mixed Electric Discharge Machining, Materials (Basel). 13 (2020) 1549. https://doi.org/10.3390/ma13071549.

[5] N. Beri, S. Maheshwari, C. Sharma, A. Kumar, Technological Advancement in Electrical Discharge Machining with Powder Metallurgy Processed Electrodes: A

Review, Mater. Manuf. Process. 25 (2010) 1186–1197. https://doi.org/10.1080/10426 914.2010.512647.

[6] N. Ekmekci, B. Ekmekci, Electrical Discharge Machining of Ti6Al4V in Hydroxyapatite Powder Mixed Dielectric Liquid, Mater. Manuf. Process. 31 (2016) 1663–1670. https://doi.org/10.1080/10426914.2015.1090591.

[7] M. Kunieda, K. Yanatori, Study on debris movement in EDM gap, Int. J. Electr. Mach. 2 (1997) 43–49.

[8] W. Zhao, Q. Meng, Z. Wang, The application of research on powder mixed EDM in rough machining, J. Mater. Process. Technol. 129 (2002) 30–33. https://doi.org/ 10.1016/S0924-0136(02)00570-8.

[9] B. Ekmekci, H. Yaşar, N. Ekmekci, A Discharge Separation Model for Powder Mixed Electrical Discharge Machining, J. Manuf. Sci. Eng. 138(8) (2016) 081006 (9 pages). https://doi.org/10.1115/1.4033042.

[10] K. Surani, Effect of Powder Mixed Electrical Discharge Machining (PMEDM) on Machining Parameters of Various Materials with Different Powders: A Review, Int. J. Res. Appl. Sci. Eng. Technol. 8 (2020) 614–630. https://doi.org/10.22214/ijra set.2020.32571.

[11] N.A.J. Hosni, M.A. Lajis, Multi-response optimization of the machining characteristics in electrical discharge machining (EDM) using span-20 surfactant and chromium (Cr) powder mixed, Materwiss. Werksttech. 50 (2019) 329–335. https://doi.org/10.1002/ mawe.201800204.

[12] S. Tripathy, D.K. Tripathy, Optimization of Process Parameters in Powder-Mixed EDM, in: 2018: pp. 239–283. https://doi.org/10.1007/978-981-10-8767-7_10.

[13] V.V. Reddy, A. Kumar, P.M. Valli, C.S. Reddy, Influence of surfactant and graphite powder concentration on electrical discharge machining of PH17-4 stainless steel, J. Brazilian Soc. Mech. Sci. Eng. 37 (2015) 641–655. https://doi.org/10.1007/s40 430-014-0193-4.

[14] S. Kumar Sahu, B. Dey, S. Datta, Selection of appropriate powder-mixed dielectric media (kerosene and used transformer oil) for desired EDM performance on Inconel 718 super alloys, Mater. Today Proc. 18 (2019) 4111–4119. https://doi.org/10.1016/ j.matpr.2019.07.355.

[15] S. Chakraborty, V. Dey, S.K. Ghosh, A review on the use of dielectric fluids and their effects in electrical discharge machining characteristics, Precis. Eng. 40 (2015) 1–6. https://doi.org/10.1016/j.precisioneng.2014.11.003.

[16] B. Singaravel, K.C. Shekar, G.G. Reddy, S. Deva Prasad, Performance Analysis of Vegetable Oil as Dielectric Fluid in Electric Discharge Machining Process of Inconel 800, Mater. Sci. Forum. 978 (2020) 77–83. https://doi.org/10.4028/www.scientific. net/MSF.978.77.

[17] A. Erden, D. Temel, Investigation on the Use of Water As a Dielectric Liquid in E.D.M., in: Proc. Twenty-Second Int. Mach. Tool Des. Res. Conf., Macmillan Education UK, London, 1982: pp. 437–440. https://doi.org/10.1007/978-1-349-06281-2_54.

[18] W. König, L. Jörres, Aqueous Solutions of Organic Compounds as Dielectrics for EDM Sinking, CIRP Ann. 36 (1987) 105–109. https://doi.org/10.1016/S0007-8506(07)62564-5.

[19] T. Masuzawa, Machining Characteristics of E.D.M. Using Water As a Dielectric Fluid, in: Proc. Twenty-Second Int. Mach. Tool Des. Res. Conf., Macmillan Education UK, London, 1982: pp. 441–447. https://doi.org/10.1007/978-1-349-06281-2_55.

[20] S.L. Chen, B.H. Yan, F.Y. Huang, Influence of kerosene and distilled water as dielectrics on the electric discharge machining characteristics of Ti–6A1–4V, J. Mater. Process. Technol. 87 (1999) 107–111. https://doi.org/10.1016/S0924-0136(98)00340-9.

[21] F.N. Leão, I.R. Pashby, A review on the use of environmentally-friendly dielectric fluids in electrical discharge machining, J. Mater. Process. Technol. 149 (2004) 341–346. https://doi.org/10.1016/j.jmatprotec.2003.10.043.

[22] Y. Liu, R. Ji, Y. Zhang, H. Zhang, Investigation of emulsion for die sinking EDM, Int. J. Adv. Manuf. Technol. 47 (2010) 403–409. https://doi.org/10.1007/s00170-009-2209-2.

[23] Y. Zhang, Y. Liu, R. Ji, B. Cai, Y. Shen, Sinking EDM in water-in-oil emulsion, Int. J. Adv. Manuf. Technol. 65 (2013) 705–716. https://doi.org/10.1007/s00170-012-4210-4.

[24] M. Kunieda, S. Furuoya, N. Taniguchi, Improvement of EDM Efficiency by Supplying Oxygen Gas into Gap, CIRP Ann. 40 (1991) 215–218. https://doi.org/10.1016/S0007-8506(07)61971-4.

[25] M. Kunieda, M. Yoshida, N. Taniguchi, Electrical Discharge Machining in Gas, CIRP Ann. 46 (1997) 143–146. https://doi.org/10.1016/S0007-8506(07)60794-X.

[26] M. Kunleda, Y. Miyoshi, T. Takaya, N. Nakajima, Y. ZhanBo, M. Yoshida, High Speed 3D Milling by Dry EDM, CIRP Ann. 52 (2003) 147–150. https://doi.org/10.1016/S0007-8506(07)60552-6.

[27] Z. Yu, T. Jun, K. Masanori, Dry electrical discharge machining of cemented carbide, J. Mater. Process. Technol. 149 (2004) 353–357. https://doi.org/10.1016/j.jmatprotec.2003.10.044.

[28] M.L. Jeswani, Electrical discharge machining in distilled water, Wear. 72 (1981) 81–88. https://doi.org/10.1016/0043-1648(81)90285-4.

[29] M.G. Xu, J.H. Zhang, Y. Li, Q.H. Zhang, S.F. Ren, Material removal mechanisms of cemented carbides machined by ultrasonic vibration assisted EDM in gas medium, J. Mater. Process. Technol. 209 (2009) 1742–1746. https://doi.org/10.1016/j.jmatprotec.2008.04.031.

[30] M. Kunieda, T. Takaya, S. Nakano, Improvement of Dry EDM Characteristics Using Piezoelectric Actuator, CIRP Ann. 53 (2004) 183–186. https://doi.org/10.1016/S0007-8506(07)60674-X.

[31] S. Joshi, P. Govindan, A. Malshe, K. Rajurkar, Experimental characterization of dry EDM performed in a pulsating magnetic field, CIRP Ann. 60 (2011) 239–242. https://doi.org/10.1016/j.cirp.2011.03.114.

[32] Y.-F. Tzeng, C.-Y. Lee, Effects of Powder Characteristics on Electrodischarge Machining Efficiency, Int. J. Adv. Manuf. Technol. 17 (2001) 586–592. https://doi.org/10.1007/s001700170142.

[33] T. Yih-fong, C. Fu-chen, Investigation into some surface characteristics of electrical discharge machined SKD-11 using powder-suspension dielectric oil, J. Mater. Process. Technol. 170 (2005) 385–391. https://doi.org/10.1016/j.jmatprotec.2005.06.006.

[34] M.P. Jahan, M. Rahman, Y.S. Wong, Study on the nano-powder-mixed sinking and milling micro-EDM of WC-Co, Int. J. Adv. Manuf. Technol. 53 (2011) 167–180. https://doi.org/10.1007/s00170-010-2826-9.

[35] N.A.J. Hosni, M.A. Lajis, Experimental investigation and economic analysis of surfactant (Span-20) in powder mixed electrical discharge machining (PMEDM) of AISI D2 hardened steel, Mach. Sci. Technol. 24 (2020) 398–424. https://doi.org/10.1080/10910344.2019.1698609.

[36] Y. Uno, A. Okada, S. Cetin, Surface modification of EDMed surface with powder mixed fluid, in: 2nd Int. Conf. Des. Prod. Dies Molds, 2001: p. 154.

[37] A. Al-Khazraji, S.A. Amin, S.M. Ali, The effect of SiC powder mixing electrical discharge machining on white layer thickness, heat flux and fatigue life of AISI D2 die steel, Eng. Sci. Technol. an Int. J. 19 (2016) 1400–1415. https://doi.org/10.1016/j.jestch.2016.01.014.

[38] G. Kibria, B.R. Sarkar, B.B. Pradhan, B. Bhattacharyya, Comparative study of different dielectrics for micro-EDM performance during microhole machining of Ti-6Al-4V alloy, Int. J. Adv. Manuf. Technol. 48 (2010) 557–570. https://doi.org/10.1007/s00170-009-2298-y.

[39] S. Mohanty, V. Kumar, R. Tyagi, S. Kumar, B. Bhushan, A.K. Das, A.R. Dixit, Surface alloying using tungsten disulphide powder mixed in dielectric in micro-EDM on Ti6Al4V, IOP Conf. Ser. Mater. Sci. Eng. 377 (2018) 012040. https://doi.org/10.1088/1757-899X/377/1/012040.

[40] R. Tyagi, K. Darmalingam, V. Shankar Patel, A. Kumar Das, A. Mandal, Deposition of WS2 and Cu nanopowder coating using EDC process and its analysis, Mater. Today Proc. 18 (2019) 5170–5176. https://doi.org/10.1016/j.matpr.2019.07.515.

[41] N. Gosai, A. Joshi, Effect of Powder Concentration in EDM Process with Powder-Mixed Dielectric (PMD-EDM), Appl. Mech. Mater. 813–814 (2015) 304–308. https://doi.org/10.4028/www.scientific.net/AMM.813-814.304.

[42] F. Klocke, D. Lung, G. Antonoglou, D. Thomaidis, The effects of powder suspended dielectrics on the thermal influenced zone by electrodischarge machining with small discharge energies, J. Mater. Process. Technol. 149 (2004) 191–197. https://doi.org/10.1016/j.jmatprotec.2003.10.036.

[43] H.K. Kansal, S. Singh, P. Kumar, Parametric optimization of powder mixed electrical discharge machining by response surface methodology, J. Mater. Process. Technol. 169 (2005) 427–436. https://doi.org/10.1016/j.jmatprotec.2005.03.028.

[44] H.K. Kansal, S. Singh, P. Kumar, Effect of Silicon Powder Mixed EDM on Machining Rate of AISI D2 Die Steel, J. Manuf. Process. 9 (2007) 13–22. https://doi.org/10.1016/S1526-6125(07)70104-4.

[45] N. Mohri, N. Saito, M. Higashi, N. Kinoshita, A New Process of Finish Machining on Free Surface by EDM Methods, CIRP Ann. 40 (1991) 207–210. https://doi.org/10.1016/S0007-8506(07)61969-6.

[46] P. Peças, E. Henriques, Effect of the powder concentration and dielectric flow in the surface morphology in electrical discharge machining with powder-mixed dielectric (PMD-EDM), Int. J. Adv. Manuf. Technol. 37 (2008) 1120–1132. https://doi.org/10.1007/s00170-007-1061-5.

[47] P. Peças, E. Henriques, Influence of silicon powder-mixed dielectric on conventional electrical discharge machining, Int. J. Mach. Tools Manuf. 43 (2003) 1465–1471. https://doi.org/10.1016/S0890-6955(03)00169-X.

[48] B. Ekmekci, H. Yasar, N. Ekmekci, A Discharge Separation Model for Powder Mixed Electrical Discharge Machining, J. Manuf. Sci. Eng. 138 (2016) 081006. https://doi.org/10.1115/1.4033042.

[49] T.T. Öpöz, H. Yaşar, N. Ekmekci, B. Ekmekci, Particle migration and surface modification on Ti6Al4V in SiC powder mixed electrical discharge machining, J. Manuf. Process. 31 (2018) 744–758. https://doi.org/10.1016/j.jmapro.2018.01.002.

[50] H. Yaşar, B. Ekmekci, Ti-6Al-4V Surfaces in SiC Powder Mixed Electrical Discharge Machining, Adv. Mater. Res. 856 (2013) 226–230. https://doi.org/10.4028/www.scientific.net/AMR.856.226.

[51] H.-M. Chow, L.-D. Yang, C.-T. Lin, Y.-F. Chen, The use of SiC powder in water as dielectric for micro-slit EDM machining, J. Mater. Process. Technol. 195 (2008) 160–170. https://doi.org/10.1016/j.jmatprotec.2007.04.130.

[52] Rival, Electrical Discharge Machining of Titanium Alloy Using Copper Tungsten Electrode with SiC Powder Suspension Dielectric Fluid, 2005. https://scholar.google.co.in/scholar?hl=en&as_sdt=0%2C5&q=Electrical+Discharge+Machining+of+Titanium+Alloy+Using+Copper+Tungsten+Electrode+with+SiC+Powder+Suspension+Dielectric&btnG= (accessed June 10, 2019).

[53] H.-M. Chow, B.-H. Yan, F.-Y. Huang, J.-C. Hung, Study of added powder in kerosene for the micro-slit machining of titanium alloy using electro-discharge machining, J. Mater. Process. Technol. 101 (2000) 95–103. https://doi.org/10.1016/S0924-0136(99)00458-6.

[54] Y.C. Lin, B.H. Yan, Y.S. Chang, Machining characteristics of titanium alloy (Ti–6Al–4V) using a combination process of EDM with USM, J. Mater. Process. Technol. 104 (2000) 171–177. https://doi.org/10.1016/S0924-0136(00)00539-2.

[55] A. Shard, D. Shikha, V. Gupta, M.P. Garg, Effect of B4C abrasive mixed into dielectric fluid on electrical discharge machining, J. Brazilian Soc. Mech. Sci. Eng. 40 (2018) 554. https://doi.org/10.1007/s40430-018-1474-0.

[56] M. Kolli, A. Kumar, Effect of Boron Carbide Powder Mixed into Dielectric Fluid on Electrical Discharge Machining of Titanium Alloy, Procedia Mater. Sci. 5 (2014) 1957–1965. https://doi.org/10.1016/j.mspro.2014.07.528.

[57] Y. Uno, A. Okada, S. Cetin, Surface Modification of EDMed Surface with Powder Mixed Fluid, in: 2nd Int. Conf. Des. Prod. Dies Molds, 2001: pp. 86–92. http://citeseerx.ist.psu.edu/viewdoc/download?doi=10.1.1.492.6325&rep=rep1&type=pdf (accessed June 10, 2019).

[58] M. Shabgard, B. Khosrozadeh, Investigation of carbon nanotube added dielectric on the surface characteristics and machining performance of Ti–6Al–4V alloy in EDM process, J. Manuf. Process. 25 (2017) 212–219. https://doi.org/10.1016/j.jmapro.2016.11.016.

[59] T. Opoz, H. Yasar, M. Murphy, N. Ekmekci, B. Ekmekci, Ti6Al4V Surface Modification by Hydroxyapatite Powder Mixed Electrical Discharge Machining for Medical Application, Int. J. Adv. Eng. Pure Sci. (2019). https://doi.org/10.7240/jeps.450383.

[60] M.L. Jeswani, Effect of the addition of graphite powder to kerosene used as the dielectric fluid in electrical discharge machining, Wear. 70 (1981) 133–139. https://doi.org/10.1016/0043-1648(81)90148-4.

[61] C. Çogun, B. Özerkan, T. Karaçay, An experimental investigation on the effect of powder mixed dielectric on machining performance in electric discharge machining, Proc. Inst. Mech. Eng. Part B J. Eng. Manuf. 220 (2006) 1035–1050. https://doi.org/10.1243/09544054JEM320.

[62] G. Setia Prihandana, T. Sriani, M. Mahardika, Improvement of machining time in micro-EDM with workpiece vibration and graphite powder mixed in dielectric fluid, 2012. http://nopr.niscair.res.in/bitstream/123456789/15815/1/IJEMS 19%286%29 375-378.pdf (accessed June 10, 2019).

[63] E. Unses, C. Cogun, Improvement of Electric Discharge Machining (EDM) Performance of Ti-6Al-4V Alloy with Added Graphite Powder to Dielectric, Strojniški Vestn. – J. Mech. Eng. 61 (2015) 409–418. https://doi.org/10.5545/sv-jme.2015.2460.

[64] M. Kolli, A. Kumar, Effect of dielectric fluid with surfactant and graphite powder on Electrical Discharge Machining of titanium alloy using Taguchi method, Eng. Sci. Technol. an Int. J. 18 (2015) 524–535. https://doi.org/10.1016/j.jestch.2015.03.009.

[65] F.L. Amorim, V.A. Dalcin, P. Soares, L.A. Mendes, Surface modification of tool steel by electrical discharge machining with molybdenum powder mixed in dielectric fluid, Int. J. Adv. Manuf. Technol. 91 (2017) 341–350. https://doi.org/10.1007/s00170-016-9678-x.

[66] G.S. Prihandana, M. Mahardika, M. Hamdi, K. Mitsui, Effect of micro MoS2 powder mixed dielectric fluid on surface quality and material removal rate in micro-EDM processes, Trans. Mater. Res. Soc. Japan. 34 (2009) 329–332. https://doi.org/10.14723/tmrsj.34.329.

[67] R. Toshimitsu, A. Okada, R. Kitada, Y. Okamoto, Improvement in Surface Characteristics by EDM with Chromium Powder Mixed Fluid, Procedia CIRP. 42 (2016) 231–235. https://doi.org/10.1016/j.procir.2016.02.277.

[68] K. Ojha, R.K. Garg, K.K. Singh, Experimental Investigation and Modeling of PMEDM Process with Chromium Powder Suspended Dielectric, Int. J. Appl. Sci. Eng. 9 (2011) 65–81.

[69] K. Ojha, R.K. Garg, K.K. Singh, Parametric Optimization of PMEDM Process using Chromium Powder Mixed Dielectric and Triangular Shape Electrodes, J. Miner. Mater. Charact. Eng. 10 (2011) 1087–1102.

[70] K. Ojha, R.K. Garg, K.K. Singh, Effect of chromium powder suspended dielectric on surface roughness in PMEDM process, Tribol. - Mater. Surfaces Interfaces. 5 (2011) 165–171. https://doi.org/10.1179/1751584X11Y.0000000021.

[71] H. Marashi, A.A.D. Sarhan, M. Hamdi, Employing Ti nano-powder dielectric to enhance surface characteristics in electrical discharge machining of AISI D2 steel, Appl. Surf. Sci. 357 (2015) 892–907. https://doi.org/10.1016/j.apsusc.2015.09.105.

[72] P. Janmanee, A. Muttamara, Surface modification of tungsten carbide by electrical discharge coating (EDC) using a titanium powder suspension, Appl. Surf. Sci. 258 (2012) 7255–7265. https://doi.org/10.1016/j.apsusc.2012.03.054.

[73] K. Furutania, A. Saneto, H. Takezawa, N. Mohri, H. Miyake, Accretion of titanium carbide by electrical discharge machining with powder suspended in working fluid, Precis. Eng. 25 (2001) 138–144. https://doi.org/10.1016/S0141-6359(00)00068-4.

[74] S.-L. Chen, M.-H. Lin, G.-X. Huang, C.-C. Wang, Research of the recast layer on implant surface modified by micro-current electrical discharge machining using deionized water mixed with titanium powder as dielectric solvent, Appl. Surf. Sci. 311 (2014) 47–53. https://doi.org/10.1016/j.apsusc.2014.04.204.

[75] H. Baseri, S. Sadeghian, Effects of nanopowder TiO2-mixed dielectric and rotary tool on EDM, Int. J. Adv. Manuf. Technol. 83 (2016) 519–528. https://doi.org/10.1007/s00 170-015-7579-z.

[76] Y.-F. Chen, Y.-C. Lin, Surface modifications of Al–Zn–Mg alloy using combined EDM with ultrasonic machining and addition of TiC particles into the dielectric, J. Mater. Process. Technol. 209 (2009) 4343–4350. https://doi.org/10.1016/j.jmatpro tec.2008.11.013.

[77] A.A. Khan, M.B. Ndaliman, Z.M. Zain, M.F. Jamaludin, U. Patthi, Surface Modification Using Electric Discharge Machining (EDM) with Powder Addition, Appl. Mech. Mater. 110–116 (2011) 725–733. https://doi.org/10.4028/www.scienti fic.net/AMM.110-116.725.

[78] B. Singh, J. Kumar, S. Kumar, Influences of Process Parameters on MRR Improvement in Simple and Powder-Mixed EDM of AA6061/10%SiC Composite, Mater. Manuf. Process. 30 (2015) 303–312. https://doi.org/10.1080/10426914.2014.930888.

[79] S. Kumar, U. Batra, Surface modification of die steel materials by EDM method using tungsten powder-mixed dielectric, J. Manuf. Process. 14 (2012) 35–40. https://doi. org/10.1016/J.JMAPRO.2011.09.002.

[80] S. Mohanty, V. Kumar, A. Kumar Das, A.R. Dixit, Surface modification of Ti-alloy by micro-electrical discharge process using tungsten disulphide powder suspension, J. Manuf. Process. 37 (2019) 28–41. https://doi.org/10.1016/j.jmapro.2018.11.007.

[81] K. Ojha, R.K. Garg, K.K. Singh, An investigation into the effect of nickel micro powder suspended dielectric and varying triangular shape electrodes on EDM performance measures of EN-19 steel, Int. J. Mechatronics Manuf. Syst. 5 (2012) 66. https://doi.org/10.1504/IJMMS.2012.046144.

[82] Y. Uno, A. Okada, T. Yamada, Y. Hayashi, Y. Tabuchi, Surface Integrity in EDM of Aluminum Bronze with Nickel Powder Mixed Fluid., J. Japan Soc. Electr. Mach. Eng. 32 (1998) 24–31. https://doi.org/10.2526/jseme.32.70_24.

[83] W. Zhao, Q. Meng, Z. Wang, The application of research on powder mixed EDM in rough machining, J. Mater. Process. Technol. 129 (2002) 30–33. https://doi.org/10.1016/S0924-0136(02)00570-8.

[84] A.M. Abdul-Rani, A.M. Nanimina, T.L. Ginta, Surface Morphology and Corrosion Behavior in Nano PMEDM, Key Eng. Mater. 724 (2016) 61–65. https://doi.org/10.4028/www.scientific.net/KEM.724.61.

[85] B. Yan, Y. Lin, F. Huang, C. Wang, Surface Modification of SKD 61 during EDM with Metal Powder in the Dielectric, Mater. Trans. 42 (2001) 2597–2604. https://doi.org/10.2320/matertrans.42.2597.

[86] A. Molinetti, F.L. Amorim, P.C. Soares, T. Czelusniak, Surface modification of AISI H13 tool steel with silicon or manganese powders mixed to the dielectric in electrical discharge machining process, Int. J. Adv. Manuf. Technol. 83 (2016) 1057–1068. https://doi.org/10.1007/s00170-015-7613-1.

[87] M.A. Tawfiq, A.S. Hameed, Effect of Powder Concentration in PMEDM on Surface Roughness for Different Die steel Types, Int. J. Curr. Eng. Technol. 5 (2015) 3323–3329.

[88] M.A. Razak, A.M.A. Rani, N.M. Saad, G. Littlefair, A.A. Aliyu, Controlling corrosion rate of Magnesium alloy using powder mixed electrical discharge machining, IOP Conf. Ser. Mater. Sci. Eng. 344 (2018) 012010. https://doi.org/10.1088/1757-899X/344/1/012010.

[89] N.S. Khundrakpam, K. Som, S. Amandeep, G.S. Brar, Study and Analysis of Zinc PMEDM Process Parameters on MRR, Int. J. Emerg. Technol. Adv. Eng. 4 (2004) 541–546.

[90] K. Paswan, A. Pramanik, S. Chattopadhyaya, Machining performance of Inconel 718 using graphene nanofluid in EDM, Mater. Manuf. Process. 35 (2020) 33–42. https://doi.org/10.1080/10426914.2020.1711924.

[91] V. Srinivasa Sai, K. Gnana Sundari, P. Gangadhara Rao, B. Surekha, Improvement of Machining Characteristics by EDM with Graphite Powder-Mixed Dielectric Medium, (2019) 41–48. https://doi.org/10.1007/978-981-13-6374-0_6.

[92] Y. Lamichhane, G. Singh, A.S. Bhui, P. Mukhiya, P. Kumar, B. Thapa, Surface modification of 316L SS with HAp nano-particles using PMEDM for enhanced Biocompatibility, Mater. Today Proc. 15 (2019) 336–343. https://doi.org/10.1016/j.matpr.2019.05.014.

[93] A.P. Tiwary, B.B. Pradhan, B. Bhattacharyya, Influence of various metal powder mixed dielectric on micro-EDM characteristics of Ti-6Al-4V, Mater. Manuf. Process. 34 (2019) 1103–1119. https://doi.org/10.1080/10426914.2019.1628265.

[94] S. Sundriyal, R.S. Walia, Vipin, M. Tyagi, Investigation on surface finish in powder mixed near dry electric discharge machining method, Mater. Today Proc. 25 (2020) 804–809. https://doi.org/10.1016/j.matpr.2019.09.031.

[95] B. Surekha, T. Sree Lakshmi, H. Jena, P. Samal, Response surface modelling and application of fuzzy grey relational analysis to optimise the multi response characteristics of EN-19 machined using powder mixed EDM, Aust. J. Mech. Eng. 19 (2021) 19–29. https://doi.org/10.1080/14484846.2018.1564527.

[96] M. V. Kavade, S.S. Mohite, D.R. Unaune, Application of metal powder to improve metal removal rate in Electric Discharge Machining, Mater. Today Proc. 16 (2019) 398–404. https://doi.org/10.1016/j.matpr.2019.05.107.

[97] R. S., M.P. Jenarthanan, B.K. A.S., Experimental investigation of powder-mixed electric discharge machining of AISI P20 steel using different powders and tool materials, Multidiscip. Model. Mater. Struct. 14 (2018) 549–566. https://doi.org/10.1108/MMMS-04-2017-0025.

[98] M. Shahbazi Dastjerdi, A. Mokhtarian, P. Saraeian, The effect of alumina powder in dielectric on electrical discharge machining parameters of aluminum composite A413-Al2O3 by the Taguchi method, the signal-to-noise analysis and the total normalized quality loss, Int. J. Mech. Mater. Eng. 15 (2020) 5. https://doi.org/10.1186/s40712-020-00117-z.

[99] V. Kumar, A. Kumar, S. Kumar, N.K. Singh, Comparative Study of Powder Mixed EDM and Conventional EDM Using Response Surface Methodology, Mater. Today Proc. 5 (2018) 18089–18094. https://doi.org/10.1016/j.matpr.2018.06.143.

[100] D.R. Sahu, A. Kumar, B.K. Roy, A. Mandal, Parametric Investigation into Alumina Nanopowder Mixed EDM of Inconel 825 Alloy Using RSM, (2019) 175–184. https://doi.org/10.1007/978-981-13-6412-9_16.

[101] V.T. Le, T.L. Banh, X.T. Tran, N. Thi Hong Minh, Surface Modification Process by Electrical Discharge Machining with Tungsten Carbide Powder Mixing in Kerosene Fluid, Appl. Mech. Mater. 889 (2019) 115–122. https://doi.org/10.4028/www.scientific.net/AMM.889.115.

[102] S. Mohanty, B. Bhushan, A.K. Das, A.R. Dixit, Production of hard and lubricating surfaces on miniature components through micro-EDM process, Int. J. Adv. Manuf. Technol. 105 (2019) 1983–2000. https://doi.org/10.1007/s00170-019-04380-z.

[103] C. Prakash, S. Singh, M. Singh, P. Antil, A.A.A. Aliyu, A.M. Abdul-Rani, S.S. Sidhu, Multi-objective Optimization of MWCNT Mixed Electric Discharge Machining of Al–30SiCp MMC Using Particle Swarm Optimization (2018) 145–164. https://doi.org/10.1007/978-981-13-2417-8_7.

[104] M.A. Abbas, M.A. Lajis, D.R. Abbas, O.M. Merzah, M.H. Kadhim, A.A. Shamran, Influence of additive materials on the roughness of AISI D2 steel in electrical discharge machining (EDM) environment, Materwiss. Werksttech. 51 (2020) 719–724. https://doi.org/10.1002/mawe.201900243.

[105] M.S. Shunmugam, P.K. Philip, A. Gangadhar, Improvement of wear resistance by EDM with tungsten carbide P/M electrode, Wear. 171 (1994) 1–5. https://doi.org/10.1016/0043-1648(94)90340-9.

[106] R. Tyagi, A.K. Das, A. Mandal, Electrical discharge coating using WS2 and Cu powder mixture for solid lubrication and enhanced tribological performance, Tribol. Int. 120 (2018) 80–92. https://doi.org/10.1016/j.triboint.2017.12.023.

[107] Z. Wang, Y. Fang, P. Wu, W. Zhao, K. Cheng, Surface modification process by electrical discharge machining with a Ti powder green compact electrode, J. Mater. Process. Technol. 129 (2002) 139–142. https://doi.org/10.1016/S0924-0136(02)00597-6.

[108] J.W. Murray, N. Ahmed, T. Yuzawa, T. Nakagawa, S. Sarugaku, D. Saito, A.T. Clare, Dry-sliding wear and hardness of thick electrical discharge coatings and laser clads, Tribol. Int. 150 (2020) 106392. https://doi.org/10.1016/j.triboint.2020.106392.

[109] J.T. Philip, D. Kumar, S.N. Joshi, J. Mathew, B. Kuriachen, Monitoring of EDM parameters to develop tribo-adaptive Ti6Al4V surfaces through accretion of alloyed matrix, Ind. Lubr. Tribol. 72 (2019) 291–297. https://doi.org/10.1108/ILT-07-2019-0257.

[110] J.T. Philip, D. Kumar, J. Mathew, B. Kuriachen, Wear Characteristic Evaluation of Electrical Discharge Machined Ti6Al4V Surfaces at Dry Sliding Conditions, Trans. Indian Inst. Met. 72 (2019) 2839–2849. https://doi.org/10.1007/s12666-019-01760-7.

[111] J.T. Philip, D. Kumar, J. Mathew, B. Kuriachen, Sliding Behavior of Secondary Phase SiC Embedded Alloyed Layer Doped Ti6Al4V Surfaces Ensuing Electro Discharge Machining, (2020) 163–172. https://doi.org/10.1007/978-981-15-0054-1_17.

[112] B.K. Panda, S. Kumar, Impact of Powder-mixed Electrical Discharge Machining on Surface Hardness of AISI D3 Die Steel, IEEE 10th Int. Conf. Mech. Aerosp. Eng., IEEE, (2019) 218–222. https://doi.org/10.1109/ICMAE.2019.8880966.

[113] J.T. Philip, D. Kumar, J. Mathew, B. Kuriachen, Tribological investigations of wear resistant layers developed through EDA and WEDA techniques on Ti6Al4V surfaces: Part I – Ambient temperature, Wear. (2020) 458–459203409. https://doi.org/10.1016/j.wear.2020.203409.

[114] J.T. Philip, D. Kumar, J. Mathew, B. Kuriachen, Tribological investigations of wear resistant layers developed through EDA and WEDA techniques on Ti6Al4V surfaces: Part II – High temperature, Wear. (2021) 466–467 203540. https://doi.org/10.1016/j.wear.2020.203540.

[115] J.T. Philip, D. Kumar, J. Mathew, B. Kuriachen, Experimental Investigations on the Tribological Performance of Electric Discharge Alloyed Ti–6Al–4V at 200–600 °C, J. Tribol. 142 (2020). https://doi.org/10.1115/1.4046016.

[116] S. Jithin, U. V. Bhandarkar, S.S. Joshi, Establishing EDM as a Method for Inducing Hydrophobicity on SS 304 Surfaces (2019) pp. 731–740. https://doi.org/10.1007/978-981-32-9425-7_66.

13 Hybrid Micro-EDM

Deepak G. Dilip*,¹ and Jose Mathew²
¹Mechanical Engineering Department, Mar Baselios College
of Engineering and Technology, Trivandrum, Kerala, India
²Department of Mechanical Engineering, National Institute
of Technology Calicut, Kerala, India
*Corresponding author

CONTENTS

13.1   INTRODUCTION

As renowned physicist and Nobel laureate Prof. Richard Feynman said in his lecture, 'There is plenty of room in the bottom', if we can find techniques to make our products smaller, extensive advancements for mankind are possible [1]. Countless research have been conducted to reduce the size of products. In order to accomplish that, the manufacturing processes must be scaled down to make miniaturized products. However, this is easier said than done. The biggest hurdle for a manufacturing engineer is to control the 'unit removal'. Unit removal is the material removed by a manufacturing process in one pass of the machining action [2]. In order to accomplish miniaturization, the unit removal of each process need to be such that small features within permissible dimensional tolerance can be produced. The second hurdle to the path of miniaturization is the availability of machine tool drives that are able to provide these small movements to obtain the required unit removal.

DOI: 10.1201/9781003202301-13

283

Another hurdle that engineers have in achieving miniaturization is the availability of adequate measurement techniques, which gives confidence that the required dimension and surface finish are attained. In short, we could say that the process, machine tool and metrology are the pre-requisites for miniaturization. Numerous work have been done by researchers to overcome these hurdles and this has resulted in rapid development in the machine tool and metrology industry. Today, we have processes that can manufacture features in the nanometer range and measurement techniques capable of measuring sub-nanometer features. Electric discharge machining (EDM), with its versatility and ability to machine any electrically conductive material, is a process on which numerous research has been conducted to attain effective miniaturization. As a result, micro-EDM was developed. It is the scaled down variant of EDM capable of producing components with dimensions going as low as 1 μm. Micro-EDM has all the advantages of EDM but many of the disadvantages too. In order to improve its efficacy, various hybrid machining techniques have been developed. More discussion on micro-EDM and its various hybrid machining techniques takes place in the following sections.

13.2 MICRO-EDM

13.2.1 THE BASIC PRINCIPLE OF MICRO-EDM

Micro-EDM works on the same basic principle of thermoelectric spark erosion as EDM [3]. The sparking phenomenon during micro-EDM can be separated into three important phases, namely, ignition, discharge and cooling off/interval phase [4]. Figure 13.1 illustrates the material removal phenomenon in micro-EDM. Initially the tool and workpiece are brought closer with dielectric fluid between them and an electric field applied across them. From Figure 13.1, it is clear that the discharge occurs between the closest point of the workpiece and tool. As the gap decreases, the energy column strength increases which culminates in the breaking down of the dielectric fluid, thus resulting in sparking. The material removal occurs via melting and vaporization. The sparking results in the production of debris which is partly removed by the flushing of dielectric, whereas the remaining might get re-solidified onto the surface. The net outcome is that after each discharge a small crater is formed on the tool and workpiece surface. This happens during the interval phase where zero voltage is applied across the electrodes. In order to have stable material removal, the pulse duration and pulse interval should be carefully selected [5].

13.2.2 MACRO VERSUS MICRO-EDM

Micro-EDM follows the same working principle as EDM. However, it cannot merely be considered as a scaled down model of EDM. There are some major differences which make micro-EDM unique. The most important one is the material removal per unit pulse. Unlike EDM, the primary focus of the micro-EDM process is to give the most accurate surface and dimensional characteristics. In order to achieve that,

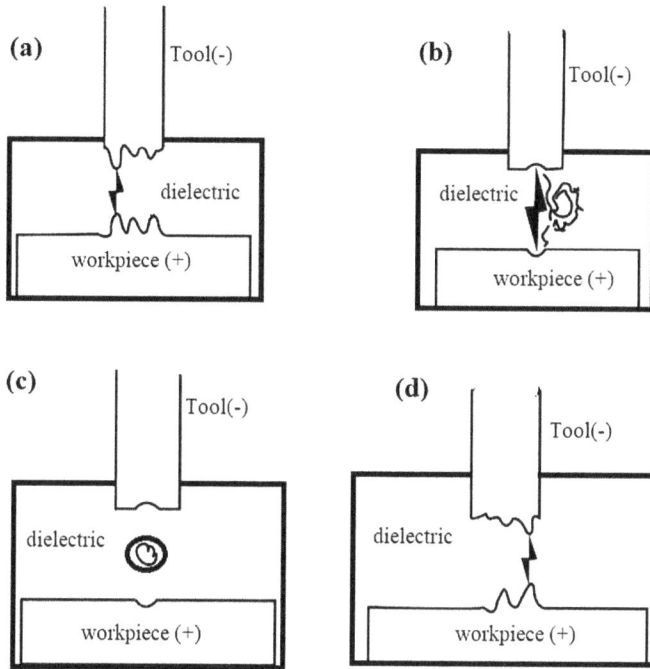

FIGURE 13.1 (a) Spark occurrence at the closest point, (b) melting and vaporization, (c) debris removal by dielectric, and (d) occurrence of next spark.

the energy per pulse and pulse duration need to be controlled. Hence, the major difference between EDM and micro-EDM comes from the difference in the ranges for energy pulse duration, which is in the range of a few milliseconds for EDM and between 50 ns to 100 µs for micro-EDM. The energy per pulse is in the order of micro joules for micro-EDM, whereas it is in the order of joules for EDM. The power supply is an important aspect to be considered in the EDM process. Generally, for EDM, a transistor-based power supply is used. However, due to the time delay in transistor-based circuits, RC-based capacitor circuits with very low capacitor ratings are used in micro-EDM. As a result of greatly reduced energy per pulse, the surface roughness is very much less in micro-EDM (less than 1 µm) compared to EDM (3–30 µm) [6].

Another important difference that is to be considered is the effect of tool size. The tool size is in the order of a few millimeters in the case of EDM. As a result, the force exerted by the dielectric fluid on the tool won't cause any significant deflection and can be neglected. However, since the tool size in micro-EDM comes in the range of only a few microns, the force exerted by the dielectric fluid can result in deflection of the tool and improper machining. Hence, the dielectric flow rate has to be chosen carefully while using micro-EDM [7]. From Figure 13.2, it can be seen that the tool dimension in micro-EDM is comparable to the grain size of the workpiece. Hence, the issues due to size effects become prominent in micro-EDM. The grain

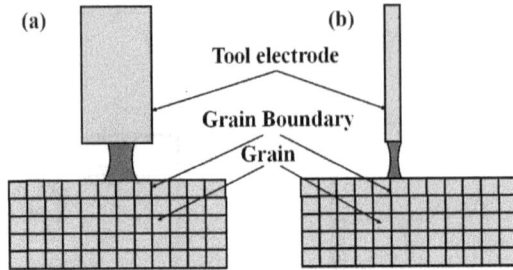

FIGURE 13.2   Schematic diagram of (a) macro and (b) micro EDM.

TABLE 13.1
EDM vs Micro – EDM

Parameters	EDM	Micro-EDM
Pulse energy per spark	Order of joule	Order of micro-joule
Pulse duration	Few milliseconds	50 ns–100 µs
Power source	Transistor-base	RC based
Surface roughness	3–30 µm	Less than 1 µm
Tool deflection due to force exerted by dielectric fluid flow	Negligible	Significant
Effect of grain size and orientation	Negligible	Significant

orientation during machining thus becomes an important factor that determines the machining performance of micro-EDM [8]. A typical example in this regard is that micro-EDM drilling will give a different performance when drilled along the grain boundary compared to when drilled within the grain [9]. Thus, it can be seen that although both EDM and micro-EDM follow the same principles of machining, there are many factors that make them distinct. The differences discussed are summarized in Table 13.1.

13.3   MICRO-EDM VARIANTS

Micro-EDM is an independent manufacturing methodology used for the machining of difficult-to-machine materials. The process, however, is classified into four major variants based on the unique strategies for machining applied by the manufacturer according to need. The major variants are discussed below.

13.3.1   Die-sinking Micro-EDM

Die-sinking micro-EDM is the simplest strategy applied in micro-EDM machining. Here, a stationary tool of any electrically conductive material/shape is made to move vertically downwards towards the workpiece. No tool rotation is imparted and tool

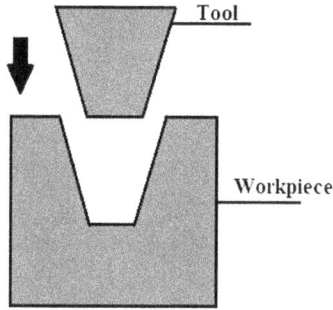

FIGURE 13.3   Die sinking micro-EDM.

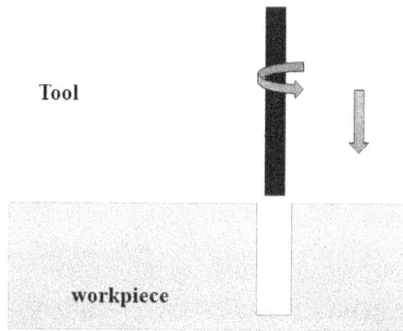

FIGURE 13.4   Micro-EDM drilling.

motion is only along the z-axis. A mirror-image of the tool will be printed on the work-piece surface. The depth of the feature is determined by controlling the tool movement in the z-axis [10]. Major applications include micro gears, micro injection molds, and the like. A schematic diagram of die-sinking micro-EDM is given in Figure 13.3.

13.3.2   MICRO-EDM DRILLING

Micro-EDM drilling is synonymous with the conventional drilling operation. Here, the tool is given a rotary motion along with motion in the z-axis. Usually, cylindrical tools are used for the purpose of micro-EDM drilling. The tool rotation assists the dielectric in efficient removal of the debris from the spark zone, thus, facilitating deep hole drilling. Major applications include micro holes for fuel injection nozzles, cooling holes in turbines, and the like [11]. The schematic of the process is depicted in Figure 13.4.

13.3.3   MICRO-EDM MILLING

Micro-EDM milling is synonymous with the conventional milling operation. It is much more complicated than micro-EDM drilling or the die sinking operations, due

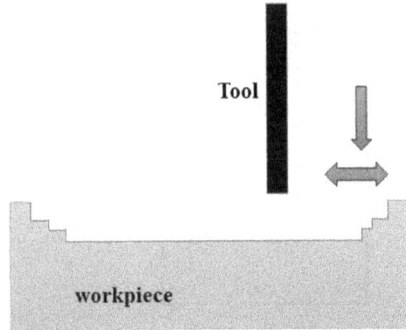

FIGURE 13.5   Micro-EDM milling.

to the simultaneous control of multiple axes in order to produce the required feature. Here, the electrode is made to move along the programmed tool path over the workpiece to produce complex features. The major disadvantages include a very small MRR and a very high tool wear rate (TWR). This makes it a viable option only for small scale applications. The major applications include micro-fluidic channels, micro-pillars, and the like [12]. A schematic diagram of the process is shown in Figure 13.5.

13.3.4   MICRO-WIRE EDM

In micro-wire EDM, the tool electrode is replaced by a travelling wire of diameter less than 100 μm. The continuously travelling wire helps in making sharp features, but issues like wire breakage, vibration, and so forth, act as a deterrent to applying this in mass production [13]. A schematic of the micro-wire EDM process is shown in Figure 13.6.

The micro-EDM process has proven to be advantageous in machining difficult to machine materials. Although a comparatively cheaper process, the very low MRR, high TWR, overcut, and so forth [11], prevent the process being used on a large scale in manufacturing.

In order to overcome the difficulties associated with micro-EDM, researchers have attempted various technologies combining micro-EDM with other machining techniques. These hybrid processes have been shown to improve overall machining performance. A brief discussion on the different hybrid machining techniques associated with micro-EDM is given below.

13.4   HYBRID MICRO-EDM

Researchers all around the globe have been trying to improve the performance characteristics of micro-EDM. Combining micro-EDM with other machining methodologies can be done in two ways, sequential and simultaneous. In the sequential process, micro-EDM is followed by another machining process which may act as a roughing/finishing operation and enhance the machining performance, whereas in the simultaneous approach, the technique is incorporated into the micro-EDM process,

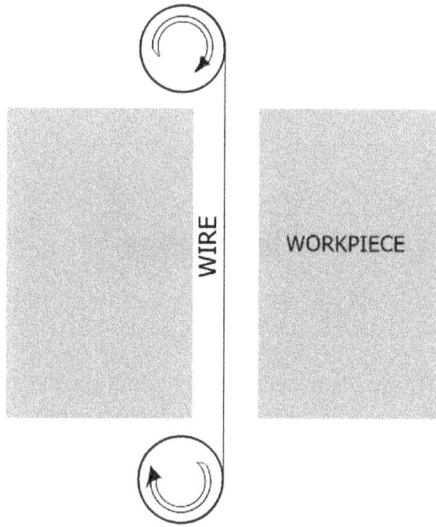

FIGURE 13.6   Micro-wire EDM.

thereby improving the machining capability of EDM. A typical example of a sequential process involves EDM followed by laser machining whereas processes such as powder mixed micro-EDM, vibration assisted micro-EDM, electro chemical discharge machining, and the like, come under the category of simultaneous processes. These hybrid micro-EDM techniques will be discussed further.

13.4.1   VIBRATION ASSISTED MICRO-EDM (VA-MEDM)

The applicability of using ultrasonic vibration to drill deep holes using EDM was studied in early 1990's and it was observed that by using ultrasonic vibration in EDM, the machining quality and efficiency could be improved. There are two important ways in which vibration applied to the workpiece/tool/dielectric can improve the overall machining performance of micro-EDM/EDM. They are as follows:

Direct contribution to material removal: The cavitation effect, occurring as a result of ultrasonic vibration can directly remove the material. The high stresses associated with the formation and collapse of the cavitation bubbles followed by sudden pressure drops in the discharge zone cause material removal by fracture. During this time, material removal also takes place due to the sparking phenomenon in micro-EDM. As a result, the overall MRR is enhanced.

Assistance in dielectric flushing: The vibratory motion provided by the ultrasonic vibration assists the dielectric fluid in removing the material from the inter-electrode gap (IEG). This results in minimization of arcing and better material removal. As a result of this the overall machining conditions are improved and better surface characteristics can be obtained [13].

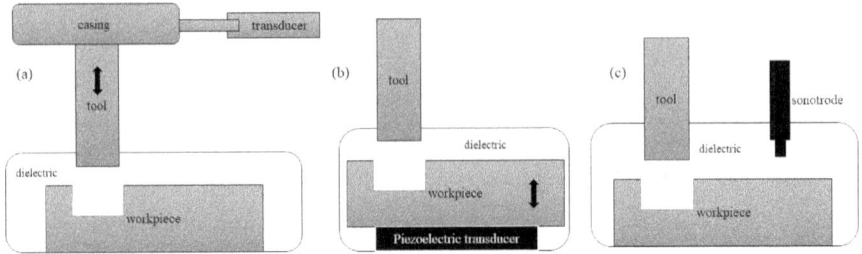

FIGURE 13.7   Vibration given to (a) tool, (b) workpiece, and (c) dielectric in micro-EDM.

Thus, improved MRR and better machining conditions are obtained by incorporating vibration assisted micro-EDM. Scientists have tried out various techniques for imparting vibration to the micro-EDM process. The first way is similar to ultrasonic machining, where ultrasonic vibration is imparted to the process by vibrating the tool. In the second method, the workpiece is vibrated in the z-axis to impart the required motion and finally in the third technique the dielectric fluid is vibrated with the help of a sonotrode to get the required movements. The schematic diagrams of all the processes are given in Figure 13.7.

In VA-MEDM, two major factors that have to be considered are the vibration frequency and amplitude of the oscillation. Selecting the optimum values is critical to the successful implementation of the process. Very high frequency or amplitude usually results in an increase in tool-workpiece contact, reduced pulse on/off time, and so forth, which will negatively affect the process. Studies have shown that providing vibration in the high/low frequency range can have positive influences on the performance measures and surface quality in features created through micro-EDM. It has its major application in high aspect ratio hole drilling. Some of the major studies done in this area are given in Table 13.2.

13.4.2   Micro-Electrochemical Discharge Machining (MECDM)

Micro-Electrochemical Discharge Machining (MECDM) combines two major micromachining processes, micro-ECM and micro-EDM. Micro-ECM works on the principle of anodic dissolution and is known for negligible tool wear and very high surface quality with significantly less surface distortions as compared to micro-EDM where significant tool wear, a heat affected zone on the workpiece, and the like, occurs. However, micro-EDM has a higher MRR. So, combining these two processes appears to be a good way to incorporate the advantages of both the processes.

The mechanism associated with MECDM is very complex. Numerous studies have been done to study the exact course of action which results in electrochemical spark machining. The mechanism may be explained as follows. The tool and workpiece are immersed in an electrolyte of very low concentration. Once electric current is passed through it, the electro chemical machining starts to result in anodic dissolution and hydrogen bubble formation at the cathode. The hydrogen bubbles start accumulating at the cathode, creating a hindrance to the smooth electron movement.

TABLE 13.2
Research work on vibration assisted micro-EDM

Authors	Findings
Sundaram et al. [13]	Studied the process parameters of VA-MEDM in ultrasonic frequency based on Taguchi method and found that the increase in the machining time significantly increases the tool wear and VA-MEDM resulted in a reduced machining time which in turn resulted in reduced tool wear.
Wansheng et al. [21]	Incorporated ultrasonic vibration of frequency 20 kHz and amplitude 2 μm in the dielectric while machining Titanium alloy and was able to achieve significant improvement in MRR and tool wear.
Meena and Azad [22]	Studied the effect of frequency, voltage, current and pulse width on the overcut, TWR and MRR during micro-hole drilling using Grey Relational Grade (GRG) and Analysis of Variance (ANOVA). They concluded that the parameter voltage played the most prominent role.
Das et al. [23]	Conducted micro-EDM drilling with ultrasonic vibration given to workpiece and observed that the recast layer thickness and micro-hardness reduced drastically when the workpiece is vibrated.
Endo et al. [24]	Conducted VA-MEDM using piezoelectric transducer in an attempt to remove debris from the IEG between a tool electrode and workpiece. Experiments were conducted with both parallel as well as perpendicular vibration to the tool, and they observed that the overall machining time, and the process stability got improved as a result of the introduction of tool vibration
Huang et al. [25]	Provided ultrasonic vibration in micro-hole drilling on Nitinol. He observed that the MRR efficiency improved by almost 60% without an appreciable change in the tool electrode wear. It was attributed to the improvement in flushing efficiency caused due to the strong stirring effect provided by the ultrasonic frequency.
Singh et al. [26]	Did a comparative study to analyze the machining performance in terms of taper angle (TA), TWR, MRR and overcut (OC) for different pulse off time, gap current and pulse on-time and found that hole drilling with VA-MEDM was better in producing micro-holes with lower TWR, higher MRR, reduced OC, better TA and smoother surface.
Gao and Liu [27]	Conducted VA-MEDM to study the MRR for both copper and stainless steel workpiece and observed that using vibration assisted drilling higher aspect ratios could be achieved.
Zhang et al. [28]	Studied the effect of vibrating the workpiece under dry EDM condition. A forced gas supply was provided through the hollow copper electrode onto the vibrating AISI 1045 steel. The obtained material removal was almost twice the MRR corresponding to dry EDM

(*continued*)

TABLE 13.2 (Continued)
Research work on vibration assisted micro-EDM

Authors	Findings
Li et al. [29]	Made micro-hole and linear groove textures on the rake face of cemented-carbide cutting tools using VA-MEDM. They observed that due to the combined action of the flowing working fluid and high-frequency vibration, cleaner textured regions with negligible micro cracks, surface burns and molten material accumulation around the cavities.
Goiogana et al. [30]	Introduced application of ultrasonic vibration in a pulsed mode for EDM finishing operations. They observed 14% and 20% decrease in the Ra value at the center and along the edges respectively for pulsed vibration assisted EDM. It was deduced that a homogeneous distribution of debris in the machining zone was achieved only through pulsed vibration which resulted in better surface finish. But too much vibration frequency or no debris in the gap occurring due to non-pulsed workpiece vibration culminated in increased surface roughness.
Xing et al. [31]	The VA-MEDM was compared with the non-ultrasonic VA-MEDM machining for a large depth-to-diameter ratio, when the amplitude of vibration was 6 μm, relative tool wear rate (RTWR) reduced by 65.8%, MRR increased by 2.4 times, overcut (OC) reduced by 32% and taper angle (θ) was reduced by 73%.
Jahan et al. [32]	Conducted deep hole drilling using VA-MEDM at low frequency on WC. The theoretical study and the associated experimental results obtained showed an improvement in machining stability, decreased EWR and increased MRR by the application of low frequency vibration. The workpiece vibration assisted in debris removal from IEG. The short-circuiting and arcing is considerably reduced which improves the overall machining stability.
Jahan et al. [33]	Developed a low frequency vibration device for VA-MEDM and concluded that amplitude of 1.5 μm and frequency of 750 Hz gave the best results. The overcut and taper ratio were observed to be very less in VA-MEDM when compared to micro-EDM without workpiece vibration
Prihandana et al. [34]	Presented an attempt to use low-frequency vibration on the stainless steel (SS304) workpiece during EDM process. The low frequency vibration resulted in an increased MRR and a decreased surface roughness and TWR. A 23% improvement in MRR was obtained.
Mahardika et al. [35]	Conducted experiments to study and optimize the machining parameters of vibrating tool electrode on micro-EDM of Poly Crystalline Diamond(PCD). The factors were capacitance, charge voltage and vibration of the tool electrode. The results showed that a 66.48% increase in MRR was observed without an increase in tool electrode wear and surface roughness.
Lee et al. [36]	Conducted studies on low frequency (10–70 Hz) range in single and multi-hole micro-EDM drilling on Cu and SS304. A 60% reduction in machining time was obtained for single hole machining with workpiece vibration at 60 Hz.

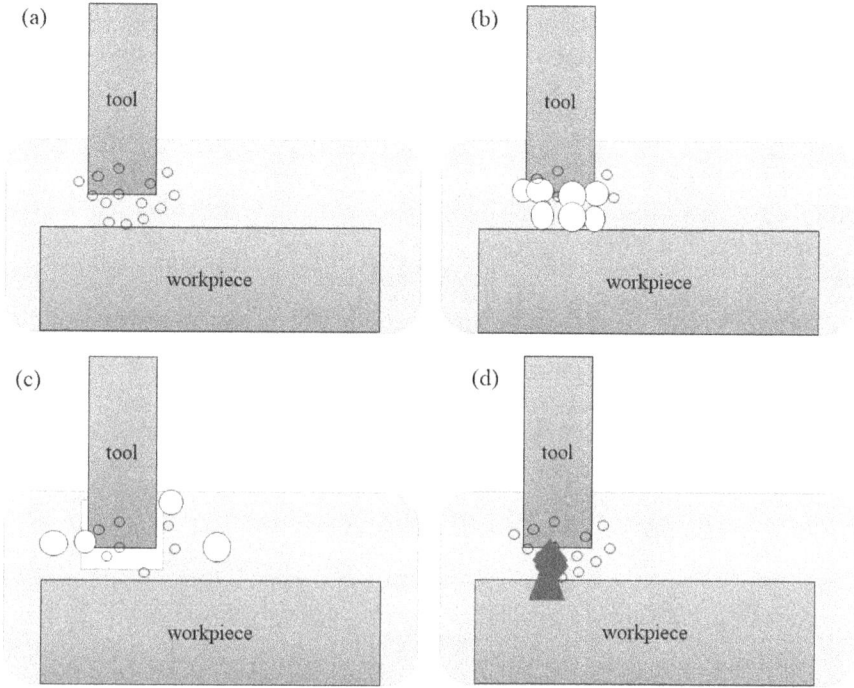

FIGURE 13.8 Micro-ECDM process: (a) bubble formation, (b) bubble expansion, (c) formation of gas film, and (d) spark at critical voltage.

This gap increases the resistance to current flow, resulting in ohmic heating. This ohmic heating in turn results in vaporizing the electrolyte in the tool vicinity, further resulting in vapor bubble formation. On increasing the voltage, these bubbles merge with each other, forming larger bubbles which exist between the tool and workpiece. On further increase of voltage, the vapor bubbles form a gas film around the electrode resulting in a sudden drop of current and an increase of current density inside the gas film. This results in the ionization of the gas and leads to the formation of a plasma channel, which in turn results in the electric discharge. Once the electric discharge occurs, the entire process is repeated and machining is accomplished [7]. A schematic diagram of the MECDM process with its various stages is given in Figure 13.8.

The major factors determining the micro-ECDM process include the supply voltage, inter-electrode gap, electrolyte type and the concentration of the electrolyte. Higher voltages result in increasing the density and thickness of the bubbles formed. The usual range is between 30–60V. Alkaline electrolytes of lower concentration are preferred for getting the ideal thin gas film layer around the tool. The inter-electrode gap should be as small as possible to ensure proper sparking. A general range observed is between 5–25 μm [14].

One of the biggest advantages of the ECDM process is its ability to machine non-conducting materials. The biggest application of the micro-ECDM process is in making micro-grooves, micro-holes and other similar features on conducting as well as on non-conducting materials. As a result, significant studies have been conducted on its machinability in materials like quartz, ceramic, and the like. The milling and drilling operations using micro-ECDM have been studied extensively and some of the findings are given in Table 13.3.

TABLE 13.3
Research on Micro–ECDM process

Authors	Findings+
Bhattacharya et al. [37]	Applied voltage of 80 V, and 25% NaOH electrolyte solution gave the optimal dimensional accuracy and moderately high machining rate for machining electrically non-conductive aluminium oxide ceramic. The tool tip shape also plays a major role in controlling the spark generation in ECDM.
Sarkar and Bhattacharya [38]	In their study of the performance of different power circuits it was seen that minimum TWR and maximum MRR can be achieved for an RL circuit at 50 V, NaOH concentration of 30 wt % and 40 mm IEG and LC circuit at 60V, NaOH concentration of 20 wt % and IEG of 20 mm respectively. Minimum Heat Affected Zone (HAZ) and minimum Radial Overcut (ROC) thickness have been obtained for RC circuit at 50 V and 10 wt % NaOH and IEG 40 mm. Hence, from an accuracy and surface finish viewpoint RC circuits are found to be favorable.
Kolhekar and Sundaram [39]	Different electrolyte concentration were studied and it was observed that at higher concentrations the energy of the spark becomes significantly higher resulting in poor surface characteristics whereas the chemical etching mechanism dominates at low concentration resulting in better surface finish and negligible heat affected zone.
Cao et al. [40]	Lower voltage and small immersion depth is suitable for producing high aspect ratio features using ECDM. Micro-holes with a diameter of 60 μm and a depth of 150 μm can be achieved with a pulse voltage of 30 V and a pulse on/off time ratio of 1ms/1ms. In ECDM milling, a pulse voltage of 23 V gave an Ra of 0.099 μm. Also in micro-milling of machined layer depth should be kept minimum in order to avoid crack formation on the workpiece and tool breakage.
Huang et al. [41]	The electrode wear was studied for the micro-hole drilling of steel using ECDM and it was obtained that the machining voltage is the most significant factor affecting electrode wear followed by tool rotation speed and tool electrode diameter. A micro hole of 0.4 μm diameter was drilled with a machining voltage of 12 V, the tool electrode rotating speed of 42,000 rpm and feed velocity of 2 μm/s.
Dong et al. [42]	From the detailed study on the effects of pulse width and peak current on the electrode wear, dimensional accuracy, taper angle and machine time, a value of 10-μs pulse width and 0.34-A peak current can be considered an ideal parameter for high speed electrochemical discharge drilling to machine micro-holes on C17200 beryllium copper alloy.
Zhang et al. [43]	Interior fluid flushing at high pressure has proven to be advantageous in enhancing the MRR, reducing the taper and improving the machining conditions

TABLE 13.3 (Continued)
Research on Micro–ECDM process

Authors	Findings+
Nguyen et al. [44]	Spark energy is the major contributor in the material removal from non-conducting materials. It was seen that tool feed rate plays a major role in improving micro-ECDM. The surface quality obtained when the tool feed rate was more than the MRR was extremely poor and can only be used for rough cutting. For good surface finish and accuracy, the tool feed rate should be significantly below the MRR.
Han et al. [45]	A 2D contour cutting technique using a textured tool for cutting a glass of 0.4 mm thickness using micro-ECDM was achieved. The best surface finish of 0.1 µm was achieved with a DC voltage of 28 V, reactive tool length of 1 mm and tool feed rate of 90 µm/min.

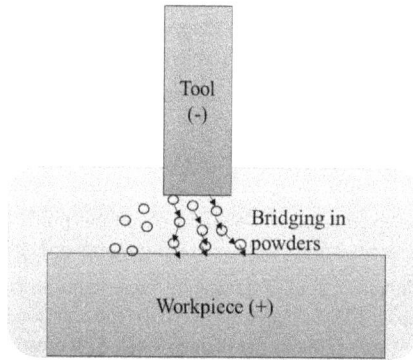

FIGURE 13.9   Schematic of PMMEDM.

13.4.3   POWDER MIXED MICRO-EDM

Another major hybrid micro-EDM machining methodology is powder mixed micro-EDM (PMMEDM), where conductive/semi-conductive powder particles are added to the dielectric before machining. The addition of powder particles changes the whole mechanism of the process. As seen from Figure 13.9, when an electric field is applied, the powder particles align themselves as chain like structures in the IEG, reducing the electrical resistance of the zone. As a result, alternate pathways/bridges are created for the spark to reach the workpiece surface. There are many significant improvements due to the addition of the powder. Firstly, it results in an increase in the inter-electrode gap for sparking. Since, the overall conductivity is increased, the energy required and time to reach the optimum inter-electrode gap is reduced. This will, in turn result in an increased MRR. Secondly, the formation of the powders in the form of chain like structures in the machining zone results in uniform distribution of the spark energy. Hence, a single pathway is replaced by multiple pathways, each sharing the energy. This results in a reduction of the spark intensity on the workpiece, resulting in an improved surface finish [15].

Along with this, particle deposition is found on the tool and/or workpiece surface. By properly controlling the process parameters, the quantity and rate of deposition can be controlled. This opens up many avenues for the application of PMMEDM in medical implant surface modification. Many factors like powder particle size, powder concentration in the dielectric, powder material, and so forth, need to be considered for the successful implementation of PMMEDM. A brief review of the various research works done in PMMEDM is given in Table 13.4.

TABLE 13.4
Research works on PMMEDM

Authors	Findings
Wong et al. [46]	The effects of introducing powder as suspension in IEG was studied. A near mirror finish was observed. The major features of the PMEDM are uniform dispersion of the electrical discharges, shorter machining time, and stable machining.
Kansal et al. [47]	Conducted a parametric optimization of PMEDM using silicon carbide powder to check the MRR and surface roughness. They found that increasing the powder concentration had a positive effect on both the responses
Pecas and Henriques [48]	Observed that use of PMEDM conditions promoted the reduction of white layer thickness, surface roughness, crater depth and crater diameter.
Jahan et al. [49]	Studied the effect of mixing different nano-powders in dielectric during micro-EDM. They observed that the addition of electrically conducting particles reduces the dielectric strength drastically and the electrical and thermal conductivity, particle size and concentration of the powders had a significant influence in the micro-EDM machinability.
Kumar et al. [50]	They tried using aluminium powder of different size and concentration and observed that the best performance is obtained for medium sized particles (mesh 325). The wear ratio and surface roughness improved by 80% and 17% respectively for EDM using medium sized particles while machining Inconel 718.
Tan and Yeo [51]	Studied the recast layer thickness under SiC powder mixed micro-EDM machining and noted a 35% decrease when the powder concentration was 0.1 g/l and the pulse on time 606 ns. They also observed that the gap distance tends to increase with increasing powder concentration but becomes steady at higher concentration levels irrespective of the pulse on time
Jahan et al. [52]	Conducted micro-EDM die sinking and milling operation on cemented tungsten carbide material using graphite nano particles mixed in the dielectric fluid. There was a visible enhancement in the machining performance and surface conditions when powder was added in the dielectric. The best surface finish was obtained for PMMEDM milling with a 0.2 g/l powder concentration at 60 V.

TABLE 13.4 (Continued)
Research works on PMMEDM

Authors	Findings
Kumar et al. [53]	Studied the effect of cryogenically cooled copper electrodes in EDM machining under graphite powder mixed conditions. A significant improvement was observed for powder mixed dielectric when the machining performance measures were TWR and wear ratio.
Kumar [54]	Applied Carbon Nano Tube (CNT) to improve surface finish in EDM. The experimental results indicated that the peak current and CNT's concentration were the most influential variables. The obtained experimental results also showed significantly better performance for Nano Powder Mixed EDM (NPMEDM) over EDM.
Kuriachen and Mathew [55]	Studied the effects of capacitance, voltage, and silicon carbide powder concentration in dielectric on TWR and MRR. From the results, it was recommended to use capacitance of 0.1 µF, powder concentration of 5 g/L, and voltage of 115 V for achieving the optimum TWR and MRR. It was also concluded that PMMEDM can also be used for surface coating based on the Si deposits observed on the workpiece using the EDS analysis
Kumar and Batra [56]	Studied the applicability of PMEDM as a surface modification technique by machining three materials, viz., OHNS die steel, D2 die steel and H13 die steel with tungsten as the powder media. Tungsten concentration of 3.25% observed on the workpiece surface and an increase in micro-hardness by over 100% validated the claim that PMEDM can be used as a surface modification technique.
Ekmekci and Ekmekci [57]	Conducted PMEDM by using Hydroxyapatite (HA) powder mixed in deionized water dielectric. The surfaces were studied using SEM, EDS and optical microscopy. High concentration of HA deposits on the workpiece was seen. Thus, PMEDM can be seen as a practical alternative for producing biocompatible interfaces or coatings for medical applications.
Opoz et al. [58]	Calculated the quantity of SiC powder material migration onto the workpiece Ti-6Al-4V during PMEDM. They noticed that the discharge current negatively affected the material deposition. For more material deposition the ideal condition would be low discharge energy and high powder concentration in dielectric.
Prihandana et al. [59]	Examined the effect of ultrasonic vibration in PMMEDM studies and observed that there was drastic improvement in the MRR when vibration was given to workpiece while being machined in the powder mixed dielectric media.
Mughal et al. [60]	Using Micro-EDM with SiC powder mixed dielectric on Ti ($\alpha + \beta$) ELI medical alloy the osseointegration was improved making it more suitable to be used as implants

13.4.4 OTHER HYBRID MICRO-EDM PROCESSES

Magnetic field assisted micro-EDM (MFAMEDM) is another hybrid micro-EDM process in which research has been conducted. It has been observed that applying a magnetic field to the micro-EDM process enhances its machining capability. The widely accepted theory is that the system is subjected to both centrifugal and magnetic forces when a magnetic field is applied to a micro-EDM process as shown in Figure 13.10. The vector sum of these forces acts in a direction perpendicular to the workpiece location thus pushing the debris particles from the discharge zone at the end of the pulse on time [16]. This helps in improving the machining performance in micro-EDM.

The MFAMEDM process has shown significant improvement in the drilling of high aspect ratio holes [17] and it has been shown to significantly reduce the micro hardness and recast layer thickness on surface created by EDM [18]. These features make it a promising technique warranting extensive studies.

Sequential Micro-EDM and laser micromachining is another hybrid machining technique used in micro-EDM where both laser micromachining and micro-EDM are combined. Laser micromachining is known for its swift machining time but it has poorer dimensional accuracy and results in significant thermal damage whereas micro-EDM has a much lower MRR and significantly better surface characteristics and dimensional accuracy. In this hybrid process, laser micromachining is followed by micro-EDM in such a way that rough machining is accomplished by laser micromachining and finish machining is achieved by micro-EDM. A 70% reduction in drilling time in diesel injection nozzles was observed by Li et al. [19] during sequential laser and micro-EDM. Another major application of this process is in the assembly of complex 3d microstructures. The microstructures are manufactured by micro-EDM and assembly is accomplished by welding these parts together by laser micromachining [20].

Sequential micro-turning and micro-EDM is another technique mainly used for in situ manufacture of the tool electrode for micro-EDM. Tool electrode handling is a major issue in micro-EDM. In this process, the tool is originally manufactured to the required dimension using micro-turning and the same tool without being removed

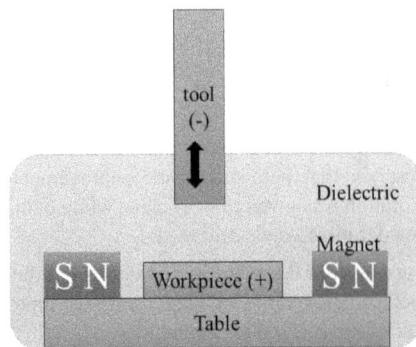

FIGURE 13.10 Schematic of magnetic field-assisted micro-EDM.

from the spindle can be used as the tool for EDM. Micromachining centres like DT110 gives you that feature. Special care needs to be taken in controlling the cutting forces during turning as a slight increase in force may result in deflection of the tool. For extremely small dimensions in the range 10–50 μm tool diameter, the micro turning is followed by electric discharge grinding to attain the required tool dimensions.

13.5   CONCLUSION

Micro-EDM is one of the most versatile machining processes. However, the inherently low MRR and comparatively lower surface finish and dimensional accuracy prevents its large scale application. In order to overcome these shortcomings, researchers have investigated various hybrid micro-EDM methodologies. Techniques such as micro-ECDM, magnetic field assisted micro-EDM and powder mixed micro-EDM give a much superior surface finish and dimensional accuracy, whereas the processes like vibration assisted micro-EDM improve the MRR. Sequential hybrid processes like sequential laser and micro-EDM are applied under specific conditions and have been shown to favourably assist micro-EDM performance. In short, hybrid micro-EDM processes help in enhancing the machining capability of micro-EDM. The selection of the hybrid process should be made according to the application. Although these processes have been found to be advantageous, a considerable amount of work is to be done in order to apply them as large scale manufacturing methodologies.

REFERENCES

[1] Feynman, R. P. (1960). There's Plenty of Room at the Bottom. *Engineering and Science*, 23 (5), 22–36.

[2] Jain, V. K. (2014). Introduction to Micromachining, *Narosa Publishing house*.

[3] Jahan, M. P., Rahman, M., & Wong, Y. S. (2011). A review on the conventional and micro-electrodischarge machining of tungsten carbide. *International Journal of Machine Tools and Manufacture*, 51(12), 837–858.

[4] B.M. Schumacher. (2004). After 60 years of EDM the discharge process remains still disputed. *Journal of Materials Processing Technology*, 149, 376–381.

[5] Allen, D.M., & Lecheheb, A. (1996) Microelectro-discharge machining of ink jet nozzles: optimum selection of material and machining parameters. *Journal of Materials Processing Technology,* 58, 53–66.

[6] Sidpara, A.M., & Malayath, G. (2019). Micro Electro Discharge Machining: Principles and Applications (1st ed.). CRC Press. https://doi.org/10.1201/9780429464782

[7] Kumar, D., Singh, N. K., & Bajpai, V. (2020). Recent trends, opportunities and other aspects of micro-EDM for advanced manufacturing: a comprehensive review. *Journal of the Brazilian Society of Mechanical Sciences and Engineering,* 42(5). doi:10.1007/s40430-020-02296-4

[8] Liu, Q., Zhang, Q., Zhang, M., & Zhang, J. (2016). Review of size effects in micro electrical discharge machining. *Precision Engineering*, 44, 29–40.

[9] Li, J. Z., Shen, F. H., Yu, Z. Y., Natsu, W. (2013). Influence of microstructure of alloy on the machining performance of micro EDM. *Surface Coating Technology*, 228, 460–465.

[10] Maradia, U., Boccadoro, M., Stirnimann, J., Beltrami, I., Kuster, F., & Wegener, K. (2012). Die-sink EDM in meso-micro machining. *Procedia CIRP*, 1, 166–171.

[11] Dilip, D.G., Panda, S. & Mathew, J. (2020). Characterization and Parametric Optimization of Micro-hole Surfaces in Micro-EDM Drilling on Inconel 718 Superalloy Using Genetic Algorithm. *Arabian Journal for Science and Engineering*, 45, 5057–5074.

[12] Kuriachen, B., & Mathew, J. (2015). Experimental investigations into the effects of micro electric-discharge milling process parameters on processing Ti-6Al-4V, *Materials and Manufacturing Processes*, 30(8), 983–990.

[13] Sundaram, M. M., Pavalarajan, G. B., & Rajurkar, K. P. (2007). A study on process parameters of ultrasonic assisted Micro EDM based on Taguchi method. *Journal of Materials Engineering and Performance*, 17(2), 210–215.

[14] Kumar, N., Mandal, N., & Das, A. K. (2020). Micro-machining through electrochemical discharge processes: a review, *Materials and Manufacturing Processes*, DOI: 10.1080/10426914.2020.1711922

[15] Kansal, H. K., Singh, S., & Kumar, P. (2007). Technology and research developments in powder mixed electric discharge machining (PMEDM). *Journal of Materials Processing Technology*, 184(1–3), 32–41.

[16] Bains, P. S., Sidhu, S. S., & Payal, H. S. (2017). Magnetic Field Assisted EDM: New Horizons for Improved Surface Properties. *Silicon, 10*(4), 1275–1282.

[17] Yeo, S. H., Murali, M., & Cheah, H. T. (2004). Magnetic field assisted micro electrodischarge machining. *Journal of Micromechanics and Microengineering*, 14(11), 1526–1529. doi:10.1088/0960-1317/14/11/013

[18] Rashef Mahbub, M., Perveen, A., & Jahan, M. P. (2018). Sequential Micro-EDM. Micro-Electrical Discharge Machining Processes, 209–229. doi:10.1007/978-981-13-3074-2_10

[19] Li, L., Diver, C., Atkinson, J., Giedl-Wagner, R., & Helml, H. J. (2006). Sequential Laser and EDM Micro-drilling for Next Generation Fuel Injection Nozzle Manufacture. *CIRP Annals,* 55(1), 179–182.

[20] Kuo, C. L., Huang, J. D., & Liang, H. Y. (2003). Fabrication of 3D metal microstructures using a hybrid process of micro-EDM and laser assembly. *The International Journal of Advanced Manufacturing Technology*, 21(10–11), 796–800.

[21] Wansheng, Z., Zhenlong, W., Shichun, D., Guanxin, C., & Hongyu, W. (2002). Ultrasonic and electric discharge machining to deep and small hole on titanium alloy. *Journal of Materials Processing Technology*, 120(1–3), 101–106.

[22] Meena, V. K., & Azad, M. S., (2012) Grey relational analysis of Micro-EDM machining of Ti-6Al-4V alloy. *Materials and Manufacturing Processes*, 27(9), 973–977.

[23] Das, A. K., Kumar, P., Sethi, A., Singh, P. K., & Hussain, M. (2016). Influence of process parameters on the surface integrity of micro-holes of SS304 obtained by micro-EDM. Journal of the Brazilian Society of Mechanical Sciences and Engineering, 38(7), 2029–2037.

[24] Endo, T., Tsujimoto, T., & Mitsui, K. (2008). Study of vibration-assisted micro-EDM – The effect of vibration on machining time and stability of discharge. Precision Engineering, 32(4), 269–277.

[25] Huang, H., Zhang, H., Zhou, L., & Zheng, H. Y., (2003). Ultrasonic vibration assisted electrodischarge machining of microholes in Nitinol. *Journal of Micromechanics and Microengineering*, 13(5), 693–700.

[26] Singh, P., Yadava, V., & Narayan, A. (2017). Comparison of machining performance of hole sinking micro-EDM without and with ultrasonic vibration on titanium alloy. *International Journal of Precision Technology*, 7(2–4), 205–221.

[27] Gao, C., & Liu Z. (2003). A study of ultrasonically aided micro-electrical-discharge machining by the application of workpiece vibration, *Journal of Materials Processing Technology*, 139(1–3), 226–228.

[28] Zhang, Q. H., Du, R., Zhang, J. H., & Zhang, Q. B. (2006) An investigation of ultrasonic assisted electrical discharge machining in gas. *International Journal of Machine Tools and Manufacture*, 46 (12–13), 1582–1588.

[29] Li, Y., Deng, J., Chai, Y., & Fan, W. (2015). Surface textures on cemented carbide cutting tools by micro EDM assisted with high-frequency vibration. *The International Journal of Advanced Manufacturing Technology*, 82 (9–12), 2157–2165.

[30] Goiogana, M., Sarasua, J. A., Ramos, J. M., Echavarri, L., & Cascn, I. (2006). Pulsed ultrasonic assisted electrical discharge machining for finishing operations, *International Journal of Machine Tools and Manufacture,* 109, 87–93.

[31] Xing, Q., Yao, Z., & Zhang, Q. (2020). Effects of processing parameters on processing performances of ultrasonic vibration-assisted micro-EDM. *The International Journal of Advanced Manufacturing Technology*. Doi:10.1007/s00170-020-06357-9

[32] Jahan, M. P., Saleh, T., Rahman, M., & Wong, Y. S. (2010) Development, modelling, and experimental investigation of low frequency workpiece vibration-assisted Micro-EDM of tungsten carbide. *Journal of Manufacturing Science and Engineering*, 132(5), 054503-1–054503-8.

[33] Jahan, M. P., Wong, Y. S., & Rahman, M. (2012). Evaluation of the effectiveness of low frequency workpiece vibration in deep-hole micro-EDM drilling of tungsten carbide. *Journal of Manufacturing Processes*, 14(3), 343–359.

[34] Prihandana, G. S., Mahardika, M., Hamdi, M., & Mitsui, K. (2011). Effect of low-frequency vibration on workpiece in EDM processes. *Journal of Mechanical Science and Technology,* 25(5), 1231–1234.

[35] Mahardika, M., Prihandana, G. S., Endo, T., Tsujimoto, T., Matsumoto, N., Arifvianto, B., & Mitsui, K. (2012). The parameters evaluation and optimization of polycrystalline diamond micro electro discharge machining assisted by electrode tool vibration, *The International Journal of Advanced Manufacturing Technology*, 60(9–12), 985–993.

[36] Lee, P. A., Kim, Y., & Kim, B. H. (2015) Effect of low frequency vibration on micro EDM drilling. *International Journal of Precision Engineering and Manufacturing*, 16(13), 2617–2622.

[37] Bhattacharyya, B., Doloi, B. N., & Sorkhel, S. K. (1999). Experimental Investigations into Electrochemical Discharge Machining (ECDM) of Non-Conductive Ceramic Materials. *Journal of Materials Processing Technology*, 95(1–3), 145–154.

[38] Sarkar, B. R., Doloi, B., & Bhattacharyya, B. (2009). Investigation into the Influences of the Power Circuit on the Micro-Electrochemical Discharge Machining Process. *Proceedings of the Institution of Mechanical Engineers, Part B: Journal of Engineering Manufacture*, 223(2), 133–144.

[39] Kolhekar, K. R., & Sundaram, M. (2016). A Study on the Effect of Electrolyte Concentration on Surface Integrity in Micro Electrochemical Discharge Machining. *Procedia CIRP*, 45, 355–358.

[40] Cao, X. D., Kim, B. H., & Chu, C. N. (2009). Micro-Structuring of Glass with Features Less than 100 Mm by Electrochemical Discharge Machining. *Precision Engineering*, 33(4), 459–465.

[41] Huang, S.F., Liu, Y., Li, J., Hu, H.X., & Sun, L.Y. (2014). Electrochemical Discharge Machining Micro-Hole in Stainless Steel with Tool Electrode High-Speed Rotating. *Materials and Manufacturing Processes*. 29(5), 634–637.

[42] Dong, S., Wang, Z., & Wang, Y. (2017). High-Speed Electrochemical Discharge Drilling (HSECDD) for Micro-Holes on C17200 Beryllium Copper Alloy in Deionized Water. *International Journal of Advanced Manufacturing Technology*, 88(1–4), 827–835.

[43] Zhang, Y., Xu, Z., Xing, J., Zhu, D. (2015). Enhanced Machining Performance of Micro Holes Using Electrochemical Discharge Machining with Super-High-Pressure Interior Flushing. *International Journal of Electrochemical Science*, 10, 8465–8483.

[44] Nguyen, K. H., Lee, P. A., & Kim, B. H. (2015). Experimental Investigation of ECDM for Fabricating Micro Structures of Quartz. *International Journal of Precision Engineering and Manufacturing*, 16, 5–12.

[45] Han, M. S., Min, B. K., & Lee, S. J. (2011). Micro – Electrochemical Discharge Cutting of Glass using a Surface Textured Tool. *CIRP Journal of Manufacturing Science and Technology*, 4(4), 362–369.

[46] Wong, Y., Lim, L., Rahuman, I., & Tee, W. (1997). Near-mirror-finish phenomenon in EDM using powder-mixed dielectric. *Journal of Materials Processing Technology*, 79(1–3), 30–40.

[47] Kansal, H. K., Singh, S., & Kumar, P. (2005). Parametric optimization of powder mixed electrical discharge machining by response surface methodology. *Journal of Materials Processing Technology*, 169(3), 427–436.

[48] Pecas, P., & Henriques, E. (2008). Electrical discharge machining using simple and powder mixed dielectric: The effect of the electrode area in the surface roughness and topography. *Journal of Materials Processing Technology*, 200(1–3), 250–258.

[49] Jahan, M. P., Rahman, M., & Wong, Y. S. (2010). Modelling and experimental investigation on the effect of nanopowder-mixed dielectric in micro-electrodischarge machining of tungsten carbide. *Proceedings of the Institution of Mechanical Engineers, Part B: Journal of Engineering Manufacture*, 224(11), 1725–1739.

[50] Kumar, A., Maheshwari, S., Sharma, C., & Beri, N. (2011). Analysis of machining characteristics in additive mixed electric discharge machining of nickel-based super alloy Inconel 718. *Materials and Manufacturing Processes*, 26(8), 1011–1018.

[51] Tan, P. C., & Yeo, S. H., (2011). Investigation of recast layers generated by a powder-mixed dielectric micro electrical discharge machining process. *Proceedings of the Institution of Mechanical Engineers, Part B: Journal of Engineering Manufacture*, 225(7), 1051–1062.

[52] Jahan, M. P., Rahman, M., & Wong, Y. S., (2011) Study on the nano-powder-mixed sinking and milling micro-EDM of WC-Co. *The International Journal of Advanced Manufacturing Technology*, 53(1–4), 167–180.

[53] Kumar, A., Maheshwari, S., Sharma, C., & Beri, N. (2012). Machining efficiency evaluation of cryogenically treated copper electrode in additive mixed EDM. *Materials and Manufacturing Processes*, 27(10), 1051–1058.

[54] Kumar, H. (2014). Development of mirror like surface characteristics using nano powder mixed electric discharge machining (NPMEDM). *The International Journal of Advanced Manufacturing Technology*, 76 (1–4), 105–113.

[55] Kuriachen, B., & Mathew, J. (2016). Effect of powder mixed dielectric on material removal and surface modification in micro-electric discharge machining of Ti-6Al-4V. *Materials and Manufacturing Processes*, 31(4), 439–446.

[56] Kumar, S., & Batra, U. (2012). Surface modification of die steel materials by EDM method using tungsten powder-mixed dielectric. *Journal of Manufacturing Processes*, 14(1), 35–40.

[57] Ekmekci. N., & Ekmekci, B. (2015). Electrical discharge machining of Ti-6Al-4V in hydroxyapatite powder mixed dielectric liquid. *Materials and Manufacturing Processes*, 31(13), 1663–670.

[58] Opoz, T. T., Yaar, H., Ekmekci, N., & Ekmekci, B. (2018). Particle migration and surface modification on Ti6Al4V in SiC powder mixed electrical discharge machining. *Journal of Manufacturing Processes*, 31, 744–758.

[59] Prihandana, G. S., Mahardika, M., Hamdi, M., Wong, Y. S., & Mitsui, K. (2009). Effect of micropowder suspension and ultrasonic vibration of dielectric fluid in micro-EDM processes – Taguchi approach. *International Journal of Machine Tools and Manufacture*, 49(12–13), 1035–1041.

[60] Mughal, M. P., Farooq, M. U., Mia, J. M. M., Shareef, M., Javed, M., Jamil, M., & Pruncu, C. I. (2021). Surface modification for osseointegration of Ti6Al4V ELI using powder mixed sinking EDM. *Journal of the Mechanical Behavior of Biomedical Materials*, 113, 104145, https://doi.org/10.1016/j.jmbbm.2020.104145

14 Modeling and Optimization of EDM-Based Hybrid Machining Processes

Sanghamitra Das, Shrikrishna Nandkishor Joshi*
and Uday Shanker Dixit
Department of Mechanical Engineering
Indian Institute of Technology Guwahati, Assam, India
* Corresponding author

CONTENTS

DOI: 10.1201/9781003202301-14

14.1 INTRODUCTION

The electric discharge machining (EDM) process is one of the most commonly used unconventional machining processes that uses spark discharges generated between two electrodes to remove material by melting and evaporation. EDM has a wide variety of applications in the modern industrialized era, such as the manufacturing of dies, automobile and aerospace components, and medical instruments. However, there are certain limitations of EDM, for example, its applicability to only electrically conductive materials, a low material removal rate, high tool electrode wear, and high energy consumption. To overcome these limitations, the EDM process mechanism is combined with other processes either simultaneously or sequentially. Combined processes are known as hybrid EDM (HEDM) processes. Research has shown that HEDM processes show greater benefits and better process performance in comparison to the individual constitutional processes, in terms of better machining efficiency, high material removal rate, and lower tool wear rate [1]. The mechanism that is coupled with EDM can be a conventional process like material removal with a wedge-shaped tool or an unconventional process like electrochemical machining, ultrasonic machining or laser machining. HEDM processes can be broadly classified into two categories: (a) primary HEDM and (b) secondary HEDM. In primary HEDM, both the constituent processes contribute simultaneously to the material removal mechanism like in electric discharge abrasive grinding (EDAG) and electrochemical discharge machining (ECDM) [2]. On the other hand, in secondary HEDM processes, the material removal mechanism is dominated by the EDM spark discharges while the other process assists in improving the efficiency of the system, like in ultrasonic-assisted EDM (UAEDM), laser-assisted EDM (LAEDM) and magnetic field-assisted EDM (MFEDM) [2]. Figure 14.1 presents an overview of some of the typical hybrid EDM processes.

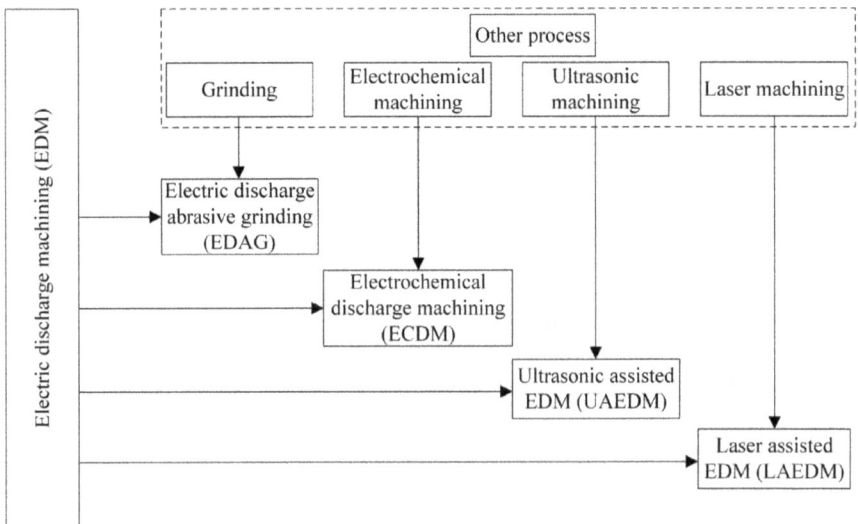

FIGURE 14.1   An overview of some typical hybrid EDM processes.

To have a better understanding of the working principle of hybrid EDM processes and for better machining efficiency, modeling and optimization of these processes are critical in the modern industrial sector. Modeling is a very useful tool for unfolding the complex mechanism of HEDM processes. Optimizing the machining parameters after the development of the model is very useful for improving machining performance. Researchers have developed various techniques for modeling and optimization of the HEDM processes, which will be discussed in detail in the following sections.

14.2   THE PHYSICS OF TYPICAL HYBRID EDM PROCESSES

To have a better understanding of hybrid EDM processes, the physical mechanism of different hybrid EDM processes are explained. Different theories have been established to evaluate the material removal rate and for the performance analysis of these mechanisms. The following sections discuss in detail, the working principles of different hybrid EDM mechanisms.

14.2.1   Electric Discharge Abrasive Grinding (EDAG)

Conventional grinding yields poor surface quality during the machining of hard materials and requires frequent dressing of the grinding wheel due to clogging of material. This limitation is overcome by the EDM process but it has the disadvantage of low material removal rate. Electric discharge abrasive grinding (EDAG) combines both these mechanisms to overcome the limitations of low surface quality during grinding and low material removal rate during EDM operation. The working principle of the EDAG process is very much similar to conventional EDM except for the fact that a metal-bonded abrasive grinding wheel acts as the cathode during the EDAG phenomenon. Figure 14.2 depicts the schematic diagram of the EDAG process. Spark discharges are generated between the metal bonded abrasive grit wheel and the workpiece, which are both submerged in the dielectric fluid. The embedded abrasive grains in the wheel perform the grinding process while the sparks are produced between the metal bond and the workpiece [3]. The heat produced due to spark discharges softens the workpiece material thus facilitating material removal by the abrasive grains. The protrusion heights of these grains play a very important role in determining the machining efficiency and quality of products [4]. In-process dressing of the grinding wheel occurs through the continuous discharges that expose fresh grains of the grinding wheel to the workpiece and this makes it easier to machine advanced materials like metal matrix composites.

14.2.2   Electrochemical Discharge Machining (ECDM)

The ECDM process is a primary hybrid machining technology that combines the electric discharge machining (EDM) and electrochemical machining (ECM) processes, which is mostly used for machining nonconductive materials. The electrodes are immersed in an electrolyte solution and material removal takes place through thermal sparks through the method of melting and vaporization. Figure 14.3 depicts

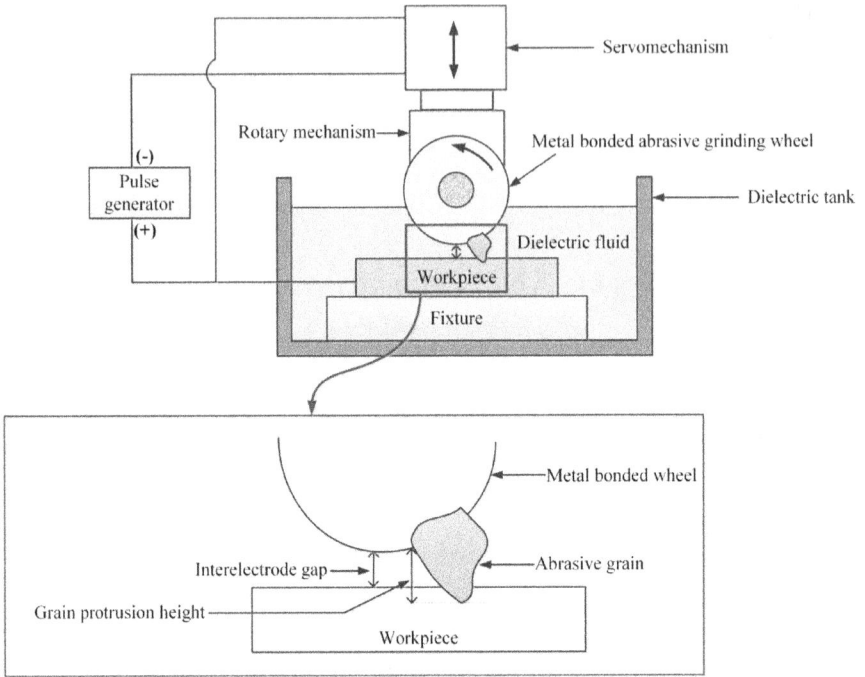

FIGURE 14.2 Schematic of EDAG process.

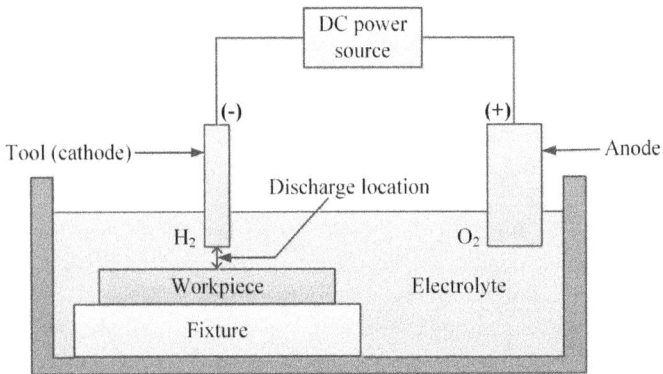

FIGURE 14.3 Schematic of ECDM process.

the schematic diagram of the ECDM process. The tool electrode acts as the cathode and a counter electrode with a large surface area acts as the anode. The workpiece is kept near the cathode tip. As a DC voltage is applied between the cathode and anode, an electrochemical reaction occurs in the electrolyte (NaOH or KOH solution), which releases hydrogen gas bubbles at the cathode and oxygen at the anode surface [5]. The hydrogen gas bubbles form a dense layer surrounding the cathode tip. When the voltage crosses the threshold limit of the insulating gas layer, a spark generates between the cathode and the electrolyte interface. Researchers have put forward different theories for the reasons behind the discharge phenomenon between the cathode and electrolyte interface. The most common theory is the "switching off" process given by Basak and Ghosh [6]. According to this theory, as the hydrogen gas is liberated at the cathode, a thick layer is formed which continues to grow in size thus forming a blanket around the tool electrode. This increases the resistance of the tool electrode interface, creating a switching off situation for a very short duration where the current drops to zero and the rate of change in current becomes very high thus producing the discharge. A fraction of the energy produced during spark formation is absorbed by the workpiece. The workpiece, which is kept in the vicinity of the tool electrode, undergoes melting and vaporization, causing material removal. Additionally, material removal from ceramic material also occurs due to thermal spalling which is a mechanical fracture because of thermal stresses generated due to steep temperature gradients during the discharge phenomenon.

14.2.3 Ultrasonic Assisted EDM (UAEDM)

UAEDM combines the mechanisms of ultrasonic machining and EDM. It is a secondary hybrid EDM process where the thermal effects of conventional EDM perform the material removal action and ultrasonic machining assists the EDM operation in improving the process performance characteristics. Figure 14.4 depicts the schematic diagram of the UAEDM process. Material removal takes place through spark discharges generated between the electrodes, coupled with ultrasonic vibration of the tool, workpiece, or dielectric. The ultrasonically vibrated machining zone facilitates cleaner production and better machining efficiency by flushing away unwanted debris

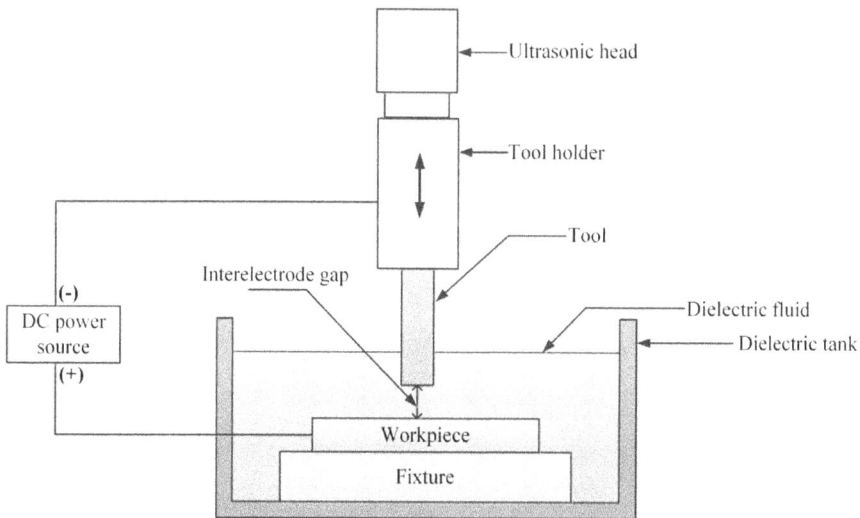

FIGURE 14.4   Schematic of UAEDM process.

generated in the inter-electrode gap [7]. Improper flushing of debris in conventional EDM produces unwanted arcs, which causes an abrupt increase in thermal load thus causing thermal shock to the machined workpiece. UAEDM overcomes this limitation. Research has shown that ultrasonic vibration also has a significant effect on bubble formation between the electrodes during the duration of the discharge [8].

14.2.4 LASER-ASSISTED EDM (LAEDM)

Laser material processing utilizes the energy from a coherent beam of light to remove, melt, or thermally modify materials. The application of a laser beam in machining depends on the thermo-optic interaction between the laser beam and the workpiece material. However, laser machining suffers from poor accuracy, whereas EDM offers the advantage of fine accuracy but has a low machining rate and high tool wear. Thus, both these processes are combined for optimum process efficiency. Generally, LAEDM is a sequential form of the HEDM process wherein the laser machining is followed by the EDM operation to remove the thermal defects generated by the laser during the drilling of micro holes into the workpiece [2].

14.2.5 MAGNETIC FIELD ASSISTED EDM (MFEDM)

Inefficient flushing of debris in the machining zone causes the generation of unwanted arcs, which hampers the overall quality and integrity of machined products during the EDM operation. The attachment of debris to the machined surface reduces the inter-electrode gap, which causes sudden arc formation, exposing the workpiece surface to thermal shock. The application of the magnetic field assists in efficient and faster removal of debris produced during the discharge phenomenon and thus stabilizes the machining zone [9]. The magnetic field is created by attaching magnets to an EDM setup near the machining zone. Figure 14.5 depicts the schematic diagram of the MFEDM process. The magnetic field creates a strong magnetic force in a direction perpendicular to the machining direction, which causes rapid and efficient flushing of debris in the machining zone. The magnetic field also increases the speed of moving electrons between the electrodes and increases the ionization of the plasma channel [10]. Thus, MFEDM improves the process efficiency in terms of higher material removal rate, lesser tool wear, cleaner machining zone, uniform cut, lesser formation of unwanted arcs, and so on.

14.3 MODELING APPROACHES OF THE HYBRID EDM PROCESSES

Modeling is a very useful tool in understanding the working principles and predicting the response characteristics for a particular set of conditions during hybrid EDM machining. Researchers have reported various modeling approaches to simplify the complex mechanism of the HEDM operation. An analytical approach and numerical methods are the most commonly used techniques to model these processes. Multiscale modeling is also presented as an approach to investigate the working principle of HEDM.

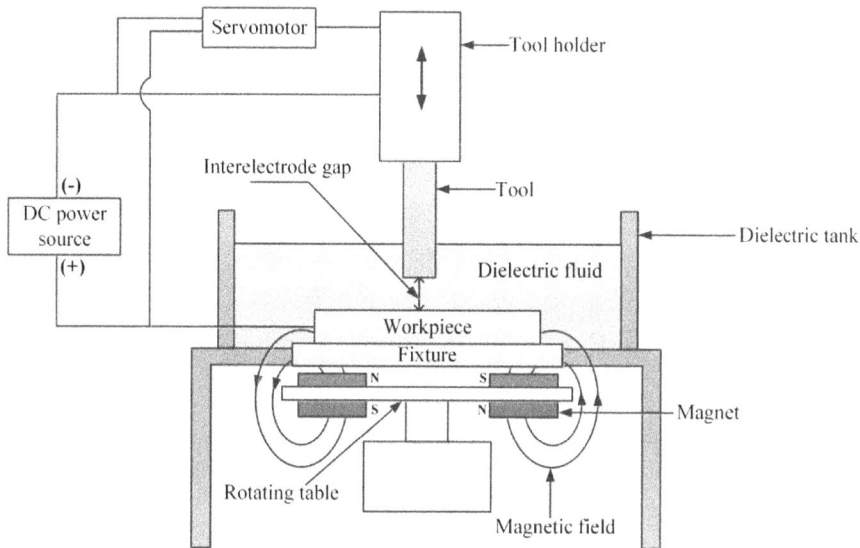

FIGURE 14.5   Schematic of MFEDM process.

14.3.1 ANALYTICAL APPROACH

An analytical model is a mathematical tool that describes a system in the form of a set of mathematical equations that specify parametric relations and their associated values as a function of time, space, and/or other system parameters. The governing equations, along with a set of initial conditions and boundary conditions, are used to solve a problem. An analytical model can be classified into (i) static, where the model parameters are independent of time, such as the mass and geometric properties of a system and (ii) dynamic, where the model values are time-dependent functions such as position as a function of time. This section presents a brief review of the analytical models of hybrid EDM processes.

Koshy et al. [4] explained the mechanism of material removal during the electro-discharge diamond grinding (EDDG) of electrically conducting hard materials. They developed an analytical model to explain the reduction in normal force and grinding power due to the thermal softening of the workpiece material in the grinding zone due to spark discharges. Basak and Ghosh [11] also developed a theoretical model to estimate the material removal rate during the ECDM operation by explaining the 'switching off' action between the tool electrode and electrolyte. The hydrogen gas liberated at the cathode surface forms a blanket around the tool-electrolyte interface. This creates a switching off action at the interface, causing the discharge phenomenon, which generates an electromotive force (EMF), E_s, given by

$$E_s = -L\frac{dI}{dt},\tag{1}$$

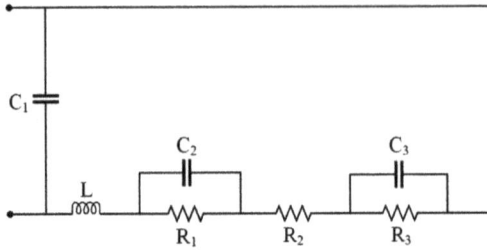

FIGURE 14.6 Equivalent circuit diagram of an ECDM setup. With permission from (Basak and Ghosh 1997) Copyright 1997 Elsevier.

where L is the inductance of the circuit and I is the instantaneous current. Figure 14.6 shows the equivalent circuit diagram of the ECDM set up.

In Figure 14.6 R_1, R_2 and R_3 are the resistances across the anode electrolyte interface, the electrolyte, and the tool electrolyte interface, respectively. The resistance R_1 is considered to be negligible because of the large anode surface area. Also, R_3 keeps on increasing with applied voltage due to bubble formation and attains the critical value R_{3c} at critical voltage V_c. Thus, the equivalent critical resistance, R_c can be expressed as

$$R_c = R_2 + R_{3C}. \tag{2}$$

The corresponding critical current I_c at the critical voltage V_c is expressed as

$$I_c = \frac{V_c}{R_c}. \tag{3}$$

A cylindrical heat source with diameter $2a$ and semi-infinite workpiece domain was considered for the analytical model. The analysis for an ECDM operation is quite similar to an EDM operation. The thermal analysis is governed by a transient heat conduction equation given by the partial differential equation

$$\frac{\partial T}{\partial t} = \alpha \nabla^2 T. \tag{4}$$

The temperature at any point $T(r, z, t)$ is determined by solving Eq. 4 with the initial condition:

$$T(r,z,t)=0, \quad t=0 \tag{5a}$$

and boundary conditions:

$$\frac{\partial T}{\partial z}\bigg|_{z=0} = 0, \text{ for } t > 0, r > a$$

$$-k\frac{\partial T}{\partial z}\bigg|_{z=0} = \frac{\lambda Q}{\pi a^2 t_d}, \text{ for } t > 0, \ 0 < r \le a \qquad (5b)$$

where k is the thermal conductivity and α is the thermal diffusivity of the workpiece material, t is the time, Q is the heat released by a single spark, t_d is the pulse duration and λ is the fraction of heat absorbed by the workpiece. The peak temperature is obtained at the end of the pulse duration along the z-axis, which is expressed as [11]

$$T(0,z,t_d) = \frac{\lambda Q}{2\pi k a t_d} \int_0^\infty J_0(\xi a) J_1(\xi a) \left[e^{-\xi z} \operatorname{erfc}\left\{ \frac{z}{2\sqrt{\alpha t_d}} - \xi\sqrt{\alpha t_d} \right\} \right] \frac{d\xi}{\xi}, \qquad (6)$$

where ξ is a dummy variable, $J_0(\xi a)$ and $J_1(\xi a)$ are Bessel's functions of zero and first order, respectively.

In the ECDM operation, spark discharges occur between the cathode and electrolyte. A part of the energy dissipated during spark formation is absorbed by the electrolyte and a part goes as latent heat for the melting of the workpiece. It was assumed that 20% of the dissipated energy was absorbed by the workpiece for the material removal process. Careful experiments are needed to ascertain the validity of this assumption.

Based on the analytical results, Basak and Ghosh [11] also provided the following empirical expression for the volume removed per spark (v):

$$v = 0.7 \times 10^{-5} Q^{1.5}, \qquad (7)$$

where Q is in joule and v is in cm^3. Further,

$$Q = \frac{1}{2} L \left(\frac{V_0}{R_c} - I_c \right)^2, \qquad (8)$$

where V_0 is the applied voltage, which is more than V_c. Considering the average value of resistance R_3, the time t_1 required to raise the current to the peak value is given by

$$t_1 = \frac{L}{R_2 + 0.5 R_{3C}} \ln\left(\frac{2R_C}{R_{3C}} \right). \qquad (9)$$

The sparking frequency f is

$$f = \frac{1}{t_1}. \tag{10}$$

The material removal rate is given by

$$M = \rho v f, \tag{11}$$

where ρ is the density of the workpiece material.

Several other authors developed analytical models for different hybrid technologies. Zhang et al. [12] predicted the material removal rate during the UAEDM operation theoretically. Mediliyegedara et al. [13] developed a real-time process controller for the ECDM operation. The tool position control system was optimized using various control algorithms to improve the material removal rate and surface finish during the operation. Kumar et al. [14] utilized graph theory and a matrix approach to model the UAEDM process to have a better understanding of the interdependency of factors on the performance index.

14.3.2 Numerical Methods

A numerical method of solving a problem is adopted when the partial differential equations (PDEs) describing the system cannot be solved analytically. Some prominent numerical methods are (i) the finite element method (FEM), (ii) the finite difference method (FDM), and (iii) the boundary element method. The finite element method and finite difference methods to model hybrid EDM mechanisms are briefly introduced.

14.3.2.1 Finite Element Method (FEM)

The finite element method (FEM) is the most commonly used numerical technique for the modelling of manufacturing processes nowadays. It predicts the behavior of a system described by a set of partial differential equations. This method is used for complicated systems with complex geometry and behavior that are difficult to solve analytically. In FEM methodology, the process domain is divided into small elements called the finite elements that are interconnected at the common points, in other words, nodes, of neighboring elements, thus forming a mesh of the domain. The finite element formulation of the problem, governed by a differential equation along with its initial and boundary conditions, results in a set of algebraic equations. The dependent variables of interest, namely, the field variables, are obtained by solving the PDEs. An unknown function is approximated over the domain and the solution is obtained at the nodes. The fundamental concept of a FEM is illustrated in Figure 14.7.

Researchers have presented various studies that employ finite element analysis to model hybrid EDM processes. Wei et al. [15] developed a finite element model to predict the material removal rate during the electrochemical discharge machining of

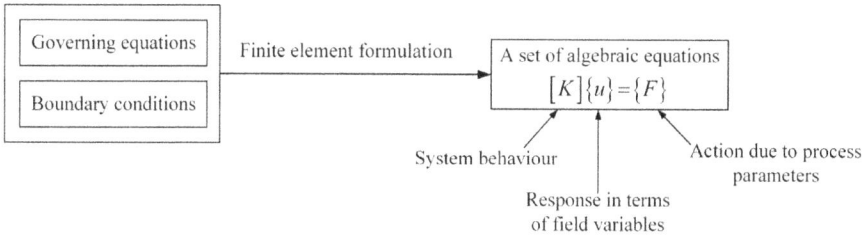

FIGURE 14.7 Fundamental approach of a FEM problem.

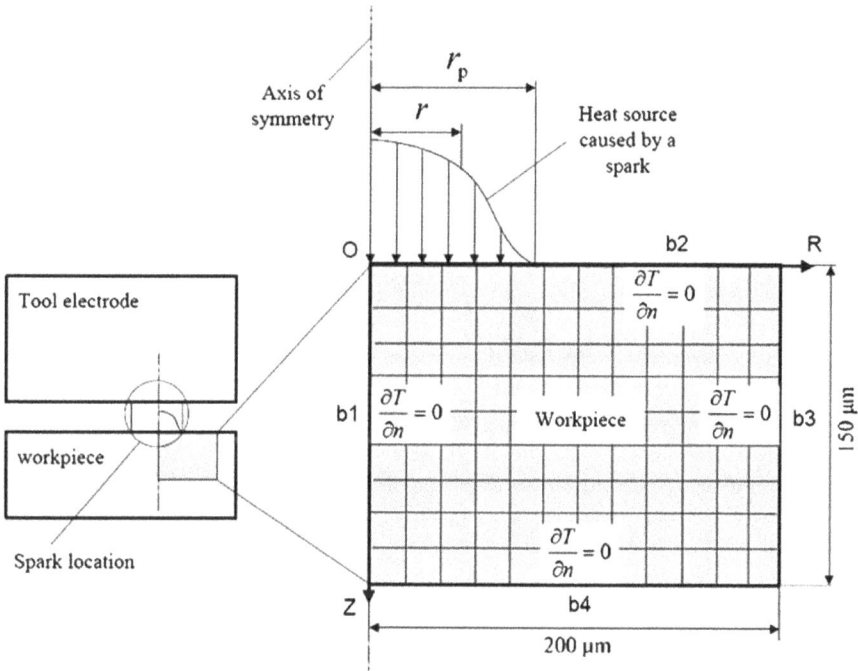

FIGURE 14.8 Schematic for a FEM model of ECDM operation. With permission from (Wei et al. 2011). Copyright 2011 Springer.

non-conductive glass in the discharge regime. Figure 14.8 depicts the axisymmetric two-dimensional domain for the FEM model along with boundary conditions. The material removal mechanism of the ECDM operation is quite similar to the EDM process except for the fact that the discharge occurs between the cathode and the electrolyte interface. The ECDM process usually machines a workpiece of very low thermal conductivity. Thus, boundaries b2, b3 and b4 can be assumed to be thermally insulated because they are far away from the spark location. The space was discretized using two-dimensional four-node PLANE 55 element of an ANSYS FEM package. A mesh size of 2 µm was selected, based on a mesh sensitivity analysis. In the ECDM process, material removal occurs through spark discharges due to melting in addition to chemical dissolution in the electrolyte. The material removal rate was computed by calculating the volume of the crater obtained after removing the elements that had attained a temperature above a critical temperature for the workpiece material. It is to be noted that FEM was used only for solving the heat conduction part. Other components of the model are solved analytically or empirically.

Jain et al. [16] employed the FEM coupled with their valve theory to predict the material removal rate during electrochemical spark machining. Bhondwe et al. [17] also developed a finite element based methodology to estimate the temperature distribution in the workpiece during ECDM. The effects of input parameters, namely, electrolyte concentration, duty factor, and energy distribution on the material removal rate were also studied. The effects of grinding time and feed on the temperature isotherm and thermal stresses generated during the EDDG were also estimated [18], [19]. Yadav and Yadava [40] modeled the temperature distribution during the EDAG operation by superimposing the results of two FEM codes, assuming a rectangular heat source for the grinding operation and a Gaussian heat flux for the EDM spark. Overall review of the literature reveals that although FEM has been used as a part of the model for hybrid EDM processes, developed models still rely on empirical results to a large extent. Researchers need to focus their attentions on developing a physics-based model that can be solved by FEM or some other analytical or numerical technique without depending on case-specific empirical models.

14.3.2.2 Finite Difference Method (FDM)

The finite difference method (FDM) is the other popular approach to solve partial differential equations. The FDM converts the derivatives of PDEs into finite difference approximations to obtain an approximate solution to the problem. The domain is divided into a finite grid containing finer elements and the equations are solved element by element over the entire region. The basic principle of the FDM methodology is to employ a Taylor series expansion for obtaining the discrete avatars of derivatives.

As an example, if $T_{i,j}$ is the temperature at any node (i, j) that is dependent on time, then $T_{i+1,j}$ at node $(i+1, j)$ can be represented as Taylor series expansion, in other words,

$$T_{i+1,j} = T_{i,j} + \left(\frac{\partial T}{\partial t}\right)_{i,j}(\Delta t) + \left(\frac{\partial^2 T}{\partial t^2}\right)_{i,j}\frac{(\Delta t)^2}{2} + \left(\frac{\partial^3 T}{\partial t^3}\right)_{i,j}\frac{(\Delta t)^3}{6} + \cdots \quad (12)$$

Assuming Δt to be very small and neglecting the higher order terms, Eq. 11 can be represented as

$$T_{i+1,j} \approx T_{i,j} + \left(\frac{\partial T}{\partial t}\right)_{i,j}(\Delta t)$$

$$\Rightarrow \left(\frac{\partial T}{\partial t}\right)_{i,j} \approx \frac{T_{i+1,j} - T_{i,j}}{\Delta t} \quad (13)$$

The neglected terms are the truncation error in the equation. Similar expressions can be obtained for higher order derivatives.

For modelling of EDM based hybrid machining processes, compared to FEM, analytical and empirical techniques, FDM has been applied sparingly. Tabar and Mobadersany [20] utilized the FDM technique and the boundary integral equation method to investigate the behavior of bubble formation and the surrounding dielectric fluid during UAEDM. The study brought out the importance of ultrasonic vibrations in improving the performance of EDM.

14.3.3 MULTISCALE MODELING APPROACH

Multiscale modeling is applied to systems with complex behavior by tackling physical, chemical, and mechanical aspects at different length and time scales. The method is suitable for hybrid machining where different phenomena are responsible for machining. Multiscale modeling approaches for a hybrid process allow the coupling of different parts of a system into a single solution. There are different types of coupling (a) manual coupling, where the output of a part of model is manually input to another part of the model at a different scale, (b) loose coupling, where the global model and all the sub models are solved separately with different sets of equations and (c) tight coupling, where all the sub models are integrated and solved by a single set of equations. Another classification of modelling approaches can be in the following way: (a) sequential modeling, where a macroscopic model is defined first by constitutive relations that are derived from a different scale model, (b) concurrent modeling, where all the constitutive relations of the system on different scales are coupled concurrently into a single comprehensive solution. Wang and Zhang [21] briefly demonstrated the hybrid approach to multiscale modeling technologies. The correlation between the models on different scales is developed by the coupling of different conditions. Zeng and Qin [22] explained the application of multiscale modeling to hybrid machining processes and they also discussed the various modeling strategies. Table 14.1 depicts the different modeling methodologies adopted at various time and length scales.

14.4 FORMULATION OF OPTIMIZATION PROBLEMS FOR HYBRID EDM PROCESSES

The modern manufacturing industry demands good quality products with minimum production time and cost. The complex mechanism of hybrid machining processes involves a large number of uncontrollable input conditions and their correlations, which need to be optimized in order to attain the desired machining efficiency and maximum productivity. The selection of process parameters to obtain optimum output performance is a difficult task due to the complicated mechanism of the hybrid EDM process. Thus, it is necessary to employ a suitable optimization technique to optimize the machining conditions for the best response characteristics.

An optimization problem is formulated by maximizing or minimizing a real-valued function within a given set. The function to be maximized or minimized is called the

TABLE 14.1
Representative multiscale modeling methodologies at different length and time scales

Length scale (m)

Time scale (s)	10⁻¹²	10⁻⁹	10⁻⁶	10⁻³
10⁻¹⁵	Quantum scale (density functional theory)			
10⁻⁹		Atomistic scale (molecular dynamics simulation)		
10⁻⁶			Coarse grained (dissipative particle dynamics)	
10⁰				Continuum (finite element method)

objective function of the problem. The type of objective function is a strong deciding factor for choosing a particular optimization algorithm. The objective function may be continuous or discrete, differentiable or non-differentiable, linear or non-linear. There are many constraints in any optimization problem that are expressed in the form of mathematical equations. There can be more than one objective required to be fulfilled simultaneously, in which case the problem is called a multi-objective problem. A typical optimization problem with k design or decision variables (denoted by k-dimensional vector x) is expressed as

$$\text{Maximize } f_1(x), f_2(x), \ldots\ldots\ldots, f_l(x),$$
$$\text{subject to}$$
$$g_i(x) \leq 0 \quad i = 1, 2, \ldots\ldots\ldots, m,$$
$$h_i(x) = 0 \quad i = 1, 2, \ldots\ldots\ldots, n,$$
$$l_{bi} \leq x_i \leq u_{bi} \quad i = 1, 2, \ldots\ldots\ldots, k. \tag{14}$$

Here, $f_i(x)$ is the ith objective function out of the total l objective functions, $g_i(x)$ is the ith inequality constraint out of the total m inequality constraints, $h_i(x)$ is the ith equality constraint out of the total n equality constraints. The upper and lower bounds of the design variable x_i are denoted by u_{bi} and l_{bi}, respectively.

The value of the objective function is dependent on the values of the design variables. The optimum solution is obtained by maximizing or minimizing the objective functions over different possible values of x, without violating constraints. It is to be noted that minimization of an objective function is equal to maximization of

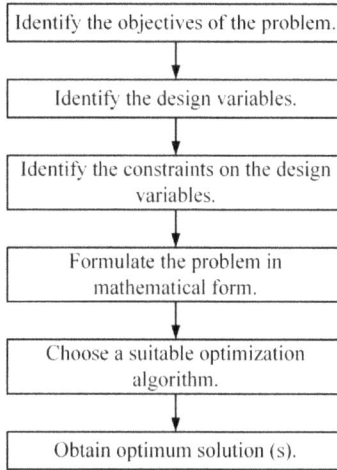

FIGURE 14.9 Flowchart for an optimization problem.

its additive or multiplicative inverse. The identification of decision variables, defining objective functions and formulating constraint relations are important subtasks of an optimization problem. The constraints in an optimization problem are certain limitations that are posed on the design variables. Constraints can be hard constraints, which must be satisfied, or soft constraints, which may be relaxed to some extent if there is a large improvement in the solution. After defining the objective function and the constraints in linguistic forms, they are converted to mathematical forms.

A single objective function results in a single optimal solution. However, in real world engineering, the optimization problems are mostly multi-objective, which results in a set of optimal solutions instead of a single value. For multi-objective problems, obtaining Pareto optimal solutions is the most appropriate. A Pareto optimal solution is the part of a Pareto-optimal solution-set, in which no solution dominates the other solutions. This means that in a Pareto-optimal set none of the solutions is better than any other solution in all aspects; one solution can be good in some aspects, while another solution can be good in other aspects. Figure 14.9 depicts the optimization procedure.

Mandal et al. [23] employed multi-objective optimization to optimize the set of process parameters for best machining performance during the EDM operation. Two response parameters were considered to decide the machining efficiency, namely, material removal rate (*MRR*) and tool wear (*TW*). Maximum *MRR* and minimum *TW* were the objectives of the optimization problem. Thus, this multi-objective problem with two objective functions of contradicting nature was formulated as

$$Objective\ 1 = \frac{1}{MRR}$$
$$Objective\ 2 = TW$$
$$Z = \text{Minimize}\left(Objective\ 1, Objective\ 2\right) \tag{15}$$

An artificial neural network (ANN) model was first developed, which was trained according to the backward propagation algorithm and utilized in determining the objective functions. Similarly Bhattacharya et al. [24] optimized the electrochemical discharge machining input conditions during micro-channel fabrication of glass. The desired objectives, namely, maximum material removal rate (*MRR*), machining depth (*MD*), minimum overcut (*OC*), and heat-affected zone (*HAZ*) were obtained using a multi-objective problem as

$$\text{Minimize } F\left(X_1, X_2, X_3\right) = \frac{1}{MRR} + \frac{1}{MD} + OC + HAZ,$$

Subject to

$$50 \le X_1 \le 60, \quad 15 \le X_2 \le 30, \quad 30 \le X_3 \le 40, \tag{16}$$

where X_1, X_2, X_3 are the input parameters representing voltage, electrolyte concentration, and inter-electrode gap, respectively. Here, a multi-objective optimization problem has been converted to a single-objective problem by assigning equal weights to all the objectives.

14.5   OPTIMIZATION TECHNIQUES FOR HYBRID EDM PROCESSES

Any process needs to be optimized for better performance without compromising the time and cost. Thus, several optimization methods have been introduced to solve modern day engineering problems according to the type of objective functions and design variables used. Hybrid EDM processes are complex mechanisms, that involve the interaction of several control variables. Thus, the usage of a suitable optimization algorithm is essential to improve machining efficiency and obtain optimal results. The following sections discuss in detail different optimization techniques used for hybrid EDM processes.

14.5.1   CLASSICAL OPTIMIZATION METHODS

Classical optimization methods are analytical methods to find the optimum solution of a problem with the help of differential equations. These methods form the basis of different optimization algorithms. The classical optimization techniques are suitable in finding the optimum solutions of unconstrained continuous and differentiable functions. Some techniques can handle constraints also. These methods have certain limitations when the objective function is discontinuous and/or non-differentiable. The problems solved by classical optimization methods fall under three major categories,

- functions with one variable,
- unconstrained multivariable functions,
- constrained multivariable functions with both equality and inequality constraints.

14.5.2 RESPONSE SURFACE METHODOLOGY (RSM)

Response surface methodology (RSM) is a statistical method that establishes a relationship between input variables and response characteristics of a process mechanism. It is an effective tool for planning, improving and optimizing a process, which saves time and resources by reducing the number of experiments and producing efficient results. The influence of various input factors and their mutual interactions on the process performance of a system is evaluated based on a set of experimental designs. The data representing the experimental set is collected using suitable designs of experiments (DOE) such as factorial design (FD), central composite design (CCD), and orthogonal array (OA). CCD is the most commonly used method for the RSM technique, which is useful in creating a second order response model. It is a suitable method for real world multi-objective problems due to its capability for dealing with multiple variables and their interrelations. RSM usually develops three-dimensional plots or contour plots, which graphically depict the variation of an output variable with the input factors or, a combination of factors. A typical relation between two inputs and one output may be of the form:

$$y = f(x_1, x_2) + e, \tag{17}$$

where y is a response variable dependent on two variables x_1 and x_2, and e is the error.

Several reported studies employed the RSM technique to model the complex mechanism of hybrid EDM processes. Ghoreishi and Atkinson [25] studied the vibratory effects of UAEDM on the material removal rate, surface finish, and tool wear rate using the RSM technique. The effects of electrode rotation combined with ultrasonic vibration on various cutting regimes were also investigated. Jain et al. [26] established the advantages of the electrochemical spark abrasive drilling of alumina and borosilicate glass over conventional machining methods in terms of enhanced material removal rate and increased machining depth using response surface equations. A similar mathematical model was developed by Sarkar et al. [27] to evaluate the effects of input conditions like applied voltage, electrolyte concentration, inter-electrode gap, and so forth, on the machining rate during the ECDM operation on silicon nitride ceramics. RSM results indicated that applied voltage is the most significant factor in influencing the material removal rate and heat affected zone of the workpiece. An RSM technique was further utilized to establish a correlation between the input and response variables during a hybrid arrangement of EDM and the end milling of AISI 1045 steel alloy [28]. Paul Hiremath [29] also optimized the input conditions for increased material removal rate and reduced tool wear rate during the ECDM of non-conducting materials. A similar attempt was made to eliminate the burrs at the exit of a drilled hole during dry UAEDM of CFRP composites [30]. The burr removal rate (BRR) was modeled as a function of input parameters and it was found that capacitance has the highest contribution to BRR followed by pulse duration and ultrasonic vibration amplitude.

14.5.3 Artificial Neural Network (ANN)

Artificial neural network (ANN) is a soft computing technique, analogous to the biological human neuron system. ANN mimics the architecture, processing, functioning, adaptability and learning capability of a biological neural system. The biological neural system consists of a huge number of biological cells, called neurons for processing and transferring information throughout the entire network. A biological neuron consists of a cell body or soma, dendrites, and an axon. The axon is a long fiber that acts as transmission lines and carries information between the neurons. Dendrites receive signals from the neurons through the axon. A neuron transmits its information to the neighboring neurons through a connection, called a synapse. Along the same lines, the primary element of an artificial neural network is an artificial neuron, which is a simple mathematical tool. A neuron is the fundamental information processing unit of the neural network. Synapses connect the neurons to the input signals as well as the neurons of the next layer. The first model of an artificial neuron was proposed by McCulloch and Pitts and is called *Threshold Logic Unit* (TLU) or a *Linear* Threshold *Gate* [31]. A neuron computes the sum of n real-valued weighted input signals, x_i followed by a threshold operation such that if the value of the sum is greater or equal than a threshold θ, then the output y of the unit is 1, otherwise, it is -1. Mathematically, it can be expressed as

$$
y(x) = \begin{cases} 1 & \text{if } \sum_{i=1}^{n} w_i\, x_i + b - \theta \geq 0 \\ -1 & \text{if } \sum_{i=1}^{n} w_i\, x_i + b - \theta < 0 \end{cases} \tag{18}
$$

where w_i is called the 'synaptic weight' or simply 'weight' analogous to the biological synapses and b is the bias associated with the neuron. The weighted sum in Eq. 16 is called the neuron activation and the function $y(x)$ is called the activation function.

When information is transferred between neurons, the weights are applied to it. The information is processed in several layers. The first layer is called the input layer, where the inputs are taken from the external world. The last layer is the output layer, which provides the output from the neural network system to the external world. The other layers are called the hidden layers. The information travelling from layer to layer is weighted by assigning appropriate weights. The weights define the influence of the input on the response. For each neuron, an activation function is used to convert the weighted inputs coming into the neuron to the output emanating from the neuron. Figure 14.10 represents the basic architecture of an ANN model with n input neurons, one hidden layer, and 3 output neurons.

Just as a biological neural system learns by exemplars from the environment, ANN models learn from a training set of data. A learning algorithm in an ANN comprises of updating the network architecture and weighted connections between the neurons to improve performance. The main goal of learning is to achieve optimized weight

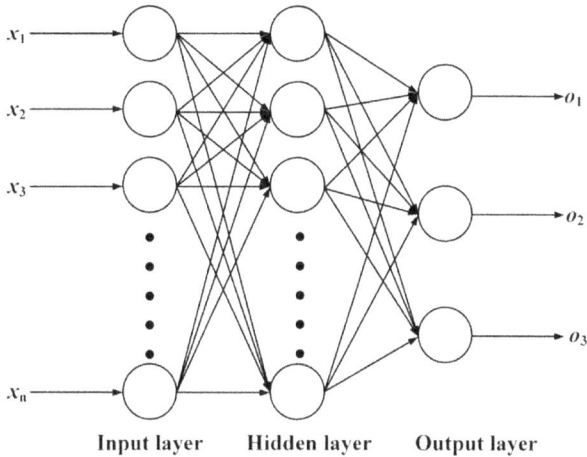

FIGURE 14.10   Basic structure of an ANN model.

values and biases based on the adaptation or training algorithm. Different learning methods are adopted by a set of training examples. The network utilizes a learning algorithm to teach the system how to behave accordingly with different input signals. There are various learning methods like supervised learning, unsupervised learning, and reinforced learning. Supervised learning basically means learning with a 'teacher' where each input is provided with a target output value and the weights are updated accordingly to obtain the desired output. Unsupervised learning or learning without a 'teacher' focuses on learning from the data set itself without any prior knowledge and trains the network according to correlations or patterns in the input data. In reinforcement learning, a 'teacher' is available but only indicates the correctness of the network outputs. The information provided is useful in the learning process of the neural network.

ANN optimization involves updating the weight functions of the network such that it minimizes the prediction error. The trained neural network with the optimized weight values produces the output within the desired accuracy. Back propagation is the most commonly used technique for training ANN architecture, but the conventional back propagation method suffers from the danger of getting trapped with the local optimum solutions. Search techniques like genetic algorithms are useful in finding the global optimum for minimizing error. Once an ANN is trained, it can be used instead of objective functions for the quick computation of objective functions.

A number of researchers have developed ANN models for a better process performance prediction of hybrid EDM mechanisms. Tsai and Wang [32] compared the performance of six different neural network models for surface roughness predictions during the EDM operation on different workpieces at different electrode polarities. Similar methods were also adopted to model hybrid EDM processes. Mediliyegedara et al. [33] developed a pulse classification system utilizing a feed-forward neural network during ECDM operation. The network had been trained to stimulate for achieving optimal solutions. Pulse classification can help in optimizing and controlling the process. Similarly, Kumar and Choudhury [34] established a three-layer feed-forward ANN model to predict the surface quality and wheel wear rate during EDAG operation. Leyva-Bravo et al. [35] estimated the MRR during ECDM of a HSS workpiece using an ANN model. Back propagation neural network and radial basis function (RBF) network modeling were employed. An RBF neural network has only one hidden layer and takes less computational time in training. The model was comprised of an input layer with three neurons representing three input variables, a hidden layer, and an output layer with one neuron representing one response variable. For the back propagation model, hyperbolic tangent activation functions were utilized and it was able to estimate the MRR values with an accuracy of 89.50 %, while the RBF model predicted the MRR with a better accuracy, namely, 97.25 %. Antil [36] employed a multilayer perceptron feed-forward neural network modeling for optimization of parameters during the ECDM of SiC reinforced metal matrix composites. The input layer was comprised of four neurons representing the input parameters, namely, voltage, electrolyte concentration, inter-electrode gap, and duty factor, a hidden layer, two output neurons representing MRR and taper.

14.5.4 Evolutionary Optimization Techniques

Evolutionary optimization (EO) methods are a metaheuristic approach for solving the optimization problem. These methods mimic the natural survival strategies to pick the best solution from the domain space. Nature's method of survival of the fittest (most adaptable to the environment) and creating more such individuals is replicated in computing algorithms. EO provides an approximate optimum solution.

The most common EO algorithm is the genetic algorithm (GA). The concept of the GA was first given by John Holland at the University of Michigan in the 1960s. This method can be used for both local and global optimization problems. The genetic algorithm is a computing technique based on natural selection and evolution by an iterative procedure [37]. In this technique, there are a number of available solutions to the problem, which constitute the population. GA initiates the optimization technique by randomly choosing solutions from the population. The solutions then recombine and mutate, creating new and better solutions for several iterations. The process continues until a stopping criterion is reached. Binary-coded GA functions on string structures, which continually evolve and create a new set of strings by using the survival of the fittest strategy. The following sequence of steps is followed in a genetic algorithm: defining the objective functions and constraints, initialization of a population of solutions, evaluating the fitness of the solutions, the creation of better solutions, and finally converging to one or more optimal solutions. To create a better set of solutions after every iteration, the random population is controlled by three operators: reproduction, crossover and mutation [38]. The reproduction operator creates more copies of the better solutions in a probabilistic manner for the next iteration. The crossover operator creates offspring from two parent-members with a motivation to find a fitter value. The mutation operator makes a random change with a base of the values of very small probability. The process of reproduction, crossover and mutation keeps going until a converged population is obtained. Convergence is often assumed when the average fitness (based on the values of the objective function) no longer improves in subsequent iterations. A real-coded version of GA has also been developed, in which there is no need of converting a real design variable to a binary form.

Particle swarm optimization (PSO) is another population based evolutionary optimization method, which was first proposed by Kennedy and Eberhart in 1995 [39]. The basic working principle of this method is based on the social behavior of birds or fish in a group. The information or data is exchanged between different particles in the group, called the swarm. Animals, birds or fish moving in a group exchange information, which is highly beneficial for their survival. Each member of the group traverses towards a point, which according to its own intelligence is better suited for its survival.

In PSO, each particle in the swarm evaluates the objective function at its current location. Every particle then determines its movement through the search space and continues to search for the optimum solution through generations (iterations). Each ith particle updates the coordinates of its own best solution (p_i) achieved so far by the particle. The current global best (p_g) solution, in other words, the best solution of all the swarm members, is also recorded. The coordinates of the particle in the swarm are updated and stored according to the following relations [38]

$$u_{i+1} = wu_i + c_1 r_1 (p_i - x_i) + c_2 r_2 (p_g - x_i), \tag{19}$$

$$x_{i+1} = x_i + u_{i+1}. \tag{20}$$

In Eq. (19), u_i denotes the displacement at the ith iteration, w the inertia factor, c_1 the self-confidence factor and c_2 the swarm-confidence factors. The numbers r_1 and r_2 are random numbers between 0 and 1. In Eq. (19), x_i denotes the position of the particle at the ith iteration. It is to be noted that most of the researchers use the term velocity (with symbol v) for displacement, which is not consistent with the physics. The new position of the particle is obtained by adding displacement to the current position and not by adding velocity. Although this matter is trivial from a computational point of view, it is better to use conceptually correct terms. The optimal solution is obtained after many iterations. In many implementations of PSO, Eq. (19) is written in somewhat modified form.

Yadav et al. [41] predicted the MRR and surface roughness during electric discharge diamond face grinding and obtained an optimum combination of input conditions, namely, wheel speed, pulse current, pulse on-time and duty factor, by utilizing non-dominated sorting genetic algorithm II (NSGA II) of GA. Antil and Singh [42] utilized the genetic algorithm technique for the optimization of input parameters, namely, discharge voltage, electrolyte concentration and duty factor during the ECDM of a polymer matrix composite. Multi-objective optimization for UAEDM was performed by Kumaran et al. [43] using grey fuzzy logic and the GA technique to establish the optimal set of process parameters (capacitance, pulse on-time, ultrasonic voltage) for reduced tool wear rate and high burr removal rate. The application of the Bees algorithm and other evolutionary methods was comprehensively explained by

Antil et al. [44] for maximizing the MRR during the ECDM of silicon carbide particle/glass fiber-reinforced polymer matrix composite.

14.5.5 HYBRID APPROACH OF ANN-GA, RSM-GA

The complex mechanism of hybrid processes can be optimized using a suitable hybrid approach in order to gain the benefits and overcome the limitations of both the constituent optimization methods. Literature has reported a few studies, which employ the hybrid methodology of ANN-GA or RSM-GA for optimizing hybrid EDM operations. Mandal et al. [23] employed a back propagation neural network with NSGA II to model the complex mechanism of the EDM process. Yadav and Yadava [41] adopted a combined hybrid model of ANN-NSGA II to optimize the input parameters, namely, current, pulse on-time, pulse off-time, wheel speed, and grit number for better material removal rate and surface finish during the EDAG process. Shrivastava and Dubey [45] employed a coupled methodology of ANN, GA and grey relational analysis (GRA) to optimize the electric discharge grinding phenomenon. An ANN model using RBF network during ECDM operation of silicon nitride ceramics was also developed [46]. The developed network was trained using GA and PSO methods. Maity et al. [47] coupled the RSM technique with the firefly algorithm to optimize the process performance of the ECDM operation. The firefly algorithm is a nature inspired methodology that can solve both single-objective and multi-objective optimization problems in a very small time scale with efficient and better results. The coupled PSO-RBF neural network model provided better optimization results as compared to the GA-RBF model due to better flexibility and better balance provided by the swarm. The hybrid approach of ANN-NSGA II optimization was also utilized by Unune et al. [48] during abrasive mixed electro discharge diamond grinding of Monel K-500.

14.6 CHALLENGING ISSUES

An extensive amount of research has been conducted so far on hybrid EDM mechanisms, as well as on various modeling and optimization techniques. However, a lot of work remains to be done in the area of modelling and the optimization of hybrid processes. Often the models make several simplifying assumptions and use a lot of empirical data. A fully computational model appears to be a distant dream. The following observations reveal the challenges in the area of modelling and optimization of the processes:

- The finite element method (FEM) has been widely used to model EDM processes as well as hybrid EDM processes; however, the complex hybrid EDM mechanisms could not be modeled efficiently using only FEM. There has been very scant work on the other numerical methods like finite difference method (FDM) and boundary element technique. There have been some efforts to use molecular dynamics simulation for modelling of EDM [49–51]. However, MDS takes a lot of computational time and hardly any work has been done in the area of the modelling of hybrid machining processes.

- It is essential to develop a coupled model for hybrid EDM processes incorporating all the thermal, structural and chemical aspects of the mechanism. For example, for the ECDM operation, adequate research has not been carried out to model the material removal mechanism through chemical dissolution during the electrochemical reaction.
- A number of ANN models have been developed for optimizing the process variables to achieve desired response characteristics, but the ANN model looks like a black-box. Moreover, it needs a lot of data for training. If ANN is combined with a physics-based model, then it can be an effective tool.
- Research in the field of optimization is at an early stage. For effective optimization, a good model of the process needs to be available.
- Limited study has been conducted on ceramics and metal matrix composites (MMCs). In many cases, the material behavior itself is not clear. There is a need to give more emphasis to inverse techniques for obtaining the material properties.

14.7   SOME DIRECTIONS FOR RESEARCH

Various modeling and optimization methods for hybrid EDM processes have been discussed in this chapter. However, there is further scope of improvement in this field to satisfy the ever-increasing demands of the modern-day manufacturing industry. Based on the information presented above and the challenges faced during modeling, some directions for future research are as follows:

- There is a need to focus on multi-objective optimization rather than focusing on only one objective. Most of the studies have concentrated on the MRR, tool wear rate and surface roughness; however, there are other response characteristics like geometrical accuracy and change in microstructure, that need to be investigated.
- Different materials have been explored using hybrid EDM mechanisms but there has been limited study on ceramics or metal matrix composites. The conventional machining of ceramics is difficult due to their high hardness, thus the machinability of ceramics and metal matrix composites can be explored using hybrid EDM techniques.
- The difference between microstructure evolution during the EDM process and the hybrid EDM process can be another direction for research. The modeling of microstructure evolution is crucial to observe the change in structure and achieve a better quality of products.
- Multiscale modeling of hybrid EDM could be an interesting research area. MDS can be an effective tool for understanding the sub-micron level phenomena, however, the requirement for huge computational time discourages its use. Reducing the computational time by employing suitable techniques such as parallel computing could be an interesting research area.
- There have been some reports on the experimental studies of LAEDM [52], [53], and MFEDM [9], [10], [54]. Bains et al. [55] have carried out semi-empirical

modelling of MFEDM for the machining of metal matrix composites. There is a huge scope for the research in the area of modelling and optimization of LAEDM and MFEDM.

14.8 CONCLUSION

Hybrid EDM processes provide better performance than an EDM process alone. Hybrid processes provide better product quality and a higher machining rate. However, the mechanisms involved in hybrid machining are complex and less explored. There is a great research opportunity in this field. This chapter has provided only exposure to the modelling and optimization issues in EDM-based machining processes. The discussion has included electric discharge abrasive grinding, electrochemical discharge machining, ultrasonic-assisted EDM, laser-assisted EDM and magnetic field assisted EDM. Among the numerical methods of modeling, the FEM technique is the most widely used. However, it has still not been applied for the modelling of LAEDM and MFEDM processes. The hybrid approaches to modelling and evolutionary optimization techniques have the potential to enhance the overall productivity of a HEDM system. There is a lot of scope in the modelling and optimization of HEDM as a part of digital manufacturing.

REFERENCES

[1] Saxena K. K., Bellotti M., Qian J. et al. 2018. "Overview of Hybrid Machining Processes." In *Hybrid Machining*, 21–41.

[2] Shrivastava P.K., Dubey A.K. 2014. "Electrical Discharge Machining-Based Hybrid Machining Processes: A Review." *Proceedings of the Institution of Mechanical Engineers, Part B: Journal of Engineering Manufacture* 228 (6): 799–825. doi:10.1177/0954405413508939.

[3] Yadava V., Jain V.K., Dixit P.M. 2002. "Temperature Distribution during Electro-Discharge Abrasive Grinding." *Machining Science and Technology* 6 (1): 97–127. doi:10.1081/MST-120003188.

[4] Koshy P., Jain V.K., Lal G.K. 1996. "Mechanism of Material Removal in Electrical Discharge Diamond Grinding." *International Journal of Machine Tools and Manufacture* 36 (10): 1173–85. doi:10.1016/0890-6955(95)00103-4.

[5] Bhattacharyya B., Doloi B.N., Sorkhel S.K. 1999. "Experimental Investigations into Electrochemical Discharge Machining (ECDM) of Non-Conductive Ceramic Materials." *Journal of Materials Processing Technology* 95 (1–3): 145–54. doi:10.1016/S0924-0136(99)00318-0.

[6] Basak I., Ghosh A. 1996. "Mechanism of Spark Generation during Electrochemical Discharge Machining: A Theoretical Model and Experimental Verification." *Journal of Materials Processing Technology* 62 (1–3): 46–53. doi:10.1016/0924-0136(95)02202-3.

[7] Goiogana M., Sarasua J.A., Ramos J.M. et al. 2016. "Pulsed Ultrasonic Assisted Electrical Discharge Machining for Finishing Operations." *International Journal of Machine Tools and Manufacture* 109. Elsevier: 87–93. doi:10.1016/j.ijmachtools.2016.07.005.

[8] Shervani-Tabar M.T., Abdullah A., Shabgard M.R. Shabgard. 2007. "Numerical and Experimental Study on the Effect of Vibration of the Tool in Ultrasonic Assisted EDM." *International Journal of Advanced Manufacturing Technology* 32 (7–8): 719–31. doi:10.1007/s00170-006-0828-4.

[9] Lin Y.C., Lee H.S. 2008. "Machining Characteristics of Magnetic Force-Assisted EDM." *International Journal of Machine Tools and Manufacture* 48 (11): 1179–86. doi:10.1016/j.ijmachtools.2008.04.004.

[10] Joshi S., Govindan P., Malshe A. et al. 2011. "Experimental Characterization of Dry EDM Performed in a Pulsating Magnetic Field." *CIRP Annals - Manufacturing Technology* 60 (1): 239–42. doi:10.1016/j.cirp.2011.03.114.

[11] Basak I., Ghosh A. 1997. "Mechanism of Material Removal in Electrochemical Discharge Machining: A Theoretical Model and Experimental Verification." *Journal of Materials Processing Technology* 71 (3): 350–59. doi:10.1016/S0924-0136(97)00097-6.

[12] Zhang, Q. 2004. "Study on Technology of Ultrasonic Vibration Aided Electrical Discharge Machining in Gas." *Journal of Materials Processing Technology* 149: 640–44. doi:10.1016/j.matprotec.2004.02.025.

[13] Mediliyegedara T.K.K.R., De Silva A.K.M., Harrison D.K. et al. 2005. "New Developments in the Process Control of the Hybrid Electro Chemical Discharge Machining (ECDM) Process." *Journal of Materials Processing Technology* 167 (2–3): 338–43. doi:10.1016/j.jmatprotec.2005.05.043.

[14] Kumar S., Grover S., Walia R.S. 2018. "Analyzing and Modeling the Performance Index of Ultrasonic Vibration Assisted EDM Using Graph Theory and Matrix Approach." *International Journal on Interactive Design and Manufacturing* 12 (1). Springer Paris: 225–42. doi:10.1007/s12008-016-0355-y.

[15] Wei C., Xu K., Ni J. 2011. "A Finite Element Based Model for Electrochemical Discharge Machining in Discharge Regime," *International Journal of Advanced Manufacturing Technology* 987–95. doi:10.1007/s00170-010-3000-0.

[16] Jain V.K., Dixit P.M., Pandey P.M. 1999. "On the Analysis of the Electrochemical Spark Machining Process." *International Journal of Machine Tools and Manufacture* 39 (1): 165–86. doi:10.1016/S0890-6955(98)00010-8.

[17] Bhondwe K.L., Yadava V., Kathiresan G. 2006. "Finite Element Prediction of Material Removal Rate Due to Electro-Chemical Spark Machining." *International Journal of Machine Tools and Manufacture* 46 (14): 1699–1706. doi:10.1016/j.ijmachtools.2005.12.005.

[18] Balaji P.S., Yadava V. 2013. "Three Dimensional Thermal Finite Element Simulation of Electro-Discharge Diamond Surface Grinding." *Simulation Modelling Practice and Theory* 35: 97–117. doi:10.1016/j.simpat.2013.03.007.

[19] Yadava V., Jain V.K., Dixit P.M. 2004. "Theoretical Analysis of Thermal Stresses in Electro-Discharge Diamond Grinding." *Machining Science and Technology* 8 (1): 119–40.

[20] Shervani-Tabar M.T., Mobadersany N. 2013. "Numerical Study of the Dielectric Liquid around an Electrical Discharge Generated Vapor Bubble in Ultrasonic Assisted EDM." *Ultrasonics* 53 (5): 943–55. doi:10.1016/j.ultras.2012.11.008.

[21] Wang C.Y., Zhang X. 2006. "Multiscale Modeling and Related Hybrid Approaches." *Current Opinion in Solid State and Materials Science* 10 (1): 2–14. doi:10.1016/j.cossms.2006.02.003.

[22] Zeng Q., Qin Y. 2018. "Multiscale Modeling of Hybrid Machining Processes." In *Hybrid Machining*, 269–98. Academic Press.

[23] Mandal D., Pal S.K., Saha P. 2007. "Modeling of Electrical Discharge Machining Process Using Back Propagation Neural Network and Multi-Objective Optimization Using Non-Dominating Sorting Genetic Algorithm-II." *Journal of Materials Processing Technology* 186 (1–3): 154–62. doi:10.1016/j.jmatprotec.2006.12.030.

[24] Bhattacharyya B., Doloi B., Sarkar B.R. et al. 2017. "Analysis on Electrochemical Discharge Machining during Micro-Channel Cutting on Glass." *International Journal of Precision Technology* 7 (1): 32. doi:10.1504/ijptech.2017.10005511.

[25] Ghoreishi M., Atkinson J. 2002. "A Comparative Experimental Study of Machining Characteristics in Vibratory, Rotary and Vibro-Rotary Electro-Discharge Machining." *Journal of Materials Processing Technology* 120 (1–3): 374–84. doi:10.1016/S0924-0136(01)01160-8.

[26] Jain V.K., Choudhury S.K., Ramesh K.M. 2002. "On the Machining of Alumina and Glass." *International Journal of Machine Tools and Manufacture* 42 (11): 1269–76. doi:10.1016/S0032-3861(02)00241-0.

[27] Sarkar B.R., Doloi B., Bhattacharyya B. 2006. "Parametric Analysis on Electrochemical Discharge Machining of Silicon Nitride Ceramics." *International Journal of Advanced Manufacturing Technology* 28 (9): 873–81. doi:10.1007/s00170-004-2448-1.

[28] Byiringiro J.B., Kim M.Y., Ko T.J. 2012. "Process Modeling of Hybrid Machining System Consisted of Electro Discharge Machining and End Milling." *International Journal of Advanced Manufacturing Technology* 61 (9–12): 1247–54. doi:10.1007/s00170-012-4089-0.

[29] Paul L., Hiremath S.S. 2013. "Response Surface Modelling of Micro Holes in Electrochemical Discharge Machining Process." *Procedia Engineering* 64: 1395–1404. doi:10.1016/j.proeng.2013.09.221.

[30] Kurniawan R., Kumaran S.T., Prabu V.A. et al. 2017. "Measurement of Burr Removal Rate and Analysis of Machining Parameters in Ultrasonic Assisted Dry EDM (US-EDM) for Deburring Drilled Holes in CFRP Composite." *Measurement* 110: 98–115. doi:10.1016/j.measurement.2017.06.008.

[31] Dixit P. M., Dixit U. S. 2008. *Metal Forming and Machining Processes. Modeling of Metal Forming and Machining Processes.* doi:10.1007/978-1-84800-189-3_1.

[32] Tsai K., Wang P. 2001. "Predictions on Surface Finish in Electrical Discharge Machining Based upon Neural Network Models" *International Journal of Machine Tools and Manufacture* 41(10): 1385–1403.

[33] Mediliyegedara T.K.K.R., Silva A.K.M. De, Harrison D.K. et al. 2004. "An Intelligent Pulse Classification System for Electro-Chemical Discharge Machining (ECDM)—a Preliminary Study" *Journal of Materials Processing Technology* 149(1–3): 499–503. doi:10.1016/j.jmatprotec.2004.04.002.

[34] Kumar S., Choudhury S.K. 2007. "Prediction of Wear and Surface Roughness in Electro-Discharge Diamond Grinding" *Journal of Materials Processing Technology* 191(1–3): 206–9. doi:10.1016/j.jmatprotec.2007.03.032.

[35] Leyva-bravo J., Chiñas-sanchez P., Hernandez-rodriguez A. et al. 2020. "Electrochemical Discharge Machining Modeling through Different Soft Computing Approaches." *The International Journal of Advanced Manufacturing Technology*, 3587–96.

[36] Antil P. 2020. "Modelling and Multi-Objective Optimization during ECDM of Silicon Carbide Reinforced Epoxy Composites." *Silicon* 12 (2): 275–88. doi:10.1007/s12633-019-00122-8.

[37] Holland, J.H. 1992. "Genetic Algorithms." *Scientific American* 267 (1): 66–73.

[38] Chandrasekaran M., Muralidhar M., Krishna C.M. et al. 2010. "Application of Soft Computing Techniques in Machining Performance Prediction and Optimization: A

Literature Review." *International Journal of Advanced Manufacturing Technology* 46 (5–8): 445–64. doi:10.1007/s00170-009-2104-x.

[39] Kennedy J. and Eberhart R. 1995. "Particle Swarm Optimization." *In: Proceedings of the IEEE International Conference on Neural Networks (ICNN'95)*. Perth, Australia. doi:10.1007/978-3-642-37846-1_3.

[40] Yadav R.N., Yadava V. 2013. "Multiobjective Optimization of Slotted Electrical Discharge Abrasive Grinding of Metal Matrix Composite Using Artificial Neural Network and Nondominated Sorting Genetic Algorithm." *Proceedings of the Institution of Mechanical Engineers, Part B: Journal of Engineering Manufacture* 227 (10): 1442–52. doi:10.1177/0954405413489294.

[41] Yadav R.N., Yadava V., Singh G.K. 2014. "Application of Non-Dominated Sorting Genetic Algorithm for Multi-Objective Optimization of Electrical Discharge Diamond Face Grinding Process" Journal of Mechanical Science and Technology 28 (6): 2299–2306. doi:10.1007/s12206-014-0520-9.

[42] Antil P., Singh S., Manna A. 2018. "Genetic Algorithm Based Optimization of ECDM Process for Polymer Matrix Composite." *Materials Science Forum* 928: 144–49.

[43] Kumaran S.T., Ko T.J., Kurniawan R. 2018. "Grey Fuzzy Optimization of Ultrasonic-Assisted EDM Process Parameters for Deburring CFRP Composites." *Measurement: Journal of the International Measurement Confederation* 123: 203–12. doi:10.1016/j.measurement.2018.03.076.

[44] Antil P., Singh S., Singh S. et al. 2019. "Metaheuristic Approach in Machinability Evaluation of Silicon Carbide Particle / Glass Fiber – Reinforced Polymer Matrix Composites during Electrochemical Discharge Machining Process" *Measurement and Control* 52: 1167–76. doi:10.1177/0020294019858216.

[45] Shrivastava P.K., Dubey A.K. 2013. "Intelligent Modeling and Multiobjective Optimization of Electric Discharge Diamond Grinding." *Materials and Manufacturing Processes* 28 (9): 1036–41. doi:10.1080/10426914.2012.700153.

[46] Pandu K.S., Surekha R.V.B. 2015. "Modeling of ECDM Micro-Drilling Process Using GA- and PSO-Trained Radial Basis Function Neural Network." *Soft Computing.* 2193–2202. doi:10.1007/s00500-014-1400-z.

[47] Maity D., Acherjee B., Kuar A. S. 2017. "Quality Improvement of Electrochemical Discharge Machining Process Using Firefly Algorithm: A Case Study." *International Journal of Swarm Intelligence* 3 (2–3): 238.

[48] Unune D.R., Nirala C.K., Mali H.S. 2018. "ANN-NSGA-II Dual Approach for Modeling and Optimization in Abrasive Mixed Electro Discharge Diamond Grinding of Monel K-500." *Engineering Science and Technology, an International Journal* 21 (3): 322–29. doi:10.1016/j.jestch.2018.04.014.

[49] Yang X., Guo J., Chen X. et al. 2011. "Molecular Dynamics Simulation of the Material Removal Mechanism in Micro-EDM." *Precision Engineering* 35 (1): 51–57. doi:10.1016/j.precisioneng.2010.09.005.

[50] Yue X., Yang X. 2017. "Molecular Dynamics Simulation of Material Removal Process and Crystal Structure Evolution in EDM with Discharge on Different Crystal Planes." *International Journal of Advanced Manufacturing Technology* 92 (9–12): 3155–65. doi:10.1007/s00170-017-0415-x.

[51] Yue X., Yang X. 2019. "Molecular Dynamics Simulation of Machining Properties of Polycrystalline Copper in Electrical Discharge Machining." *Proceedings of the Institution of Mechanical Engineers, Part B: Journal of Engineering Manufacture* 233 (2): 371–80. doi:10.1177/0954405417748187.

[52] Al-Ahmari A.M.A., Rasheed M.S., Mohammed M.K. et al. 2016. "A Hybrid Machining Process Combining Micro-EDM and Laser Beam Machining of

Nickel-Titanium-Based Shape Memory Alloy." *Materials and Manufacturing Processes* 31 (4): 447–55. doi:10.1080/10426914.2015.1019102.

[53] Antar M., Chantzis D., Marimuthu S. et al. 2016. "High Speed EDM and Laser Drilling of Aerospace Alloys." *Procedia CIRP* 42: 526–31. doi:10.1016/j.procir.2016.02.245.

[54] Naveen Anthuvan R., Krishnaraj V., Parthiban M. 2020. "Magnetic Field-Assisted Electrical Discharge Machining of Micro-Holes on Ti-6Al-4V." *Materials Today: Proceedings*. doi:10.1016/j.matpr.2020.06.153.

[55] Bains P.S., Sidhu S.S., Payal H.S. 2016. "Semi Empirical Modeling of Magnetic Field Assisted Ed Machining of Metal Matrix Composites." *Proceedings of the American Society for Composites - 31st Technical Conference*.

15 Application of EDM-Based Hybrid and Sequential Processes in Micro Manufacturing

Sebastian Skoczypiec*

Chair of Production Engineering, Faculty of Mechanical Engineering, Cracow University of Technology, Kraków, Poland

*Corresponding author

CONTENTS

15.1 INTRODUCTION

The integration of different manufacturing techniques into a single machine tool is one of the research and development trends in today's manufacturing technology. This development trend is also observed in micro-manufacturing and the aim of such a solution is to increase machining reliability, efficiency or to shorten the production chain. Technology integration might lead to the following possibilities:

- a hybrid machining process where different machining actions or phases are combined to interact with or remove material,
- combined or sequential machining, which involves at least two different machining technologies on a single machine tool or links two different machine tools in a production chain.

DOI: 10.1201/9781003202301-15

Adequate combination of thermal, mechanical and chemical interaction in material removal mechanisms allows the obtaining of a synergetic effect or economic benefits during production of highly advanced products.

Electro-discharge machining (EDM) is one of the most popular and widespread electro-physical machining methods in all branches of industry. In EDM, due to a series of electric discharges, electrical energy is changed into thermal energy. The process is conducted in a thin interelectrode gap filled with a dielectric medium. One of the electrodes is a workpiece and the second one is a tool. Electric discharges allow the removal of the material by melting and evaporation and the dielectric ensures favorable conditions for spark generation and flushes away debris of melted material. In respect of the micromachining application, one can state following features of EDM [1]:

- the process is non-contact and mechanical forces during machining are negligible, therefore it can be used for shaping thin, fragile and high-aspect ratio structures,
- there is high processing flexibility, relatively good precision and low costs,
- there is the possibility to machine electrically conductive materials regardless of their chemical structure, morphology and mechanical properties (namely, strength or hardness),
- there is very low material removal rate and occurrence of tool wear,
- resulting from thermal material removal (melting and vaporizing), there is a decreased quality of surface layer.

The main technological factors of electro discharge shaping are the rate of material removal, electrode tool wear and technological surface layer integrity. The following group of parameters affect these factors:

- the timing and electrical parameters (in other words, the polarity of the electrodes, pulse and pause time, voltage and current amplitude),
- the properties of the machined material and tool-electrode material (especially thermal conductivity, specific heat capacity, melting point and electrical conductivity),
- the composition, physical parameters and hydrodynamics of the dielectric,
- machine tool design (especially the strategy for gap regulation).

EDM can be used for shaping parts in different variants like sinking, wire-cutting, milling or drilling [2–5] and its application in micromachining is connected with scaling down the process. This includes the minimization of unit removal occurrences and appropriate adaptation of machine tools and tooling. Therefore, adaptation of EDM to micromachining needs the matching of the above mentioned electrical, non-electrical and material factors to minimize the amount of material removed in a single discharge (Figure 15.1). In EDM a single portion of machined material is a single discharge trace, in other words, a discharge crater [6]. Its size depends on discharge energy and defines the resolution of the machining and the possibility of

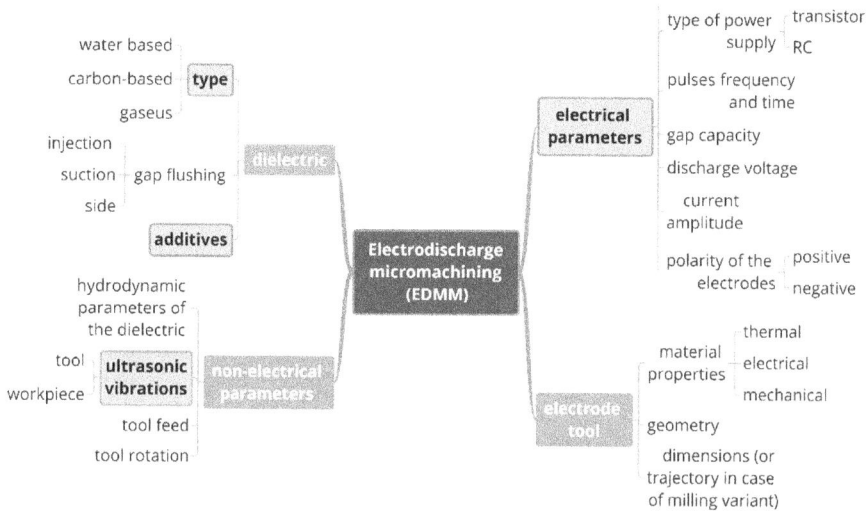

FIGURE 15.1 Different factors of electro-discharge machining responsible for its successful adoption to micromachining (factors related to make process hybrid are marked in shaded color).

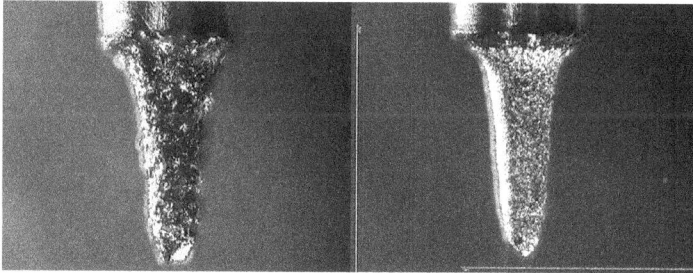

FIGURE 15.2 The comparison of EDMed shaft surface after machining with different discharge energies (left – transistor power supplier ($t_i=t_p=25$ µs, I= 1A,); right – low discarge energy (55,1 nJ) resistance–capacitance supplier.

the miniaturization of machined elements. An example of differences resulting from different discharge energies is presented in Figure 15.2.

Depending on the type of applied tool and its shape and kinematics, the following variants of electro discharge micromachining (EDMM) can be distinguished:

- wire electro discharge cutting – the electrode tool is a thin wire (even up to 20 µm), which due to wear, scrolls through the gap,
- electro discharge sinking (the reproduction of the shape of the electrode in the machined material),
- electro discharge drilling, which is applied to produce holes with circular cross-section with diameter up to 5 µm,

- electro discharge machining with a universal electrode tool where the shape of the workpiece is obtained due to reproduction of the tool trajectory (this variant is also called electro discharge milling),
- wire electro discharge grinding (wire electro discharge turning), in which a travelling wire is used as the tool electrode, a process developed to shaping very thin rods.

The minimum workpiece dimensions that it is possible to obtain in EDMM is ~5 μm (in cutting and drilling), the shape aspect ratio is in the range from 10 (in milling) to 100 (in drilling), and the minimum surface roughness is in the range from 0.05 to 0.5 μm [1]. It is worth mentioning that in respect of shaping any 3D structures (namely, microforming tools), the greatest possibilities are connected with the application of a universal electrode tool with as small as possible a diameter and appropriate motion control of electrodes (also including tool wear compensation algorithms). However, regardless of the variant of EDMM being considered, the adaptation of the process to micromachining (discussed in paragraph 2) has two main disadvantages, namely, significant tool wear and very low efficiency. Therefore, one of the EDMM development trends is in carrying out research to solve these problems. Effective solutions in this area involve combining EDMM with other machining technologies. A typical example is the abrasive electro discharge grinding process which allows machining with better efficiency (compared to EDMM) and decreased tool wear (compared to grinding) [7] or, alternatively, to apply ultrasonic vibrations to improve the stability and reliability of machining, increase process efficiency and minimize tool wear [8].

15.2   EDM ADAPTATION TO MICROMACHINING

When scaling down the geometrical dimensions of the machined part some unexpected results arise in the machining process. It is called, size effect [9] and has to be considered during development of the electro discharge micromachining process. The technological characteristics of EDMM differ from those developed for a typical EDM process. In [10], the scale effect in relation to EDM was defined as *deviations from proportional extrapolated values of micro EDM performance values that occur when down-scaling the processing parameters and/or the global or local geometrical dimensions of the electrodes.* With change of the geometrical dimensions of the machining system, factors affecting the direction of the EDM process do not scale in the same way, hence differences occur. The dimensions of electrodes, width of the interelectrode gap and the characteristic dimensions of material structure, in other words, the grain size, can be understood as geometrical dimensions. While the process parameters include parameters, which influence discharge energy (i.e. voltage and current amplitude, pulse time and pause). Its comparison for macro- and micro-EDM is presented in Table 15.1.

The scale effect in EDMM can be considered in three interrelated categories [10]:

- material microstructure, which means that characteristic dimensions for the material microstructure (in other words, grain size) do not change with

TABLE 15.1
Comparison of typical machining parameters for typical process (EDM) and process adapted to micromachining (EDMM)

	Electro discharge Machining (EDM)	Electro discharge Micromachining (EDMM)
Preferred power source	transistor power supply	RC
Preferred dielectric fluid	carbon based dielectrics, deionized water	carbon based dielectrics
Gap flushing	forced flow, internal, external	none
Electrode-tool material	copper, graphite	tungsten, tungsten carbide,
Current amplitude	0.5–400 A	0.1–10 mA
Voltage	40–400 V	60–120 V
Pulse time	10^{-6}–10^{-3} s	10^{-9}–10^{-6} s
Tool wear	< 5%	up to 100%

miniaturization of the machining system (in other words, electrode-tool, work-piece, gap size) and the material cannot be considered as homogenic and isotropic,

- change of process parameters (decreased discharge energy), which can the lead to different technological characteristics,
- thermal conductivity, which changes due to a change of timescale of the phenomena being viewed (significantly higher than normally discharge frequency).

From a technological point of view, the scale effect is responsible for the following problems during EDMM (which are negligible in macro machining):

- the relationship between material microstructure, machining efficiency and surface integrity (the volume of grain boundaries influences the melting temperature and heat conductivity, so in the case of small workpieces, the grain boundaries will affect course of physical phenomena in the gap),
- the mechanism of material removal during electro discharge machining is responsible for the occurrence of residual stresses in workpiece and electrode material. When machining system dimensions are small, it causes deformation of workpiece and influences accuracy and precision of machining. When high aspect ratio parts are machined, it even defines the limits of miniaturization [11] (an example of such effect is presented in Figure 15.3),
- the emergence of electrical forces during machining (resulting from potential difference), magnetic forces (resulting from current flow) and thermal and hydrodynamic impact (resulting from changes of temperature and flow of dielectric), are source of deformations and vibrations [12–14],
- due to the short pulse time, tool wear is significant and the negative tool polarity is to be preferred. Additionally, for high frequencies of discharge, skin effect makes current density higher near the surface of conductor, which results in

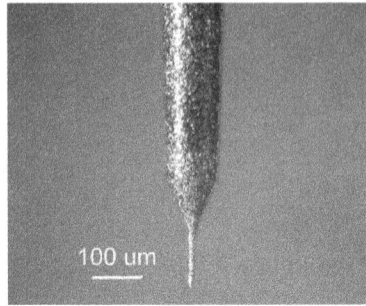

FIGURE 15.3 Example of electrical, magnetic, thermal and hydrodynamic impact on EDMM process – microtool with deformed tip (diameter 15 μm).

significantly higher wear of the edges and of the lateral surfaces of the electrode tool,

- significantly lower discharge energy ensures that a higher proportion of energy to create the plasma channel and the share of discharge energy between electrodes in EDMM is estimated at 10% to 15% (in comparison, during EDM it is in range of 25–40%). This is a reason for rapid temperature rise in the gap; therefore, the EDMM is also connected with increased gap flushing demands to keep machining stable.

15.3   HYBRID PROCESSES IN MICROMACHINING

According to CIRP definition proposed in [15],

hybrid manufacturing processes are based on the simultaneous and controlled interaction of process mechanisms and/or energy sources/tools having a significant effect on the process performance.

The classification of hybrid manufacturing processes includes the following two groups (Figure 15.4):

- processes, where a combination of different energy sources/tools takes place,
- processes, where controlled application of process mechanisms (normally done in separated processes) takes place.

The first group covers mixed processes, where two or more processes are directly involved in material removal (several process mechanisms or possibly a new machining mechanism occurs) and assisted processes, where main/primary processes are dominant and secondary processes only assist (significantly improving conditions for machining). In [16], more than 30 hybrid micromachining processes were specified and reviewed. One can state than in the micromachining field of application, assisted processes dominate (~70%) and the most effective is the application of ultrasonic vibration and laser beam to deliver enhanced energy.

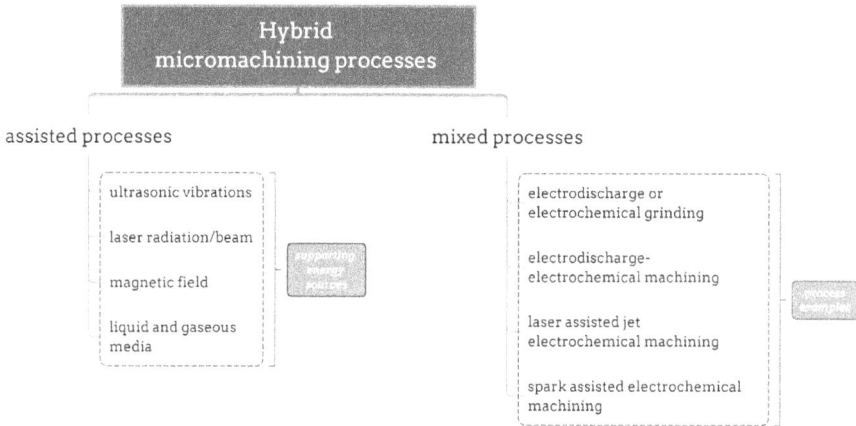

FIGURE 15.4   Classification of hybrid micromachining processes (scheme based on [16]).

In developing hybrid micromachining processes the electro discharge interaction in the majority of cases plays a leading role and is assisted by other energy sources. The most investigated and most promising example is electro discharge machining supported by ultrasonic vibrations of the electrode tool, workpiece or by introducing vibrations to the dielectric medium. The EDMM of ferromagnetic can be also take place in a magnetic field. In this case the vertical component of the magnetic field vector improves the evacuation of the debris of melted material from the interelectrode gap. The common feature of electro discharge and electrochemical micromachining is that both processes can be easily and effectively integrated as hybrid or sequential (complete or combine) machining processes. In the latter case, integration is justified by relatively high shaping accuracy (compared to other micromachining technologies).

15.4   EDM-BASED HYBRID MICROMACHINING PROCESSES

15.4.1   Ultrasonically Assisted Electro Discharge Machining

One of the most effective solutions to improve the efficiency of electro discharge micromachining is in introducing ultrasonic vibrations to the machining system. It can be achieved in the following three ways: (i) vibration of the electrode tool [17, 18], (ii) vibration of the workpiece [19–21] or (iii) by application of a vibration exciter immersed in the dielectric medium [22, 23]. The applied vibration frequency is less than 50 kHz and its amplitude does not exceed 10 μm (peak to peak), however its active interaction improves dielectric flow, favors the evacuation of debris of solidified material from the gap and causes even distribution of dielectric and erosion products in the machining area. These effects are especially noticeable during machining with low discharge energy. Therefore, application of vibrations in EDM is preferred in micromachining operation. The following benefits of such solutions can be shown to be: improvement of machining stability (especially during machining high aspect ratio, thin-walled and slim elements), a decrease of the amount of arc discharges,

(a)   (b)

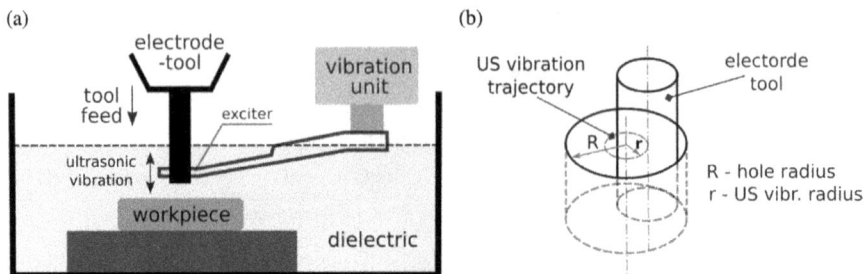

FIGURE 15.5   Schemes of variants of vibration assisted EDMM process: (a) application of special vibration exciter immersed in dielectric fluid and (b) application of ultrasonic circular vibration electrode tool (scheme according to [24]).

an increase in machining efficiency, a decrease of tool wear and an improvement of technological surface layer integrity. Due to the possibility of shaping high-aspect structures, the most common vibration-assisted variant of EDMM is drilling, especially when machined diameters are small and the application of typical technological solutions (namely, high-pressure dielectric supply through tube electrode-tools) are impossible. Considering the dimensions of the EDMM system, the most attractive solution is to apply a special vibration exciter immersed in the dielectric medium giving a direct vibration excitation of the fluid [22, 23]. This solution also simplifies the technical problems because it does not require integration into an existing structure of the machine tool and the vibration exciter acts as electrode-tool guide (schema depicted in the Figure 15.5a). In the case of electro discharge microdrilling, this solution improves machining stability, decreases machining time (average feed rate 33 times faster) and decreases the tool wear ratio.

The interesting approach of ultrasonic assistance of the microdrilling process was presented in [24], where application of ultrasonic circular vibrations of the electrode tool was proposed (schema depicted in Figure 15.5b). Application of 32.85 kHz frequency and 3 µm of eccentric tool vibrations allows the improvement of drilling stability (electrode retreat frequency decreases and solidified material debris adhesion minimizes). Circular tool vibrations also favorably affect surface morphology and decrease the hole taper angle. Results presented in [24] were supported by flow-field numerical modelling, which confirmed that the advantageous effect of vibrations are connected with significantly enlarged dielectric flow velocity and an enhanced capability of the dielectric fluid to remove debris particles from the gap. In this case, the additional tool velocity component (perpendicular to the feed) mitigates abnormal discharges and the machining process is more stable.

15.4.2   ELECTROCHEMICAL-ELECTRO DISCHARGE MICROMACHINING (ECEDMM)

The integration of electro discharge and electrochemical machining are developed for synergetic or process reasons. In the former case, it results in a typical hybrid mixed process, while in the latter case, it results in a sequential process. In the hybrid integration the material is removed by electrical discharges accompanied by electro-chemical dissolution (both processes occurring simultaneously in the gap). Depending on the type of applied power source, this process can be conducted using constant voltage (Electrochemical Arc Machining, ECAM) or pulse voltage (Electrochemical Discharge Machining, ECDM). The greatest application potential in micromachining is connected to the development of the ECDM process and therefore, further discussion focuses on this variant. Such processes take place in a low concentration electrolyte (for example, a less than 5% aqueous solution of $NaNO_3$) with pulse parameters in the following ranges: pulse voltage 10-120 V, pulse on time 10-2000 μs, pulse off time 5-500 μs. In some cases, during pulse off time, the voltage is above zero and current amplitude depends on gap conditions. The machining parameters such as electric current density, voltage amplitude, pulse on/off time and gap thickness define the mode of the material removal mechanism. Thus, machining can be carried as (i) electrochemical dissolution accompanied by electrical discharges or (ii) only by electrochemical dissolution. The share between (i) and (ii) results from the intensity of the electrochemical reaction connected with the amount of gas and from the temperature of the electrolyte. It can be stated that electrochemical-electro discharge machining is a form of electrochemical machining, which is conducted with a continuously controlled critical state in the gap. It means, that the passing electric current significantly increases the temperature of the electrolyte which starts to boil and then finally evaporates. Near one of the electrodes a gas-vapor layer is created so that the electrolyte conductivity decreases thus finally arresting the dissolution process. This occurs in areas where the gap thickness is the smallest (in these areas the current density must be higher than a critical value). The result is an increase in the electric potential, so electric discharges occur. The material begins to be removed due to the EDM mechanism. The ECEDM process can be used in various kinematic variants, however, the most common application is during drilling or sinking. However, the mechanism of simultaneous electrical discharges and electrochemical dissolution can also be applied in special applications, namely, trueing and dressing of metal bond grinding wheels [25]. It allows the attainment of grain protrusion in the range of 75–100% of the average grain diameter, improves the roundness of the grinding wheel (in comparison to conventional methods) and the application of such a tool reduces grinding forces and improves geometrical accuracy.

In comparison to EDM shaping, the benefit of ECEDM process application is in a significant increase of machining efficiency. This is especially noticeable during the machining of materials with high tensile strength and heat resistance. The effect of the electrochemical–electro discharge material removal mechanism can be obtained during machining with deionized water as a dielectric (such a working fluid can be also treated as a low concentration electrolyte). In [26], research results relating to the ECEDM of aluminum alloy-based metal matrix composite (MMCs) with silicon carbide (SiC) reinforcement were presented. Specific properties of such material, in other words, structural heterogeneity are a key challenge during machining. In many cases this is the reason of rapid tool wear, poor integrity and residual stresses in the technological surface layer. The studies presented in [26] were carried out in milling kinematics (Figure 15.6). Depending on machining parameters the process is conducted with a different predominant mechanism of material removal. When

(a)

dielectric
fluid

a

electrode
tool

L

(b)

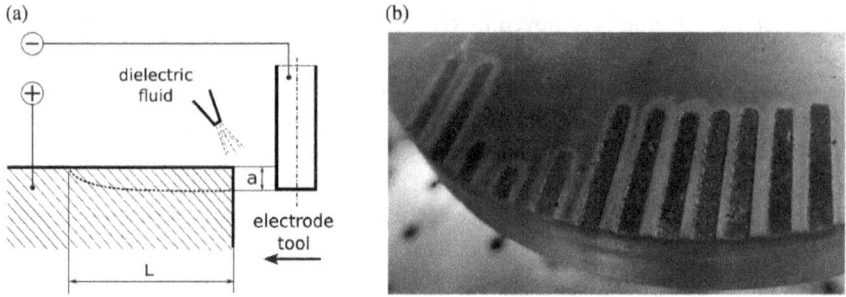

FIGURE 15.6 (a) Scheme of machining process and (b) photography of the sample with machined groves in AL-SIC composite sample; a – layer depth [26]. With permission from Elsevier.

U=30 V, I = 2,15 A, t_i=10 µs	U=10 V, I = 2,15 A, t_i=100 µs	U=30 V, I = 2,15 A, t_i=100 µs

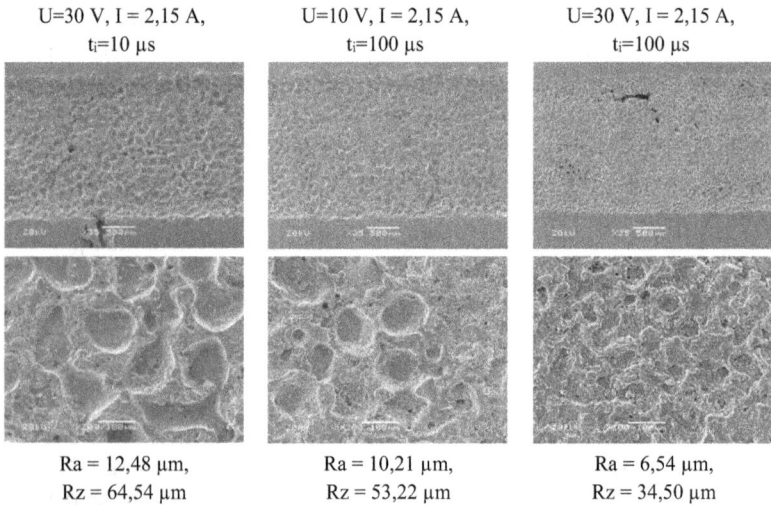

Ra = 12,48 µm, Rz = 64,54 µm	Ra = 10,21 µm, Rz = 53,22 µm	Ra = 6,54 µm, Rz = 34,50 µm

FIGURE 15.7 The SEM photographs of Al-SiC composite surface EDMed with various process parameters [26]. With permission from Elsevier.

discharge energy is low, the electrochemical dissolution plays an important role in material removal and the SiC reinforcement grains flow out much more easily from the machined surface. This is indicated by the surface integrity (Figure 15.7). When the electric current and voltage amplitude is low, machining conditions are good for the rinsing of the reinforcement grains from the machined surface, therefore, surface roughness increases. When the voltage, electric current and pulse time are high enough, the deionized water starts to boil in the gap and its conductivity decreases. The result is a decrease in the gap size, electrical discharges occur, and the material is removed by melting and evaporation as happens during the EDM process.

The published research indicates that in ECEDM, the process share of the electro discharge and of the electrochemical material removal mechanism can be regulated by adequate selection of the process parameters such as voltage, electric current amplitude and pulse duration. The possibility of choice of a dominant machining mechanism (electrical discharges or electrochemical dissolution) by changes to the process parameters while machining is one of the main advantages of ECDM technology. Through a special design of the process energy source, it can be used in the sequential integration of electrochemical and electro discharge micromachining. Such examples were described in Section 15.5.

15.4.3 Spark Assisted Electrochemical Machining (SAEM)

One of the recently developed hybrid technologies based on electrical discharges is a process based on the electrochemical discharge phenomenon. In this case the machining mechanism is based simultaneously on thermal and chemical erosion. This process is known by several names in literature, namely, *Electrochemical Arc Machining*, *Electrochemical Discharge Machining*, *Electrochemical Spark Machining*, *Discharge Machining of Non-conductors*. However, one can observe that the physical basis of this process is completely different than of that described in the previous section of the electrochemical-electro discharge machining process. In SAEM, the phenomena of the breakdown of electrolysis, when electrochemical cell voltage is increased above the critical value, is used to initiate material removal. The process is carried out in an electrochemical system consisting of tool and auxiliary electrodes connected to an external power source and with the workpiece immersed in electrolyte solution. Electrochemical reactions between tool (connected as cathode) and auxiliary electrode (connected as anode) are responsible for the formation of a gas film around the electrode tool. It creates conditions for the formation of electric discharges. The machining in the SAEM system is carried out with the voltage around 30 V - 40 V, which is optimal to maintaining the gas film around the electrode-tool. The electrical field in this area is sufficient to enable discharges between tool and electrolyte. The tool electrode is close to the workpiece surface (less than 1 mm), so that the discharges interact with the workpiece material and cause its removal mainly by [27, 28] melting, evaporation, mechanical impact (thermal stresses, micro-cracking, spalling, impact of expanding gases and moving electrolyte), high temperature dissolution and electrochemical etching. The electrode-tool feed mechanism keeps the machining constant and stable. Process control is based on gravity feed (the workpiece and tool are in contact during the whole machining process) or on keeping a constant force between tool and workpiece.

The area of application of the SAEM process are machining nonconductive materials that are difficult-to-cut and also difficult-to-machine by non-traditional methods (for example, EDM, ECM or LBM). As an example, there are hard and brittle dielectric materials like Pyrex glass, granite, quartz, ceramic (that is, Al_2O_3 based) or composite materials. The SAEM can be carried out in the operations of drilling [29], machining with a universal electrode tool [30] or wire-cutting. The size of machined shapes does not exceed a few millimeters, that is, for quartz it is in the range of 1–3 mm, for ceramic Al_2O_3 based materials it is in the range 0.1–1 mm and for glass it is in the range of 0.1–2 mm [31]. The example of holes machined in glass and granite are presented in Figure 15.8. It is worth mentioning that the application of the SAEM process is oriented to micromachining. The shaping of structures with

(a)   (b)   (c)

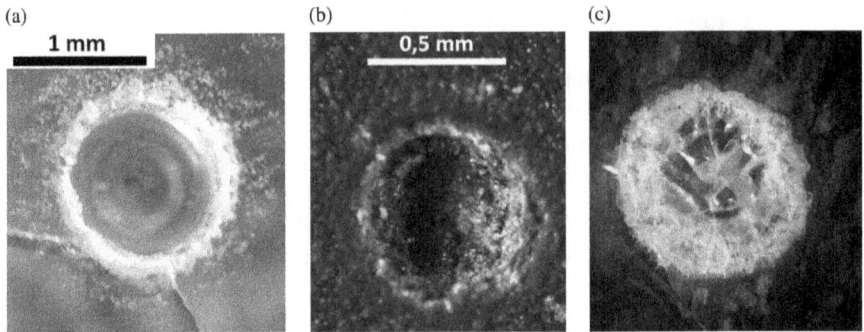

FIGURE 15.8 Examples of holes machined by SAEM process in (a and c) glass, (b) granite; and (c) the hole with micro cracks resulting from ineffective electrolyte supply.

higher dimension with SAEM is limited by the application of higher current that means a significant amount of Joule heating, which has to be effectively dissipated from the machining area. Another problem, especially when machining high aspect-ratio holes, is maintaining an effective electrolyte supply to the machining area. Lack of electrolyte causes a decrease in process efficiency and deterioration of the surface quality (micro cracks occur, example in Figure 15.8c). However, one of the main disadvantages of the SAEM process is insufficient repeatability of results, therefore research on the removal mechanism in this process are still in progress.

15.5   EDM-BASED SEQUENTIAL MICROMACHINING PROCESSES

Sequential integration of different micromachining technologies can be implemented through combined or through complete machining. In both cases at least two different machining processes are involved to obtain a highly sophisticated part or product on a single machine tool (in some cases also with two machine tools). The difference is that complete machining relates to the entire shaping of the part without a change in the fixing. One can mention the following advantages of such an approach, namely, a decrease in machining time (set-up time reduction), a reduction of cost, and quality improvement. Such advantages are attractive and popular in conventional machining, especially when cutting methods (that is, turning, milling) are combined with grinding.

In respect of EDMM one can use a combination of laser and electro discharge micromachining to solve the problem of unsatisfactory process efficiency [32]. The application of a laser source in rough machining operations allowed a decrease in machining time of 90% (in the case of drilling) and about 75% (in the case of 3D-EDMM) in comparison to 3D-EDMM. However, due to high costs connected with laser application, this sequence was not implemented in industry.

Presented in [33], the characteristics of electrochemical and electro discharge micromachining indicate a number of relevant complementary advantages. As similarities between ECMM and EDMM, one can state machining variants, tool material and shape, limitations of the workpiece material, and machine tool design and its components. Therefore, many developments in the field of combining electro discharge and electrochemical machining in sequential processes have been proposed. This research mainly focusses on applying electrochemical treatment as a finishing operation to improve the surface integrity that had been changed by prior EDM. Such combinations were successfully implemented in macro machining, in other words, for the electrochemical finishing of WEDMed parts [34], where the part left after wire cutting can be applied as the tool electrode in the electrochemical finishing operation (in an investigated case, in order to achieve a final mirror–like surface, only 20 s of finishing is necessary). In [35], the case of shaping an arc shaped neck with a thickness of 50 μm was analysed. The ED/ECMM sequence was proposed as a suitable replacement for boring and grinding operations. The application of EDM followed by pulse ECM finishing with the same electrode tool enabled accuracy to be improved (the application of EDM and ECM eliminate the workpiece deformations) without a negative impact on the surface layer quality (surface integrity after EDM is improved in the pulse ECM operation).

Electrochemical machining can be also applied to improve the surface integrity of EDMed micro parts [36–39]. Application of ECMM after EDMM allows the reduction of surface roughness from Ra = 1 μm to Ra = 0.2 μm [36]. The research was carried out in sinking and milling kinematics on the same machine tool and with the same tool clamping. This interesting approach is the application of partially deionized water as the dielectric for EDMM and as the electrolyte for the ECMM operation. It significantly decreases ECMM efficiency and forces the use of a high interelectrode voltage (in range of 100–150 V). In this case the change of machining mode from EDMM to ECMM was obtained by the decrease of capacitance of the power supply and the reduction of electrode tool feed rate. Additionally, results presented in [37] were obtained for ECMM and EDMM carried out in deionized water (resistivity ~0.5 MΩ cm). Additionally, the sequence was conducted with application of the same power supply and electrical signal with the following parameters: voltage amplitude 60 V, frequency 500 kHz, duty factor in the range of 0.25 to 0.4. During milling with a universal electrode tool (a cylinder with diameter 100 μm), the share of electrical discharges and electrochemical dissolution depends on the feed rate of the electrode tool. For a feed rate of 50 μm/s, the obtained surface is typical for EDMM (Ra = 142 nm), while for a feed rate 10 μm/s there are no signs of discharges and Ra = 22 nm. Due to the same working fluid and electrical signal the process presented in [37] can be also classified as hybrid, however a change of the machining parameters (mainly feed rate) influences the electrochemical dissolution. Similar research carried out in Cracow University of Technology confirmed this (Figure 15.9). The proposed order from EDMM to ECMM is justified by the needs of the processes (improved efficiency and surface integrity). One of the disadvantages of the ECMM finishing operation is a decrease of accuracy and rounding of the edges, therefore to avoid this, tool lateral surface insulation should be applied.

A typical example of an EC/EDMM sequence was also presented in [39]. The sequence was conducted on the Sarix machine tool with the same cylindrical electrode tool made of tungsten carbide and diameter 100 μm. In comparison to previously presented solutions, this sequence was achieved with two different working fluids (namely, kerosene as dielectric and 3 wt% $NaClO_3$+1 wt% EDTA mixture as

(a)   (b)

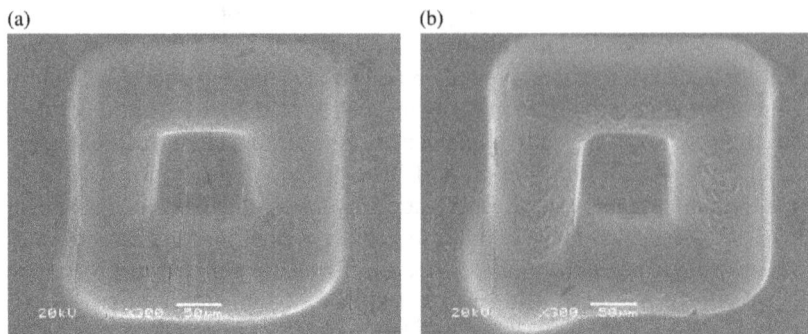

FIGURE 15.9 Examples of microcavities machined with application of resistance-capacitance power supplier in milling kinematics for different feed rates: (a) 200 μm/min (the typical surface for ECM) and (b) 250 μm/min (visible signs of discharges on the surface); machining in deionized water as electrolytes [40].

electrolyte) and separate energy sources for the ECMM and EDMM operations. The sequence consists of EDMM shaping and ECMM finishing. The removal of 15 μm of allowance in the electrochemical finishing operation made it possible to improve surface roughness from Ra = 0.707 μm to Ra = 0.143 μm. The ED/ECMM sequence is also effective in micro-hole drilling [41]. EDMM and ECMM conducted with the same electrode and on the same machine tool, albeit with a different working fluid, allows machining efficiency to be improved by a factor of about 9 with a concurrent improvement of shape accuracy and hole precision.

An interesting approach was developed by Wan and Kunieda [42, 43]. They proposed a novel method to change the machining mode between EDMM and ECMM by changing the diode polarity. This solution allows machining with the same bipolar ultrashort pulse generator in the electrolyte (2% aqueous solution of $NaNO_3$) and using the same tungsten electrode tool. Switching between EDMM and ECMM is based on the principle that a positively polarized tungsten electrode is easily oxidized (tungsten is highly susceptible to passivation). It is used in the EDMM mode for discharge ignition, while in the ECMM mode oxide film on the electrode tool surface is not formed and dissolution may occur. The capacity of the method was presented in case of shaping high aspect ratio SUS304 micro-rod with diameter of 57 μm and L/D radio up to 14. The process included the EDMM mode as rough machining (with feed 1.5 μm/s and material removal rate ~56 $μm^3$/s) and the ECMM mode to finish (feed 0,5 μm/s and material removal rate 2225 $μm^3$/s). It is worth mentioning that to achieve such results, authors use inexpensive ultra-short pulse generators, the design of which is based on an electrostatic induction feeding method. This solution is based on voltage supply to the working gap through a capacitance.

The above examples have focused on applying electrochemical treatment to improve the quality of the technological surface layer that has been changed in an earlier stage of electro discharge machining (EDMM is followed by the ECMM process). In [33], authors, both on a theoretical and experimental basis, proposed a change to this order and apply an ECMM followed by EDMM. Such a solution is focused on

(a)   (b)   (c)

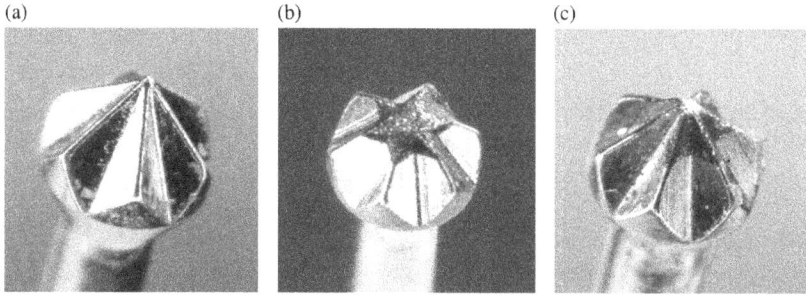

FIGURE 15.10 Photographs of the electrode tool: (a) before machining, (b) after electrodischarge machining of the cavity and (c) after properly designed EC/EDMM sequence.

the micromachining application and allows for a relevant reduction of the drawbacks and underlines the advantages of both micromachining methods. In this case the main aim of combination is in machining time reduction. The research was carried out on a specially designed, assembled and developed prototype of an electrochemical-electro discharge machine tool for micro detail manufacturing. The presented case studies of EC/EDMM sequence application indicate a possibility of efficient and with limited tool wear (compared with EDMM, Figure 15.10) and precision (compared with ECMM) 3D micro detail shaping carried out with the kinematics of sinking. Considering efficiency, for an optimal combination of both processes, ECMM should be conducted with a focus on the maximal removal rate (that is, with relatively high inter-electrode gap). In such a sequence final accuracy (possibly of up to ~5 μm) and surface quality is in relation to the thickness of allowance removed during the EDMM stage (it depends on the shape deviation after ECMM). It is worthwhile underlining that implementation of sequence EC/EDMM is much more complex, nevertheless, due to its advantages (accuracy improvement) the area of its application is in the machining of micro part prototypes or tools for microforming.

15.6 CONCLUSION

The main topic of this chapter has been electro discharge-based micromachining methods. Technologies that have be discussed may be classified as nonconventional machining processes and they have remained an area of the author's scientific interest for almost 20 years. The chapter began with a discussion of the physical and technological fundamentals of electro discharge micromachining, its adaptation to micromachining, kinematic variants and various perspectives of their application. The adaptation of the EDM process to micromachining has two main disadvantages: significant tool wear and very low efficiency. One of the research trends to overcome these problems is the combination of electro discharge interaction with other machining methods in hybrid (assisted or mixed) and sequential processes. In both of these areas noteworthy progress in recent years has been observed. In a group of assisted hybrid processes the electro discharge interaction in the majority of cases plays a leading role and the most effective and promising solution is electro discharge micromachining supported by ultrasonic vibrations. Note, also that electro discharge and electrochemical machining technologies have a number of common features as they are complementary to each other. They also have similar areas of application and can be easily integrated into a single machine tool in a hybrid or sequential process. Therefore, discussion on electro discharge and electrochemical hybrid mixed and sequential machining process conducted on a single machine tool was presented in the chapter.

The preferred application areas for EDM based micromachining hybrid and sequential processes are (1) machining of high-aspect ratio micro holes with circular or polygonal cross section; (2) manufacturing of 3D structures and tools, (in other words, micro-molds, parts of technological tooling and prototypes, high aspect ratio and thin walled structures) and (3) micro-rods for micro tools and probes; (4) machining of thin, fragile and high-aspect ratio shapes.

REFERENCES

[1] Liu K., Lauwers B., Reynaerts D. Process capabilities of micro-EDM and its applications, Int J Adv Manuf Technol, 47:11–19, 2010.

[2] Wang D., Zhao W. S., Gu L., Kang X. M., A study on micro-hole machining of polycrystalline diamond by micro-electrical discharge machining, Journal of Materials Processing Technology, 211:3–11, 2011.

[3] Cheng, X. H. Yang, Y. M. Huang, G. M. Zheng, L. Li, Helical surface creation by wire electrical discharge machining for micro tools, Robotics and Computer-Integrated Manufacturing, 30(3):287–294, 2014.

[4] F. C. Hsu, T. Y. Tai, V. N. Vo, S. Y. Chen, Y. H. Chen, The machining characteristics of polycrystalline diamond (PCD) by micro-WEDM, Procedia CIRP, 6:261–266, 2013.

[5] Yan J., Watanabe K., Aoyama T., Micro-electrical discharge machining of polycrystalline diamond using rotary cupronickel electrode, CIRP Annals - Manufacturing Technology, 63:209–212, 2014.

[6] Kunieda M., Lauwers B., Rajurkar K. P., Schumacher B.M., Advancing EDM through Fundamental Insight into the Process, CIRP Annals – Manufacturing Technology, 2005, pp. 64–87

[7] Ruszaj A., Skoczypiec S., Wyszyński D., Recent Developments in Abrasive Hybrid Manufacturing Processes, Management and Production Engineering Review, 8(2): 81–90, 2017.

[8] Iwai M., Ninomiy S., Suzuki K., Improvement of EDM properties of PCD with electrode vibrated by ultrasonic transducer, Procedia CIRP 6:146–150, 2013.

[9] Vollertsen F, Biermann D, Hansen HN, Jawahir IS, Kuzman K. Size effects in manufacturing of metallic components. CIRP Ann Manuf Technol, 58(2):566–87, 2009.

[10] Liu Q., Zhang Q., Zhang M., Zhang J. Review of size effects in micro electrical discharge machining, Precision Engineering, 44:29–40, 2016.

[11] Kawakami T., Kunieda M. Study on factors determining limits of minimum machinable size in micro EDM. CIRP Annals – Manufacturing Technology, 54(1):167–170, 2005.

[12] Karthikeyan G., Garg A.K., Ramkumar J., Dhamodaran S. A microscopic investigation of machining behavior in ED-milling process. Journal of Manufacturing Processes, 14(3):297–306, 2012.

[13] Klocke F., Garzón M., Dieckmann J., Klink A. Process force analysis on sinking-EDM electrodes for the precision manufacturing. Production Engineering, 5(2):183–190, 2011.

[14] Garzon M., Adams O., Veselovac D., Blattner M., Thiel R., Kirchheim A. High speed micro machining processes analysis for the precision manufacturing. Procedia CIRP, 1:609–614, 2012.

[15] Lauwers B., Klocke F., Klink A., Tekkaya A.E., Neugebauer R., McIntosh D. Hybrid processes in manufacturing. CIRP Annals – Manufacturing Technology, 63(2):561–583, 2014.

[16] Chavoshi S.Z., Luo X. Hybrid micro-machining processes: A review. Precision Engineering, 41:1–23, 2015.

[17] Goiogana M., Sarasua J.A., Ramos J.M. Ultrasonic assisted electrical discharge machining for high aspect ratio blind holes. Procedia CIRP, 68:81–85, 2018.

[18] Yu Z.Y., Zhang Y., Li J., Luan J., Zhao F., Guo D. High aspect ratio micro-hole drilling aided with ultrasonic vibration and planetary movement of electrode by micro-EDM. CIRP Annals – Manufacturing Technology, 58(1):213–216, 2009.

[19] Hoang K.T., Yang S.H., A study on the effect of different vibration-assisted methods in micro-WEDM. Journal of Materials Processing Technology, 213(9):1616–1622, 2013.

[20] Jahan M.P., Saleh T., Rahman M, Wong Y.S. Study of micro-EDM of tungsten carbide with workpiece vibration. Advanced Materials Research, 264–265:1056–1061, 2011.

[21] Tong H., Li Y., Wang Y. Experimental research on vibration assisted EDM of micro-structures with non-circular cross-section. Journal of Materials Processing Technology, 208(1–3):289–298, 2008.

[22] Ichikawa T., Natsu W. Study on machining characteristics of ultrasonic vibration assisted micro-EDM. Applied Mechanics and Materials, 217–219:2163–2166, 2012.

[23] Ichikawa T., Natsu W. Realization of micro-EDM under ultra-small discharge energy by applying ultrasonic vibration to machining fluid. Procedia CIRP, 6:326–331, 2013.

[24] Zhengkai Li, Jiajing Tang, Jicheng Bai, A novel micro-EDM method to improve microhole machining performances using ultrasonic circular vibration (UCV) electrode. International Journal of Mechanical Sciences, 175:105574, 2020.

[25] Singh T., Dvivedi A. Developments in electrochemical discharge machining: A review on electrochemical discharge machining, process variants and their hybrid methods. International Journal of Machine Tools and Manufacture, 105:1–13, 2016.

[26] Skoczypiec S., Bizoń W., Podolak-Lejtas A., Selected Aspects of Electrodischarge Milling of Aluminum Alloy-Based Metal Matrix Composite with SiC Reinforcement. Procedia Manufacturing, 47:795–798, 2020.

[27] Jiang B., Lan S., Ni J., Zhang Z. Experimental investigation of spark generation in electrochemical discharge machining of non-conducting materials. J. Mater. Process. Technol., 214(4):892–98, 2014.

[28] Wüthrich R., Allagui A. Building micro and nanosystems with electrochemical discharges. Electrochim. Acta., 55(27):8189–96, 2010.

[29] Hof L., Ziki J.A. Micro-hole drilling on glass substrates – a review. Micromachines, 8(2):53, 2017.

[30] Zheng Z.-P., Cheng W.-H., Huang F.-Y., Yan B.-H. 3D microstructuring of pyrex glass using the electrochemical discharge machining process. Journal of Micromechanics and Microengineering, 17(5):960, 2007.

[31] Wüthrich R., Mandin P. Electrochemical discharges – discovery and early applications. Electrochimica Acta, 54(16):4031–4035, 2009.

[32] Kim S., Kim B.H., Chung D.K., HShin H.S., Chu Ch.N. Hybrid micromachining using a nanosecond pulsed laser and micro EDM. Journal of Micromechanics and Microengineering, 20(1), 2010. Paper No. 015037.

[33] Skoczypiec S., Ruszaj A. A sequential electrochemical–electrodischarge process for micropart manufacturing. Precision Engineering, 38:680–690, 2014.

[34] Masuzawa T, Sakai S. Quick finishing of WEDM products by ECM using a mateelectrode. CIRP Ann Manuf Technol, 36:123–126, 1987.

[35] Xiaowei L, Zhixin J, Jiaqi Z, Jinchun L. A combined electrical machining process for the production of flexure hinge. J Mater Process Technol 71:373–6, 1997.

[36] Kurita T., Hattori M. A study of edm and ECM/ECM-lapping complex machining technology. International Journal of Machine Tools and Manufacture, 46(14):1804–1810, 2006.

[37] Nguyen M.D., Rahman M., Wong Y.S. Enhanced surface integrity and dimensional accuracy by simultaneous micro-ED/EC milling. CIRP Annals – Manufacturing Technology, 61:191–194, 2012.

[38] Richter C., Krah Th., Buttgenbach St. Novel 3D manufacturing method combining microelectrial discharge machining and electrochemical polishing. Microsystem Technology, 18:1109–1118, 2012.

[39] Zeng Z., Wang Y., Wang Z., Shan D., He X. A study of micro-EDM and micro-ECM combined milling for 3D metallic micro-structures. Precision Engineering, 36:500–509, 2012.

[40] Drozd, K. Manufacturing of 3D microstructures with application of electro-discharge milling in deionized water. (Unpublished master's thesis, supervisor Skoczypiec S.). Cracow University of Technology, Krakow, Poland, 2014.

[41] He XL, Wang YK, Wang ZL, Zeng ZQ. Micro-hole drilled by EDM–ECM combined processing. Key Eng Mater 562–565:52–6, 2011.

[42] Han W, Kunieda M., A novel method to switch machining mode between Micro-ECM and Micro-EDM using oxide film on surface of tungsten electrode, Precision Engineering, 56: 455–465, 2019.

[43] Han W, Kunieda M., Research of micro EDM/ECM method in same electrolyte with running wire tool electrode, Precision Engineering 70:1–14, 2021.

Index

351

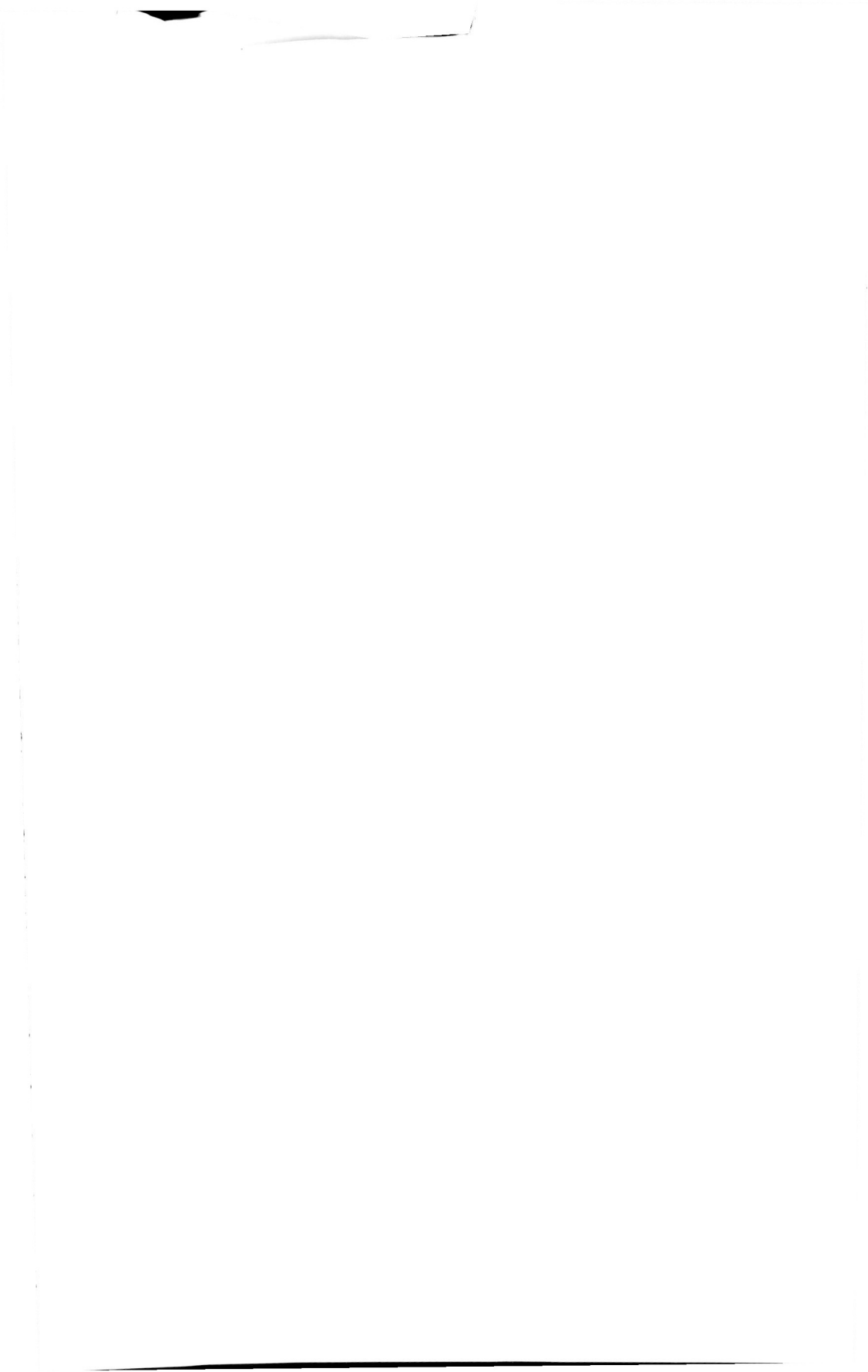

For Product Safety Concerns and Information please contact our EU
representative GPSR@taylorandfrancis.com
Taylor & Francis Verlag GmbH, Kaufingerstraße 24, 80331 München, Germany

www.ingramcontent.com/pod-product-compliance
Lightning Source LLC
Chambersburg PA
CBHW060758220326
41598CB00022B/2482

9 781032 064352